Introduction to Polymers

Introduction to Polymers

Second Edition

R.J. Young

Professor of Polymer Science and Technology
Manchester Materials Science Centre
University of Manchester and UMIST

and

P.A. Lovell

Reader in Polymer Science and Technology
Manchester Materials Science Centre
University of Manchester and UMIST

Stanley Thornes (Publishers) Ltd

First edition published in 1981 by Chapman & Hall
Second edition published in 1991 by Chapman & Hall

Reprinted in 2000 by:
Stanley Thornes (Publishers) Ltd
Delta Place
27 Bath Road
CHELTENHAM
GL53 7TH
United Kingdom

00 01 02 03 04 / 10 9 8 7 6 5 4 3 2

A catalogue record for this book is available from the British Library
ISBN 0 7487 5740 6

Typeset by Best-set Typesetters Ltd, Hong Kong
Printed in Great Britain by T.J. International Ltd, Padstow, Cornwall

Contents

Preface to the second edition

The decade that has passed since the first edition was written has seen further growth in the uses of polymers. During this time much research effort has been focussed upon the development of speciality polymers for high-performance applications, and this has served to emphasize the importance of polymer chemistry. It is partly for this reason that through the introduction of a second author in the Second Edition, the first three chapters have been reorganized, revised and expanded to give a broader and more thorough coverage of the fundamental aspects of polymer synthesis and polymer characterization. In particular, the sections upon ionic polymerization, Ziegler–Natta polymerization, copolymerization, static light scattering, phase separation, gel permeation chromatography and spectroscopy have been substantially revised and expanded. Additionally, new sections upon ring-opening polymerization, specialized methods of polymerization, dynamic light scattering and small-angle X-ray and neutron scattering have been included to give brief introductions to these topics which are of growing importance. Whilst Chapters 4 and 5 are to a large extent as they appeared in the First Edition, they also have been expanded. A more in-depth treatment of factors affecting glass transition and melting temperatures is given, and new sections have been added to introduce the important topics of thermoplastic elastomers and toughening of brittle polymers.

The approach used, and the design and structure of the book are the same as for the first edition. Thus a modern treatment has been used for presentation of much of the subject matter, and the book seeks to fuse together aspects of the chemistry, structure and mechanical properties of polymers, thereby introducing important relationships between synthesis, structure, and molecular and bulk properties. The book is designed principally for undergraduate and postgraduate students who are studying polymers, but also should be of use to scientists in industry and research who need to become familiar with the fundamentals of Polymer Science. It has been written to be, as far as is possible, self-contained with most equations fully derived and critically discussed. Nevertheless, lists of books are given at the end of each chapter for background and further reading. Together with the problems which have been included, they will enable the reader to reinforce, extend and test his or her knowledge and understanding of specific subjects.

In addition to the people and organizations who assisted in the preparation of the First Edition, the authors would like to thank Mrs Susan Brandreth and Mrs Jean Smith for typing the new manuscript. They are also grateful to Dr Frank Heatley, Dr Tony Ryan, Dr John Stanford and Dr Bob Stepto for useful comments on aspects of the new material. Finally, they would like to express their sincere gratitude to their families for the understanding and support they have shown during the writing and preparation of the new edition.

ROBERT J. YOUNG
PETER A. LOVELL

Manchester Materials Science Centre
1990

Preface to the first edition

Polymers are a group of materials made up of long covalently-bonded molecules, which include plastics and rubbers. The use of polymeric materials is increasing rapidly year by year and in many applications they are replacing conventional materials such as metals, wood and natural fibres such as cotton and wool. The book is designed principally for undergraduate and postgraduate students of Chemistry, Physics, Materials Science and Engineering who are studying polymers. An increasing number of graduates in these disciplines go on to work in polymer-based industries, often with little grounding in Polymer Science and so the book should also be of use to scientists in industry and research who need to learn about the subject.

A basic knowledge of mathematics, chemistry and physics is assumed although it has been written to be, as far as is possible, self-contained with most equations fully derived and any assumptions stated. Previous books in this field have tended to be concerned primarily with either polymer chemistry, polymer structure or mechanical properties. An attempt has been made with this book to fuse together these different aspects into one volume so that the reader has these different areas included in one book and so can appreciate the relationships that exist between the different aspects of the subject. Problems have also been given at the end of each chapter so that the reader may be able to test his or her understanding of the subject and practise the manipulation of data.

The textbook approaches the subject of polymers from a Materials Science viewpoint, being principally concerned with the relationship between structure and properties. In order to keep it down to a manageable size there have been important and deliberate omissions. Two obvious areas are those of polymer processing (e.g. moulding and fabrication) and electrical properties. These are vast areas in their own right and it is hoped that this book will give the reader sufficient grounding to go on and study these topics elsewhere.

Several aspects of the subject of polymer science have been updated compared with the normal presentation in books at this level. For example, the mechanical properties of polymers are treated from a mechanistic viewpoint rather than in terms of viscoelasticity, reflecting modern developments in the subject. However, viscoelasticity being an important aspect of polymer properties is also covered but with rather less emphasis than it has been given in the past. The presentation of some theories and experimental results has been changed from the original approach for the sake of clarity and consistency of style.

I am grateful to Professor Bill Bonfield for originally suggesting the book and for his encouragement throughout the project. I am also grateful to my other colleagues at Queen Mary College for allowing me to use some of their material and problems and to many people in the field of Polymers who have contributed micrographs. A large part of the book was written during a period of study leave at the University of the Saarland in West Germany. I would like to thank the Alexander von Humboldt Stiftung for financial support during this period. The bulk of the manuscript was typed by Mrs Rosalie Hillman and I would like to thank her for her help. Finally, my gratitude must go to my wife and family for giving me their support during the preparation of the book.

ROBERT J. YOUNG

Queen Mary College, London
1980

1 Introduction

1.1 The origins of polymer science and the polymer industry

Polymers have existed in natural form since life began and those such as DNA, RNA, proteins and polysaccharides play crucial roles in plant and animal life. From the earliest times, man has exploited naturally-occurring polymers as materials for providing clothing, decoration, shelter, tools, weapons, writing materials and other requirements. However, the origins of today's polymer industry commonly are accepted as being in the nineteenth century when important discoveries were made concerning the modification of certain natural polymers.

In 1820 Thomas Hancock discovered that when masticated (i.e. subjected repeatedly to high shear forces), natural rubber becomes more fluid making it easier to blend with additives and to mould. Some years later, in 1839, Charles Goodyear found that the elastic properties of natural rubber could be improved, and its tackiness eliminated, by heating with sulphur. Patents for this discovery were issued in 1844 to Goodyear, and slightly earlier to Hancock, who christened the process vulcanization. In 1851 Nelson Goodyear, Charles' brother, patented the vulcanization of natural rubber with large amounts of sulphur to produce a hard material more commonly known as hard rubber, ebonite or vulcanite.

Cellulose nitrate, also called nitrocellulose or gun cotton, first became prominent after Christian Schönbein prepared it in 1846. He was quick to recognize the commercial value of this material as an explosive, and within a year gun cotton was being manufactured. However, more important to the rise of the polymer industry, cellulose nitrate was found to be a hard elastic material which was soluble and could be moulded into different shapes by the application of heat and pressure. Alexander Parkes was the first to take advantage of this combination of properties and in 1862 he exhibited articles made from Parkesine, a form of plasticized cellulose nitrate. In 1870 John and Isaiah Hyatt patented a similar but more easily processed material, named celluloid, which was prepared using camphor as the plasticizer. Unlike Parkesine, celluloid was a great commercial success.

In 1892 Charles Cross, Edward Bevan and Clayton Beadle patented the 'viscose process' for dissolving and then regenerating cellulose. The process was first used to produce viscose rayon textile fibres, and subsequently for production of cellophane film.

The polymeric materials described so far are semi-synthetic since they are produced from natural polymers. Leo Baekeland's Bakelite phenol–formaldehyde resins have the distinction of being the first fully-synthetic polymers to be commercialized, their production beginning in 1910. The first synthetic rubber to be manufactured, known as methyl rubber, was produced from 2,3-dimethylbutadiene in Germany during World War I as a substitute, albeit a poor one, for natural rubber.

Although the polymer industry was now firmly established, its growth was restricted by the considerable lack of understanding of the nature of polymers. For over a century scientists had been reporting the unusual properties of polymers, and by 1920 the common belief was that they consisted of physically-associated aggregates of small molecules. Few scientists gave credence to the viewpoint so passionately believed by Hermann Staudinger, that polymers were composed of very large molecules containing long sequences of simple chemical units linked together by covalent bonds. Staudinger introduced the word 'macro-molecule' to describe polymers, and during the 1920s vigorously set about proving his hypothesis to be correct. Particularly important were his studies of the synthesis, structure and properties of polyoxymethylene and of polystyrene, the results from which left little doubt as to the validity of the macromolecular viewpoint. Staudinger's hypothesis was further sub-stantiated by the crystallographic studies of natural polymers reported by Herman Mark and Kurt Meyer, and by the classic work of Wallace Carothers on the preparation of polyamides and polyesters. Thus by the early 1930s most scientists were convinced of the macromolecular structure of polymers. During the following 20 years, work on polymers increased enormously: the first journals devoted solely to their study were published and most of the fundamental principles of *Polymer Science* were established. The theoretical and experimental work of Paul Flory was prominent in this period, and for his long and substantial contribution to Polymer Science he was awarded the Nobel Prize for Chemistry in 1974. In 1953 Staudinger had received the same accolade in recognition of his pioneering work.

Not surprisingly, as the science of macromolecules emerged, a large number of synthetic polymers went into commercial production for the first time. These include polystyrene, poly(methyl methacrylate), nylon 6.6, polyethylene, poly(vinyl chloride), styrene–butadiene rubber, sili-cones and polytetrafluoroethylene, as well as many others. From the 1950s onwards regular advances, too numerous to mention here, have continued to stimulate both scientific and industrial progress.

Whilst Polymer Science is now considered to be a mature subject, its breadth is ever increasing and there are many demanding challenges awaiting scientists who venture into this fascinating multidisciplinary science.

1.2 Basic definitions and nomenclature

Several important terms and concepts must be understood in order to discuss fully the synthesis, characterization, structure and properties of polymers. Most of these will be defined and discussed in detail in subsequent chapters. However, some are of such fundamental importance that they must be defined at the outset.

In strict terms, a *polymer* is a *substance* composed of molecules which have long sequences of one or more species of atoms or groups of atoms linked to each other by primary, usually covalent, bonds. The emphasis upon substance in this definition is to highlight that although the words polymer and *macromolecule* are used interchangeably, the latter strictly defines the molecules of which the former is composed.

Macromolecules are formed by linking together *monomer* molecules through chemical reactions, the process by which this is achieved being known as *polymerization*. For example, polymerization of ethylene yields polyethylene, a typical sample of which may contain molecules with 50 000 carbon atoms linked together in a chain. It is this long chain nature which sets polymers apart from other materials and gives rise to their characteristic properties.

1.2.1 *Skeletal structure*

The definition of macromolecules presented up to this point implies that they have a *linear* skeletal structure which may be represented by a chain with two ends. Whilst this is true for many macromolecules, there are also many with *non-linear* skeletal structures of the type shown in Fig. 1.1.

Branched polymers have side chains, or *branches*, of significant length which are bonded to the main chain at *branch points* (also known as

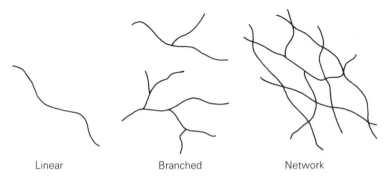

Linear   Branched   Network

Fig 1.1 *Representive skeletal structures of linear and non-linear polymers.*

junction points), and are characterized in terms of the number and size of the branches. *Network polymers* have three-dimensional structures in which each chain is connected to all others by a sequence of junction points and other chains. Such polymers are said to be *crosslinked* and are characterized by their *crosslink density*, or *degree of crosslinking*, which is related directly to the number of junction points per unit volume.

Non-linear polymers may be formed by polymerization, or can be prepared by linking together (i.e. *crosslinking*) pre-existing chains.

Variations in skeletal structure give rise to major differences in properties. For example, linear polyethylene has a melting point about 20°C higher than that of branched polyethylene. Unlike linear and branched polymers, network polymers do not melt upon heating and will not dissolve, though they may swell considerably in compatible solvents. The importance of crosslink density has already been encountered in terms of the vulcanization (i.e. sulphur-crosslinking) of natural rubber. With low crosslink densities (i.e. low levels of sulphur) the product is a flexible elastomer, whereas it is a rigid material when the crosslink density is high.

1.2.2 *Homopolymers*

The formal definition of a homopolymer is a polymer derived from one species of monomer. However, the word *homopolymer* often is used more broadly to describe polymers whose structure can be represented by multiple repetition of a single type of *repeat unit* which may contain one or more species of *monomer unit*. The latter is sometimes referred to as a structural unit.

The chemical structure of a polymer usually is represented by that of the repeat unit enclosed by brackets. Thus the hypothetical homopolymer $\sim$A—A—A—A—A—A—A—A$\sim$ is represented by $+$A$+_n$ where n is the number of repeat units linked together to form the macromolecule. Table 1.1 shows the chemical structures of some common homopolymers together with the monomers from which they are derived and some comments upon their properties and uses. It should be evident that slight differences in chemical structure can lead to very significant differences in properties.

The naming of polymers or envisaging the chemical structure of a polymer from its name is often an area of difficulty. At least in part this is because most polymers have more than one correct name, the situation being further complicated by the variety of trade-names which also are used to describe certain polymers. The approach adopted here is to use names which most clearly and simply indicate the chemical structures of the polymers under discussion.

The names given to the polymers in Table 1.1 exemplify elementary

TABLE 1.1 *Some common homopolymers*

Monomers	Polymer	Comments
(1) Ethylene $CH_2=CH_2$	Polyethylene (PE) $+CH_2-CH_2+_n$	Moulded objects, tubing, film, electrical insulation, e.g. 'Alkathene', 'Lupolen'.
(2) Propylene $CH_2=CH$ $\quad\quad\mid$ $\quad\quad CH_3$	Polypropylene (PP) $+CH_2-CH+_n$ $\quad\quad\quad\mid$ $\quad\quad\quad CH_3$	Similar uses to PE; lower density, stiffer, e.g. 'Propathene', 'Novolen'.
(3) Tetrafluoroethylene $CF_2=CF_2$	Polytetrafluoroethylene (PTFE) $+CF_2-CF_2+_n$	Mouldings, film, coatings; high temperature resistance, chemically inert, excellent electrical insulator, very low coefficient of friction; expensive, e.g. 'Teflon', 'Fluon'.
(4) Styrene $CH_2=CH$	Polystyrene (PS) $+CH_2-CH+_n$	Cheap moulded objects, e.g. 'Styron', 'Hostyren'. Modified with rubbers to improve toughness, e.g. high-impact polystyrene (HIPS) and acrylonitrile–butadiene–styrene copolymer (ABS). Expanded by volatilization of a blended blowing agent (e.g. pentane) to produce polystyrene foam.

TABLE 1.1 *continued*

Monomers	Polymer	Comments
(5) Methyl methacrylate $$CH_2{=}C\begin{array}{c}CH_3\\|\\|\\C{=}O\\|\\OCH_3\end{array}$$	Poly(methyl methacrylate) (PMMA) $$\left(CH_2{-}C\begin{array}{c}CH_3\\|\\|\\C{=}O\\|\\OCH_3\end{array}\right)_{\!\!n}$$	Transparent sheets and mouldings; used for aeroplane windows; more expensive than PS, e.g. 'Perspex', 'Diakon', 'Lucite', 'Oroglass', 'Plexiglas'.
(6) Vinyl chloride $$CH_2{=}CH\atop{\textstyle|\atop\textstyle Cl}$$	Poly(vinyl chloride) (PVC) $$\left(CH_2{-}CH\atop{\textstyle|\atop\textstyle Cl}\right)_{\!\!n}$$	Water pipes and gutters, bottles, gramophone records; plasticized to make PVC leathercloth, raincoats, flexible pipe and hose, toys, sheathing on electrical cables, e.g. 'Darvic', 'Welvic', 'Vinoflex', 'Hostalit'.
(7) Vinyl acetate $$CH_2{=}CH\begin{array}{c}\\|\\O\\|\\C{=}O\\|\\CH_3\end{array}$$	Poly(vinyl acetate) (PVA) $$\left(CH_2{-}CH\begin{array}{c}\\|\\O\\|\\C{=}O\\|\\CH_3\end{array}\right)_{\!\!n}$$	Surface coatings, adhesives, chewing gum.

TABLE 1.1 *continued*

Monomers	Polymer	Comments
(8) Ethylene glycol HO—CH₂—CH₂—OH	Poly(ethylene glycol) (PEG) $\left[CH_2-CH_2-O\right]_n$	Water-soluble packaging films, textile sizes, thickeners, e.g. 'Carbowax'.
(9) Ethylene glycol HO—CH₂—CH₂—OH and terephthalic acid HO—C(=O)—⬡—C(=O)—OH	Poly(ethylene terephthalate) (PET)* $\left[O-CH_2-CH_2-O-\underset{\parallel}{\overset{O}{C}}-⬡-\underset{\parallel}{\overset{O}{C}}\right]_n$	Textile fibres, film, bottles, e.g. 'Terylene', 'Dacron', 'Melinex', 'Mylar'.
(10) Hexamethylene diamine H₂N—(CH₂)₆—NH₂ and sebacic acid HO—C(=O)—(CH₂)₈—C(=O)—OH	Poly(hexamethylene sebacate) (nylon 6.10)* $\left[\underset{\underset{6\text{ carbons}}{\underbrace{}}}{N-(CH_2)_6-}\underset{\underset{10\text{ carbons}}{\underbrace{}}}{N-C-(CH_2)_8-C}\right]_n$ with H on each N, and O on each C	Mouldings, fibres, e.g. 'Ultramid 6.10'.

* The polymer has two monomer units in the repeat unit.

aspects of nomenclature. Thus source-based nomenclature places the prefix 'poly' before the name of the monomer, the monomer's name being contained within parentheses unless it is a simple single word. In structure-based nomenclature the prefix poly is followed in parentheses by words which describe the chemical structure of the repeat unit. This type of nomenclature is used for polymers nine and ten in Table 1.1.

1.2.3 *Copolymers*

The formal definition of a *copolymer* is a polymer derived from more than one species of monomer. However, in accordance with use of the word homopolymer, it is common practice to use a structure-based definition. Thus the word copolymer more commonly is used to describe polymers whose molecules contain two or more different types of repeat unit. Hence polymers nine and ten in Table 1.1 usually are considered to be homopolymers rather than copolymers.

There are several categories of copolymer, each being characterized by a particular form of arrangement of the repeat units along the polymer chain. For simplicity, the representation of these categories will be illustrated by copolymers containing only two different types of repeat unit (A and B).

Statistical copolymers are copolymers in which the sequential distribution of the repeat units obeys known statistical laws (e.g. Markovian). *Random copolymers* are a special type of statistical copolymer in which the distribution of repeat units is truly random (some words of caution are necessary here because older textbooks and scientific papers often use the term random copolymer to describe both random and non-random statistical copolymers). A section of a truly random copolymer is represented below

$\sim$ B—B—B—A—B—B—A—B—A—A $\sim$

Alternating copolymers have only two different types of repeat unit and these are arranged alternately along the polymer chain

$\sim$ A—B—A—B—A—B—A—B—A—B $\sim$

Statistical, random and alternating copolymers generally have properties which are intermediate to those of the corresponding homopolymers. Thus by preparing such copolymers it is possible to combine the desirable properties of the homopolymers into a single material. This is not normally possible by blending because most homopolymers are immiscible with each other.

Block copolymers are linear copolymers in which the repeat units exist only in long sequences, or *blocks*, of the same type. Two common block copolymer structures are represented below and usually are termed AB di-block and ABA tri-block copolymers

A—A—A—A—A—A—A—A—A—A—B—B—B—B—B—B—B—B—B—B

A—A—A—A—A—A—A—A—B—B—B—B—B—B—B—A—A—A—A—A—A—A—A

Graft copolymers are branched polymers in which the branches have a different chemical structure to that of the main chain. In their simplest form they consist of a main homopolymer chain with branches of a different homopolymer

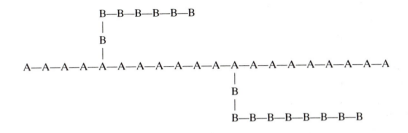

In distinct contrast to the types of copolymer described earlier, block and graft copolymers usually show properties characteristic of each of the constituent homopolymers. They also have some unique properties arising from the chemical linkage(s) between the homopolymer sequences preventing them from acting entirely independently of each other.

The current principles of nomenclature for copolymers are indicated in Table 1.2 where A and B represent source- or structure-based names for these repeat units. Thus a statistical copolymer of ethylene and propylene

TABLE 1.2 *Principles of nomenclature for copolymers*

Type of copolymer	Example of nomenclature
Unspecified	Poly(A-*co*-B)
Statistical	Poly(A-*stat*-B)
Random	Poly(A-*ran*-B)
Alternating	Poly(A-*alt*-B)
Block	PolyA-*block*-polyB
Graft*	PolyA-*graft*-polyB

* The example is for polyB branches on a polyA main chain.

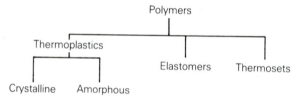

Fig. 1.2 *Classification of polymers.*

is named poly(ethylene-*stat*-propylene), and an ABA tri-block copolymer of styrene (A) and isoprene (B) is named polystyrene-*block*-polyisoprene-*block*-polystyrene. In certain cases, additional square brackets are required. For example, an alternating copolymer of styrene and maleic anhydride is named poly[styrene-*alt*-(maleic anhydride)].

1.2.4 *Classification of polymers*

The most common way of classifying polymers is outlined in Fig. 1.2 where they are first separated into three groups: *thermoplastics*, *elastomers* and *thermosets*. Thermoplastics are then further separated into those which are crystalline and those which are amorphous (i.e. non-crystalline). This method of classification has an advantage in comparison to others since it is based essentially upon the underlying molecular structure of the polymers.

Thermoplastics, often referred to just as plastics, are linear or branched polymers which can be melted upon the application of heat. They can be moulded (and remoulded) into virtually any shape using processing techniques such as injection moulding and extrusion, and now constitute by far the largest proportion of the polymers used in industry. Generally, thermoplastics do not crystallize easily upon cooling to the solid state because this requires considerable ordering of the highly coiled and entangled macromolecules present in the liquid state. Those which do crystallize invariably do not form perfectly crystalline materials but instead are *semi-crystalline* with both crystalline and amorphous regions. The crystalline phases of such polymers are characterized by their *melting temperature* (T_m). Many thermoplastics are, however, completely amorphous and incapable of crystallization, even upon annealing. Amorphous polymers (and amorphous phases of semi-crystalline polymers) are characterized by their *glass transition temperature* (T_g), the temperature at which they transform abruptly from the *glassy state* (hard) to the *rubbery state* (soft). This transition corresponds to the onset of chain motion; below T_g the polymer chains are unable to move and are 'frozen' in position. Both T_m and T_g increase with increasing chain stiffness and increasing forces of intermolecular attraction.

Elastomers are crosslinked rubbery polymers (i.e. rubbery networks) that can be stretched easily to high extensions (e.g. 3× to 10× their original dimensions) and which rapidly recover their original dimensions when the applied stress is released. This extremely important and useful property is a reflection of their molecular structure in which the network is of low crosslink density. The rubbery polymer chains become extended upon deformation but are prevented from permanent flow by the cross-links, and driven by entropy, spring back to their original positions on removal of the stress. The word rubber, often used in place of elastomer, preferably should be used for describing rubbery polymers which are not crosslinked.

Thermosets normally are rigid materials and are network polymers in which chain motion is greatly restricted by a high degree of crosslinking. As for elastomers, they are intractable once formed and degrade rather than melt upon the application of heat.

1.3 Molar mass and degree of polymerization

Many properties of polymers show a strong dependence upon the size of the polymer chains, so that it is essential to characterize their dimensions. This normally is done by measuring the *molar mass* (M) of a polymer which is simply the mass of 1 mole of the polymer and usually is quoted in units of $g\,mol^{-1}$ or $kg\,mol^{-1}$. The term 'molecular weight' is still often used instead of molar mass, but is not preferred because it can be somewhat misleading. It is really a dimensionless quantity, the relative molecular mass, rather than the weight of an individual molecule which is of course a very small quantity (e.g. $\sim 10^{-19} - \sim 10^{-18}\,g$ for most polymers). By multiplying the numerical value of molecular weight by the specific units $g\,mol^{-1}$ it can be converted into the equivalent value of molar mass. For example, a molecular weight of 100 000 is equivalent to a molar mass of $100\,000\,g\,mol^{-1}$ which in turn is equivalent to a molar mass of $100\,kg\,mol^{-1}$.

For network polymers the only meaningful molar mass is that of the polymer chains existing between junction points (i.e. *network chains*), since the molar mass of the network itself essentially is infinite.

The molar mass of a homopolymer is related to the *degree of polymerization* (x), which is the number of repeat units in the polymer chain, by the simple relation

$$M = xM_0 \tag{1.1}$$

where M_0 is the molar mass of the repeat unit. For copolymers the sum of the products xM_0 for each type of repeat unit is required to define the molar mass.

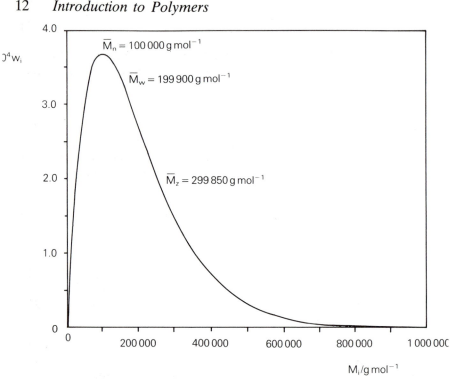

Fig. 1.3 *A typical molar mass distribution curve.*

1.3.1 *Molar mass distribution*

With very few exceptions, polymers consist of macromolecules (or network chains) with a range of molar masses. Since the molar mass changes in intervals of M_0, the distribution of molar mass is discontinuous. However, for most polymers these intervals are extremely small in comparison to the total range of molar mass and the distribution can be assumed to be continuous, as exemplified in Fig. 1.3.

1.3.2 *Molar mass averages*

Whilst a knowledge of the complete molar mass distribution is essential in many uses of polymers, it is convenient to characterize the distribution in terms of molar mass averages. These usually are defined by considering the discontinuous nature of the distribution in which the macromolecules exist in discrete fractions i containing N_i molecules of molar mass M_i.

The *number-average molar mass* $(\overline{M}_n)$ is defined as 'the sum of the products of the molar mass of each fraction multiplied by its mole fraction' i.e. $\quad \overline{M}_n = \sum X_i M_i$ (1.2)

where X_i is the mole fraction of molecules of molar mass M_i and is given by the ratio of N_i to the total number of molecules. Therefore it follows that

$$\overline{M}_n = \sum N_i M_i \bigg/ \sum N_i$$ (1.3)

showing this average to be the arithmetic mean of the molar mass distribution. It is often more convenient to use weight fractions rather than numbers of molecules. The weight fraction w_i is defined as the mass of molecules of molar mass M_i divided by the total mass of all the molecules present

i.e. $\quad w_i = N_i M_i \bigg/ \sum N_i M_i$ (1.4)

from which it can be deduced that

$$\sum (w_i/M_i) = \sum N_i \bigg/ \sum N_i M_i$$ (1.5)

Combining Equations (1.3) and (1.5) gives $\overline{M}_n$ in terms of weight fractions

$$\overline{M}_n = 1 \bigg/ \sum (w_i/M_i)$$ (1.6)

The *weight-average molar mass* $(\overline{M}_w)$ is defined as 'the sum of the products of the molar mass of each fraction multiplied by its weight fraction'

i.e. $\quad \overline{M}_w = \sum w_i M_i$ (1.7)

By combining this equation with Equation (1.4) $\overline{M}_w$ can be expressed in terms of the numbers of molecules

$$\overline{M}_w = \sum N_i M_i^2 \bigg/ \sum N_i M_i$$ (1.8)

The ratio $\overline{M}_w/\overline{M}_n$ must by definition be greater than unity for a *polydisperse* polymer and is known as the *polydispersity* or *heterogeneity index*. Its value often is used as a measure of the breadth of the molar mass distribution, though it is a poor substitute for knowledge of the complete distribution curve. Typically $\overline{M}_w/\overline{M}_n$ is in the range 1.5–2.0, though there are many polymers which have smaller or very much larger values of polydispersity index. A perfectly *monodisperse* polymer would have $\overline{M}_w/\overline{M}_n = 1.00$.

Higher molar mass averages sometimes are quoted. For example, certain methods of molar mass measurement (e.g. sedimentation equilibrium) yield the *z*-average molar mass $(\overline{M}_z)$ which is defined as follows

$$\overline{M}_z = \sum N_i M_i^3 \bigg/ \sum N_i M_i^2 = \sum w_i M_i^2 \bigg/ \sum w_i M_i \tag{1.9}$$

In addition, more complex exponent averages can be obtained (e.g. by dilute solution viscometry and sedimentation measurements).

Degree of polymerization averages are of more importance than molar mass averages in the theoretical treatment of polymers and polymerization, as will be highlighed in the subsequent chapters. For homopolymers they may be obtained simply by dividing the corresponding molar mass average by M_0. Thus the *number-average* and *weight-average degrees of polymerization* are given by

$$\overline{x}_n = \overline{M}_n / M_0 \tag{1.10}$$

and $\quad \overline{x}_w = \overline{M}_w / M_0 \tag{1.11}$

Further reading

Billmeyer, F.W. (1984), *Textbook of Polymer Science*, 3rd edn, Wiley-Interscience, New York.

Cowie, J.M.G. (1973), *Polymers: Chemistry and Physics of Modern Materials*, International Textbook Company, Aylesbury, UK.

Elias, H-G. (1987), *Mega Molecules*, Springer-Verlag, Berlin.

ICI Plastics Division (1962), *Landmarks of the Plastics Industry*, Kynoch Press, Birmingham.

Jenkins, A.D. and Loening, K.L. (1989), 'Nomenclature' in *Comprehensive Polymer Science*, Vol. 1 (ed. C. Booth and C. Price), Pergamon Press, Oxford.

Kaufman, M. (1963), *The First Century of Plastics — Celluloid and its Sequel*, The Plastics Institute, London.

Mandelkern, L. (1983), *An Introduction to Macromolecules*, 2nd edn, Springer-Verlag, New York.

Mark, H.F. (1970), *Giant Molecules*, Time-Life Books, New York.

Morawetz, H. (1985), *Polymers — The Origins and Growth of a Science*, John Wiley, New York.

Treloar, L.R.G. (1970), *Introduction to Polymer Science*, Wykeham Publications, London.

2 Synthesis

2.1 Classification of polymerization reactions

The most basic requirement for polymerization is that each molecule of monomer must be capable of being linked to two (or more) other molecules of monomer by chemical reaction, i.e. monomers must have a *functionality* of two (or higher). Given this relatively simple requirement, there are a multitude of chemical reactions and associated monomer types that can be used to effect polymerization. To discuss each of these individually would be a major task which fortunately is not necessary since it is possible to place most polymerization reactions in one of two classes, each having distinctive characteristics.

The classification used in the formative years of polymer science was due to Carothers and is based upon comparison of the molecular formula of a polymer with that of the monomer(s) from which it was formed. *Condensation polymerizations* are those which yield polymers with repeat units having fewer atoms than present in the monomers from which they are formed. This usually arises from chemical reactions which involve the elimination of a small molecule (e.g. H_2O, HCl). *Addition polymerizations* are those which yield polymers with repeat units having identical molecular formulae to those of the monomers from which they are formed. Table 1.1 (Section 1.2.2) contains examples of each class: the latter three examples are condensation polymerizations involving elimination of H_2O, whereas the others are addition polymerizations.

Carothers' method of classification was found to be unsatisfactory when it was recognized that certain condensation polymerizations have the characteristic features of typical addition polymerizations and that some addition polymerizations have features characteristic of typical condensation polymerizations. A better basis for classification is provided by considering the underlying polymerization mechanisms, of which there are two general types. Polymerizations in which the polymer chains grow step-wise by reactions that can occur between any two molecular species are known as *step-growth* polymerizations. Polymerizations in which a polymer chain grows only by reaction of monomer with a reactive end-group on the growing chain are known as *chain-growth* polymerizations, and usually require an initial reaction between the monomer and an *initiator* to start the growth of the chain. There has been a tendency in recent years to change these names to *step polymerization* and *chain polymerization*, and this practice will be used here. The essential

TABLE 2.1 *A schematic illustration of the fundamental differences in reaction mechanism between step polymerization and chain polymerization* *

Formation of	Step polymerization	Chain polymerization
Dimer	o + o → o–o	I + o → I–o I–o + o → I–o–o
Trimer	o–o + o → o–o–o	I–o–o + o → I–o–o–o
Tetramer	o–o–o + o → o–o–o–o o–o + o–o → o–o–o–o	I–o–o–o + o → I–o–o–o–o
Pentamer	o–o–o–o + o → o–o–o–o–o o–o + o–o–o → o–o–o–o–o	I–o–o–o–o + o → I–o–o–o–o–o
Hexamer	o–o–o–o–o + o → o–o–o–o–o–o o–o + o–o–o–o → o–o–o–o–o–o o–o–o + o–o–o → o–o–o–o–o–o	I–o–o–o–o–o + o → I–o–o–o–o–o–o
Heptamer	o–o–o–o–o–o + o → o–o–o–o–o–o–o o–o + o–o–o–o–o → o–o–o–o–o–o–o o–o–o + o–o–o–o → o–o–o–o–o–o–o	I–o–o–o–o–o–o + o → I–o–o–o–o–o–o–o
Octomer	o–o–o–o–o–o–o + o → o–o–o–o–o–o–o–o o–o + o–o–o–o–o–o → o–o–o–o–o–o–o–o o–o–o + o–o–o–o–o → o–o–o–o–o–o–o–o o–o–o–o + o–o–o–o → o–o–o–o–o–o–o–o	I–o–o–o–o–o–o–o + o → I–o–o–o–o–o–o–o–o

* Definition of symbols used : o, molecule of monomer; –, chemical link; I, initiator species

differences between these classes of polymerization are highlighted in Table 2.1 which illustrates for each mechanism the reactions involved in growth of the polymer chains to a degree of polymerization equal to eight. In step polymerizations the degree of polymerization increases steadily throughout the reaction, but the monomer is rapidly consumed in its early stages (e.g. when $\bar{x}_n = 10$ less than 1% of the monomer remains unreacted). By contrast, in chain polymerizations high degrees of polymerization are attained at low monomer conversions, the monomer being consumed steadily throughout the reaction.

Step Polymerization

2.2 **Linear step polymerization**

Step polymerizations involve successive reactions between pairs of mutually-reactive functional groups which initially are provided by the monomer(s). The number of functional groups present on a molecule of monomer (i.e. its functionality) is of crucial importance, as can be appreciated by considering the formation of ester linkages from the condensation reaction of carboxylic acid groups with hydroxyl groups. Acetic acid and ethyl alcohol are *monofunctional* compounds which upon reaction together yield ethyl acetate with elimination of water

$$CH_3COOH + CH_3CH_2OH \rightarrow CH_3COOCH_2CH_3 + H_2O$$

but because ethyl acetate is incapable of further reaction a polymer chain cannot form. Now consider the reaction between terephthalic acid and ethylene glycol, both of which are *difunctional*

$$HOOC-\langle O \rangle-COOH + HOCH_2CH_2OH \rightarrow$$

$$HOOC-\langle O \rangle-COOCH_2CH_2OH + H_2O$$

The product of their reaction is an ester which possesses one carboxylic acid end-group and one hydroxyl end-group (i.e. it also is difunctional). This *dimer*, therefore, can react with other molecules of terephthalic acid, ethlyene glycol or dimer leading to the formation of *difunctional trimers* or *difunctional tetramer*. Growth of linear polymer chains then proceeds via further condensation reactions in the manner indicated for step polymerization in Table 2.1. Hence *linear step polymerizations* involve reactions of difunctional monomers. If a *trifunctional* monomer were included, reaction at each of the three functional groups would lead initially to the formation of a branched polymer but ultimately to the formation of a network. For example, if terephthalic acid were reacted with glycerol, $HOCH_2CH(OH)CH_2OH$, the product would be a non-linear polyester. It follows that polymerizations involving monomers of functionality greater than two will produce non-linear polymers.

2.2.1 *Polycondensation*

Step polymerizations that involve reactions in which small molecules are eliminated are termed *polycondensations*. The formation of linear *polyesters*, as described in the previous section, is typical of these reactions and may be represented more generally by

$$nHOOC—R_1—COOH + nHO—R_2—OH \rightarrow$$
$$H\text{-}\!\!\{OOC—R_1—COO—R_2\}_n OH + (2n-1)H_2O$$

where R_1 and R_2 represent any divalent group (usually hydrocarbon). Such reactions often are referred to as $RA_2 + RB_2$ step polymerizations where R is any divalent group and A and B represent the mutually-reactive functional groups.

Polyesters may also be prepared from single monomers which contain both types of functional group, e.g. ω-hydroxy carboxylic acids

$$nHO—R—COOH \rightarrow HO\text{-}\!\!\{R—COO\}_n H + (n-1)H_2O$$

With each condensation reaction the polymer chain grows but remains an ω-hydroxy carboxylic acid and so can react further. This is an example of an ARB step polymerization. The use of monomers of this type has the advantage that, provided they are pure, an exact stoichiometric equivalence of the two functional groups is guaranteed. Very slight excesses of one monomer in a $RA_2 + RB_2$ polymerization significantly reduce the attainable degree of polymerization because the polymer chains become terminated with functional groups derived from the monomer present in excess (e.g. both end-groups are ultimately of type B if RB_2 is in excess). Since these functional groups are unreactive towards each other, further growth of the chains is not possible.

Polyamides can be prepared by polycondensations analogous to those used to prepare polyesters, the hydroxyl groups simply being replaced by amine groups, e.g.

$$nH_2N—R_1—NH_2 + nHOOC—R_2—COOH \rightarrow$$
$$H\text{-}\!\!\{NH—R_1—NHOC—R_2—CO\}_n OH + (2n-1)H_2O$$
$$nH_2N—R—CO_2H \rightarrow H\text{-}\!\!\{NH—R—CO\}_n OH + (n-1)H_2O$$

The formation of *polyethers* by dehydration of diols is one of the relatively few examples of RA_2 step polymerization

$$nHO—R—OH \rightarrow H\text{-}\!\!\{O—R\}_n OH + (n-1)H_2O$$

Preparation of *siloxanes* by hydrolysis of dichlorodialkylsilanes, e.g. dichlorodimethylsilane

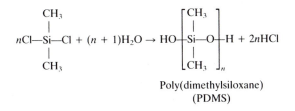

Poly(dimethylsiloxane)
(PDMS)

is unusual in that after partial hydrolysis of the monomer via

$$
n\text{Cl} \!-\! \underset{\underset{\text{CH}_3}{|}}{\overset{\overset{\text{CH}_3}{|}}{\text{Si}}} \!-\! \text{Cl} + 2\text{H}_2\text{O} \rightarrow \text{HO} \!-\! \underset{\underset{\text{CH}_3}{|}}{\overset{\overset{\text{CH}_3}{|}}{\text{Si}}} \!-\! \text{OH} + 2\text{HCl}
$$

both $RA_2 + RB_2$ polymerization

$$
\text{Cl} \!-\! \underset{\underset{\text{CH}_3}{|}}{\overset{\overset{\text{CH}_3}{|}}{\text{Si}}} \!-\! \text{Cl} + \text{HO} \!-\! \underset{\underset{\text{CH}_3}{|}}{\overset{\overset{\text{CH}_3}{|}}{\text{Si}}} \!-\! \text{OH} \rightarrow \text{Cl} \!-\! \underset{\underset{\text{CH}_3}{|}}{\overset{\overset{\text{CH}_3}{|}}{\text{Si}}} \!-\! \text{O} \!-\! \underset{\underset{\text{CH}_3}{|}}{\overset{\overset{\text{CH}_3}{|}}{\text{Si}}} \!-\! \text{OH} + \text{HCl}
$$

and RA_2 polymerization can occur

$$
\text{HO} \!-\! \underset{\underset{\text{CH}_3}{|}}{\overset{\overset{\text{CH}_3}{|}}{\text{Si}}} \!-\! \text{OH} + \text{HO} \!-\! \underset{\underset{\text{CH}_3}{|}}{\overset{\overset{\text{CH}_3}{|}}{\text{Si}}} \!-\! \text{OH} \rightarrow \text{HO} \!-\! \underset{\underset{\text{CH}_3}{|}}{\overset{\overset{\text{CH}_3}{|}}{\text{Si}}} \!-\! \text{O} \!-\! \underset{\underset{\text{CH}_3}{|}}{\overset{\overset{\text{CH}_3}{|}}{\text{Si}}} \!-\! \text{OH} + \text{H}_2\text{O}
$$

The latter reaction occurs readily and in order to control the degree of polymerization attained upon complete hydrolysis (as indicated in the general equation above), it is usual to include monofunctional chlorosilanes, e.g. chlorotrimethylsilane would lead to PDMS with unreactive end-groups

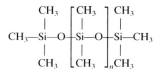

TABLE 2.2 *Some high-performance polymers prepared by polycondensation*

Polycondensation	Comments
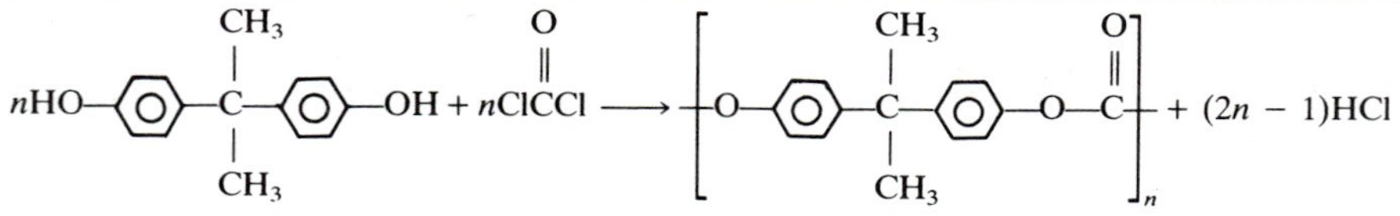 Polycarbonate (PC)	Mouldings and sheet; transparent and tough:used for safety glasses, screens and glazing, e.g. 'Lexan', 'Merlon'.
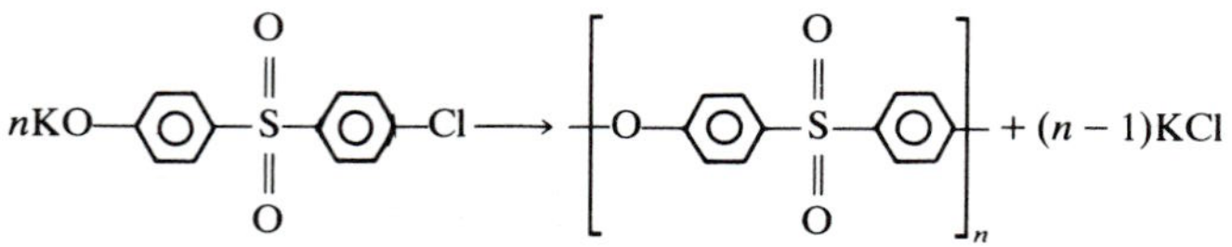 Polyethersulphone (PES)	Mouldings, coatings, membranes, e.g. 'Victrex PES'.
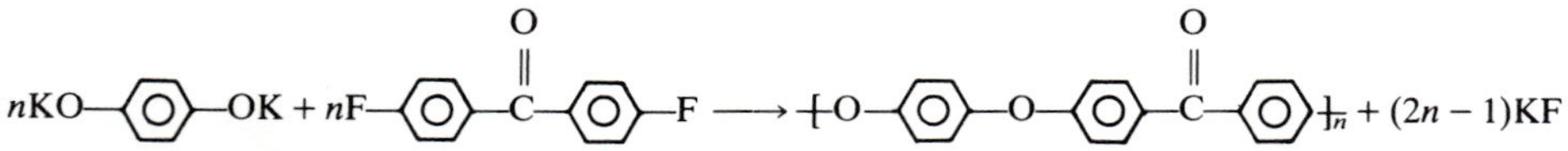 Polyetheretherketone (PEEK)	Mouldings, composites, bearings, coatings; very high continuous use temperature (260°C), e.g. 'Victrex PEEK'

$+ (2n - 1)HCl$

$+ (n - 1)KCl$

$+ (2n - 1)KF$

TABLE 2.2 continued

Polycondensation	Comments

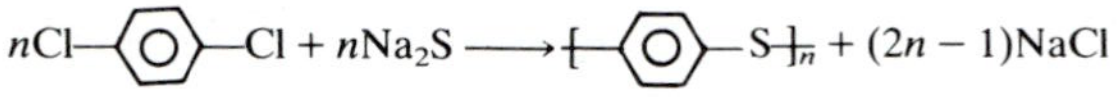

$$n\text{Cl}-\langle\bigcirc\rangle-\text{Cl} + n\text{Na}_2\text{S} \longrightarrow \left[\langle\bigcirc\rangle-\text{S}\right]_n + (2n-1)\text{NaCl}$$

Poly(phenylene sulphide) (PPS)

Mouldings, composites, coatings, e.g. 'Ryton', 'Tedur', 'Fortron'.

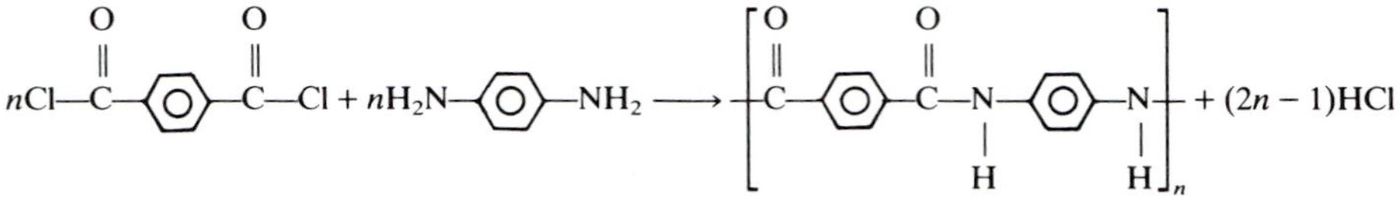

$$n\text{Cl}-\overset{\text{O}}{\underset{}{\text{C}}}-\langle\bigcirc\rangle-\overset{\text{O}}{\underset{}{\text{C}}}-\text{Cl} + n\text{H}_2\text{N}-\langle\bigcirc\rangle-\text{NH}_2 \longrightarrow \left[\overset{\text{O}}{\underset{}{\text{C}}}-\langle\bigcirc\rangle-\overset{\text{O}}{\underset{}{\text{C}}}-\underset{\text{H}}{\text{N}}-\langle\bigcirc\rangle-\underset{\text{H}}{\text{N}}\right]_n + (2n-1)\text{HCl}$$

Poly(p-phenylene terephthalamide)

High modulus fibres, e.g. 'Kevlar', 'Twaron'.

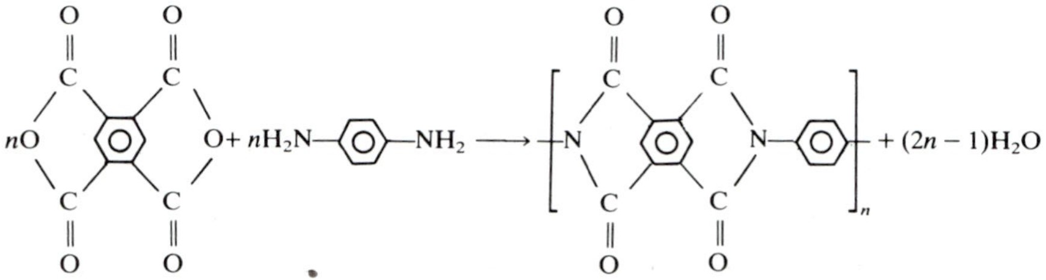

A polyimide

Films, coatings, adhesives, laminates, e.g. 'Kapton', 'Vespel'.

Many high-performance polymers are prepared by polycondensation and some specific examples are shown in Table 2.2.

2.2.2 Polyaddition

Step polymerizations in which the monomers react together without the elimination of other molecules are termed *polyadditions*. The preparation of *polyurethanes* by the $RA_2 + RB_2$ reaction of diisocyanates with diols is one of relatively few important examples of linear polyaddition

$$n O{=}C{=}N{-}R_1{-}N{=}C{=}O + n HO{-}R_2{-}OH \rightarrow$$

$$\left[\begin{array}{cc} O & O \\ \| & \| \\ {-}C{-}NH{-}R_1{-}NH{-}C{-}O{-}R_2{-}O{-} \end{array} \right]_n$$

The analogous reaction of diisocyanates with diamines yields *polyureas*

$$n O{=}C{=}N{-}R_1{-}N{=}C{=}O + n H_2N{-}R_2{-}RN_2 \rightarrow$$

$$\left[\begin{array}{cc} O & O \\ \| & \| \\ {-}C{-}NH{-}R_1{-}NH{-}C{-}NH{-}R_2{-}NH{-} \end{array} \right]_n$$

Diels–Alder reactions also have been applied to the preparation of polymers by polyaddition. A simple example is the self-reaction of cyclopentadiene

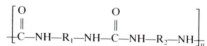

2.2.3 The principle of equal reactivity of functional groups

Chemical reactions proceed as a consequence of collisions during an encounter between mutually-reactive functional groups. At each encounter the functional groups collide repeatedly until they either diffuse apart or, far more rarely, react. Under normal circumstances, the reactivity of a functional group depends upon its collision frequency and not upon the collision frequency of the molecule to which it is attached. As molecular size increases, the rate of molecular diffusion decreases leading to larger time intervals between encounters (i.e. to fewer encounters per unit time). This effect is compensated by the greater duration of each encounter giving rise to a larger number of functional group collisions per encounter. Hence the reactivity of a functional group can be expected to be approximately independent of molecular size.

Mathematical analysis of step polymerization is simplified greatly by assuming that the intrinsic reactivity of a functional group is independent of molecular size and unaffected by reaction of the other functional group(s) in the molecule of monomer from which it is derived. This principle of equal reactivity of functional groups was proposed by Flory who demonstrated its validity for many step polymerizations by examining the kinetics of model reactions. On this basis step polymerization involves random reactions occurring between any two mutually-reactive molecular species. Intrinsically, each of the possible reactions is equally probable and their relative preponderances depend only upon the relative numbers of each type of molecular species (i.e. monomer, dimer, trimer, etc.). The assumption of equal reactivity is implicit in each of the theoretical treatments of step polymerization that follow.

2.2.4 *Carothers theory*

Carothers developed a simple method of analysis for predicting the molar mass of polymers prepared by step polymerization. He recognized that the number-average degree of polymerization *with respect to monomer units* is given by the relation

$$\bar{x}_n = \frac{N_o}{N} \tag{2.1}$$

where N_o is the number of molecules present initially and N is the number of molecules remaining after a time t of polymerization.

Assuming that there are equal numbers of functional groups, $\bar{x}_n$ can be related to the *extent of reaction p* at time t which is given by

$$p = \frac{\text{Number of functional groups that have reacted}}{\text{Number of functional groups present initially}}$$

and is the probability that any functional group present initially has reacted. Since the total number of molecules decreases by one for each pair-wise reaction between functional groups

$$p = \frac{N_o - N}{N_o} \text{ or } \frac{N_o}{N} = \frac{1}{1 - p} \tag{2.2}$$

Combining Equations (2.1) and (2.2) gives the Carothers equation

$$\bar{x}_n = \frac{1}{1 - p} \tag{2.3}$$

This equation is applicable to $RA_2 + RB_2$, ARB and RA_2 polymerizations in which there is an exact stoichiometric balance in the numbers of mutually-reactive functional groups. The equation highlights the need to attain very high extents of reaction in order to produce polymers with useful physical properties. Normally, degrees of polymerization of the order of 100 or above are required, hence demanding values of $p \geqslant 0 \cdot 99$. This clearly demonstrates the necessity for using monomers of high purity and reactions which are either highly efficient or can be forced towards completion.

The number-average molar mass $\bar{M}_n$ is related to $\bar{x}_n$ by

$$\bar{M}_n = \bar{M}_o \bar{x}_n$$

where $\bar{M}_o$ is the mean molar mass of a monomer unit and is given by

$$\bar{M}_o = \frac{\text{Molar mass of the repeat unit}}{\text{Number of monomer units in the repeat unit}}$$

Slight stoichiometric imbalances significantly limit the attainable values of $\bar{x}_n$. Consider a $RA_2 + RB_2$ polymerization in which RB_2 is present in excess. The ratio of the numbers of the different functional groups present initially is known as the *reactant ratio r*, and for linear step polymerization is always defined so that it is less than or equal to one. Thus for the reaction under consideration

$$r = \frac{N_A}{N_B} \tag{2.4}$$

where N_A and N_B are respectively the numbers of A and B functional groups present initially. Since there are two functional groups per molecule

$$N_o = \frac{N_A + N_B}{2}$$

TABLE 2.3 *Variation of number-average degree of polymerization with extent of reaction and reactant ratio*

r	$\bar{x}_n$ at				
	$p = 0.90$	$p = 0.95$	$p = 0.99$	$p = 0.999$	$p = 1.000^*$
1.000	10.0	20.0	100.0	1000.0	∞
0.999	10.0	19.8	95.3	666.8	1999.0
0.990	9.6	18.3	66.8	166.1	199.0
0.950	8.1	13.4	28.3	37.6	39.0
0.900	6.8	10.0	16.1	18.7	19.0

* As $p \to 1$, $\bar{x}_n \to \dfrac{(1+r)}{(1-r)}$

which upon substitution for N_A from Equation (2.4) gives

$$N_o = \frac{N_B(1+r)}{2} \tag{2.5}$$

It is common practice to define the extent of reaction, p, in terms of the functional groups present in minority (i.e. A groups in this case). On this basis,

$$\text{number of unreacted A groups} = N_A - pN_A$$
$$= rN_B(1-p)$$
$$\text{number of unreacted B Groups} = N_B - pN_A$$
$$= N_B(1-rp)$$

so that

$$N = \frac{rN_B(1-p) + N_B(1-rp)}{2}$$

i.e. $$N = \frac{N_B(1+r-2rp)}{2} \tag{2.6}$$

Substitution of Equations (2.5) and (2.6) into Equation (2.1) yields the more *general Carothers equation*

$$\bar{x}_n = \frac{1+r}{1+r-2rp} \tag{2.7}$$

of which Equation (2.3) is the special case for $r = 1$. Table 2.3 gives values of $\bar{x}_n$ calculated using Equation (2.7) and reveals the dramatic reduction in $\bar{x}_n$ when r is less than unity. Thus only very slight stoichiometric imbalances can be tolerated if useful polymers are to be formed. In practice, such imbalances are used to control $\bar{x}_n$.

Equation (2.7) also is applicable to reactions in which a monofunctional compound is included to control $\bar{x}_n$, e.g. $RA_2 + RB_2 + RB$ or $ARB + RB$. All that is required is to re-define the reactant ratio

$$r = \frac{N_A}{N_B + 2N_{RB}}$$

where N_A and N_B are respectively the initial numbers of A and B functional groups from the difunctional monomer(s) and N_{RB} is the number of molecules of RB present initially. The factor of 2 is required because one RB molecule has the same quantitative effect in limiting $\bar{x}_n$ as one excess RB_2 molecule.

2.2.5 *Statistical theory*

The theory of Carothers is restricted to the prediction of number-average quantities. In contrast, simple statistical analyses based upon the random nature of step polymerization allow prediction of size distributions. Such analyses were first described by Flory.

For simplicity $RA_2 + RB_2$ and ARB polymerizations in which there is exactly equivalent stoichiometry will be considered here. The first stage in the analysis is to calculate the probability $P(x)$ of existence of a molecule consisting of exactly x monomer units at time t when the extent of reaction is p. A molecule containing x monomer units is created by the formation of a sequence of $(x - 1)$ linkages. The probability that a particular sequence of linkages has formed is the product of the probabilities of forming the individual linkages. Since p is the probability that a functional group has reacted, the probability of finding a sequence of two linkages is p^2, the probability of finding a sequence of three linkages is p^3, and the probability of finding a sequence of $(x - 1)$ linkages is $p^{(x-1)}$. For a molecule to contain exactly x monomer units, the xth (i.e. last) unit must possess a terminal unreacted functional group. The probability that a functional group has not reacted is $(1 - p)$ and so

$$P(x) = (1 - p)p^{(x-1)} \tag{2.8}$$

Since $P(x)$ is the probability that a molecule chosen at random contains exactly x monomer units, it must also be the *mole fraction* of x-mers. If the total number of molecules present at time t is N, then the total number, N_x, of x-mers is given by

$$N_x = N(1 - p)p^{(x-1)} \tag{2.9}$$

Often N cannot be measured and so is eliminated by substitution of the rearranged form of Equation (2.2), $N = N_o(1 - p)$, to give

$$N_x = N_o(1 - p)^2 p^{(x-1)} \tag{2.10}$$

which is an expression for the number of molecules of degree of polymerization x in terms of the initial number of molecules, N_o, and the extent of reaction, p.

The weight fraction, w_x, of x-mers is given by

$$w_x = \frac{\text{Total mass of molecules with degree of polymerization } x}{\text{Total mass of all the molecules}}$$

Thus, neglecting end groups

$$w_x = \frac{N_x(x\bar{M}_0)}{N_o\bar{M}_0} = \frac{xN_x}{N_o} \tag{2.11}$$

Combining Equations (2.10) and (2.11) gives

$$w_x = x(1-p)^2 p^{(x-1)} \tag{2.12}$$

Equations (2.8) and (2.12) define what is known as the *Most Probable (or Shultz–Flory) Distribution*, the most important features of which are

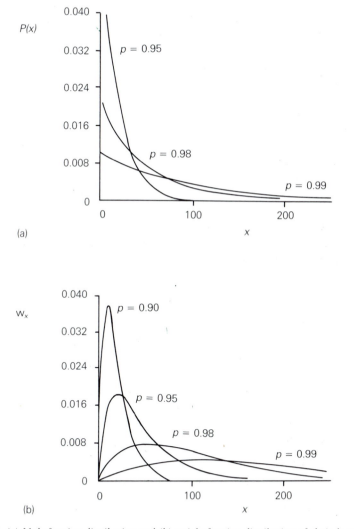

(a)

(b)

Fig. 2.1 *(a) Mole fraction distribution and (b) weight fraction distribution of chain lengths for various extents of reaction in a linear step polymerization (both sets of curves taken from Flory).*

illustrated by the plots shown in Fig. 2.1. Thus the mole fraction $P(x)$ decreases continuously as the number of monomer units in the polymer chain increases, i.e. at all extents of reaction the mole fraction of monomer is greater than that of any other species. In contrast, the weight fraction distribution shows a maximum which is very close to $\bar{x}_n$. As the extent of reaction increases, the maximum moves to higher values of x and the weight fraction of monomer becomes very small.

Knowledge of the distribution functions (i.e. $P(x)$ and w_x) enables molar mass averages to be evaluated. From Equation (1.2) of Section 1.3.2 the number-average molar mass may be written as

$$\bar{M}_n = \sum P(x)M_x$$

Recognizing that $M_x = x\bar{M}_0$ and also substituting for $P(x)$ using Equation (2.8) gives

$$\bar{M}_n = \sum x\bar{M}_0(1-p)p^{(x-1)}$$

i.e. $\bar{M}_n = \bar{M}_0(1-p)\sum xp^{(x-1)}$

Using the mathematical relation

$$\sum_{x=1}^{\infty} xp^{(x-1)} = (1-p)^{-2} \text{ for } p<1$$

the equation for $\bar{M}_n$ reduces to

$$\bar{M}_n = \frac{\bar{M}_0}{(1-p)} \tag{2.13}$$

Since $\bar{x}_n = \bar{M}_n/\bar{M}_0 = 1/(1-p)$ this equation is equivalent to the Carothers Equation (2.3), but this time it has been derived from purely statistical considerations.

Application of Equation (1.7) from Section 1.3.2 enables the weight-average molar mass to be written as

$$\bar{M}_w = \sum w_x M_x$$

Using Equation (2.12) it follows that

$$\bar{M}_w = \bar{M}_0(1-p)^2 \sum x^2 p^{(x-1)}$$

and another mathematical relation

$$\sum_{x=1}^{\infty} x^2 p^{(x-1)} = (1+p)(1-p)^{-3} \text{ for } p<1$$

leads to

$$\overline{M}_w = \overline{M}_0 \frac{(1+p)}{(1-p)} \tag{2.14}$$

and hence to the weight-average degree of polymerization

$$\bar{x}_w = \frac{(1+p)}{(1-p)} \tag{2.15}$$

The polydispersity index $\overline{M}_w/\overline{M}_n$ is then given by

$$\frac{\overline{M}_w}{\overline{M}_n} = 1 + p \tag{2.16}$$

and for most linear polymers prepared by step polymerization is close to 2 (since high values of p are required to form useful polymers).

The mole fraction and weight fraction distributions for step polymerizations in which there is a stoichiometric imbalance are similar to those just derived for the case of exactly equivalent stoichiometry. Thus all linear step polymerizations lead to essentially the same form of molar mass distribution.

Before closing this section, it must again be emphasized that the degrees of polymerization given are with respect to monomer units *and not* repeat units.

2.2.6 *Kinetics*

The assumption of equal reactivity of functional groups also greatly simplifies the kinetics of step polymerization since a single rate constant applies to each of the step-wise reactions. It is usual to define the overall rate of reaction as the rate of decrease in the concentration of one or other of the functional groups, i.e. in general terms for equimolar stoichiometry

$$\text{Rate of reaction} = -\frac{d[A]}{dt} = -\frac{d[B]}{dt}$$

Most step polymerizations involve bimolecular reactions which often are catalysed. Thus neglecting elimination products in polycondensations, the general elementary reaction is

$$\text{\Large\sim} A + B \text{\Large\sim} + \text{catalyst} \rightarrow \text{\Large\sim} AB \text{\Large\sim} + \text{catalyst}$$

and so the rate of reaction is given by

$$-\frac{d[A]}{dt} = k'[A][B][\text{Catalyst}] \tag{2.17}$$

where k' is the rate constant for the reaction. Since the concentration of a true catalyst does not change as the reaction proceeds it is usual to simplify the expression by letting $k = k'[\text{Catalyst}]$ giving

$$-\frac{d[A]}{dt} = k[A][B] \tag{2.18}$$

For equimolar stoichiometry $[A] = [B] = c$ and Equation (2.18) becomes

$$-\frac{dc}{dt} = kc^2$$

This equation may be integrated by letting $c = c_0$ at $t = 0$

$$\int_{c_0}^{c} -\frac{dc}{c^2} = \int_{0}^{t} kdt$$

and gives

$$\frac{1}{c} - \frac{1}{c_0} = kt$$

which may be rewritten in terms of the extent of reaction by recognizing that $c_0/c = N_0/N$ and applying Equation (2.2)

$$\frac{1}{(1-p)} - 1 = c_0 kt \tag{2.19}$$

This equation also applies to reactions which proceed in the absence of catalyst, though the rate constant is different and obviously does not include a term in catalyst concentration.

Certain step polymerizations are self-catalysed, i.e. one of the types of functional group also acts as a catalyst (e.g. carboxylic acid groups in a polyesterification). In the absence of an added catalyst the rate of reaction for such polymerizations is given by

$$-\frac{d[A]}{dt} = k''[A][B][A] \tag{2.20}$$

assuming that the A groups catalyse the reaction. Again letting $[A] = [B] = c$, Equation (2.20) becomes

$$-\frac{dc}{dt} = k''c^3$$

which upon integration as before gives

$$\frac{1}{c^2} - \frac{1}{c_0^2} = 2k''t$$

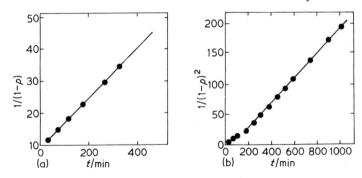

Fig. 2.2 *(a) Plot of 1/(1 − p) as a function of time for the polymerization of decamethylene glycol with adipic acid using p-toluene sulphonic acid as a catalyst at 369 K and (b) plot of 1/(1 − p)² as a function of time for the polymerization of ethylene glycol with adipic acid at 459 K (both sets of data from Flory).*

or in terms of the extent of reaction

$$\frac{1}{(1-p)^2} - 1 = 2c_0^2 k'' t \qquad (2.21)$$

Equations (2.19) and (2.21) have been derived assuming that the reverse reaction (i.e. depolymerization) is negligible. This is satisfactory for many polyadditions, but for reversible polycondensations requires the elimination product to be removed continuously as it is formed. The equations have been verified experimentally using step polymerizations that satisfy this requirement, as is shown by the polyesterification data plotted in Fig. 2.2. These results further substantiate the validity of the principle of equal reactivity of functional groups.

2.2.7 Ring formation

A complication not yet considered is the *intramolecular* reaction of terminal functional groups on the same molecule. This results in the formation of cyclic molecules (i.e. rings), e.g. in the preparation of a polyester

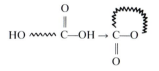

The ease of ring formation depends strongly upon the number of atoms linked together in the ring. For example, 5-, 6- and, to a lesser extent,

7-membered rings are stable and often form in preference to linear polymer. For the self-condensation of ω-hydroxy carboxylic acids $HO—(CH_2)_i—CO_2H$ when $i = 3$ only the monomeric lactone is produced

$$HO—(CH_2)_3—CO_2H \rightarrow \begin{array}{c} CH_2—CH_2 \\ | \quad\quad | \\ CH_2 \quad O \\ \diagdown \quad \diagup \\ C \\ \| \\ O \end{array} + H_2O$$

When $i = 4$ some polymer is produced in addition to the corresponding monomeric lactone, and when $i = 5$ the product is a mixture of polymer with some of the monomeric lactone.

Normally, 3- and 4-membered rings and 8- to 11-membered rings are unstable due to bond-angle strain and steric repulsions between atoms crowded into the centre of the ring respectively, and usually are not formed. Whilst 12-membered and larger rings are more stable and can form, their probability of formation decreases as the ring size increases. This is because the probability of the two ends of a single chain meeting decreases as their separation (i.e. the chain length) increases. Thus large rings rarely form.

Ring formation disturbs the form of the molar mass distribution and reduces the ultimate molar mass attainable. However, since linear polymerization is a bimolecular process and ring formation is a unimolecular process, it is possible to greatly promote the former process relative to the latter by using high monomer concentrations. This is why many step polymerizations are performed in bulk (i.e. using only monomer(s) plus catalysts).

2.2.8 Polymerization systems

The preceding sections highlight the many constraints upon the formation of high molar mass polymers by linear step polymerization. Special polymerization systems often have to be developed to overcome these constraints and are exemplified here by systems developed for the preparation of polyesters and polyamides.

Ester interchange (or *transesterification*) reactions commonly are employed in the production of polyesters, the most important example being the preparation of poly(ethylene terephthalate). The direct polyesterification reaction of terephthalic acid with ethylene glycol indicated in Table 1.1 (Section 1.2.2) is complicated by the high melting point of terephthalic acid (in fact it sublimes at 573 K before melting) and its low solubility. Thus poly(ethylene terephthalate) is prepared in a two-stage process. The first stage involves formation of bis(2-hydroxyethyl)terephthalate either by reaction of dimethylterephthalate with an excess of ethylene glycol (i.e. via ester interchange)

$$CH_3OOC-\langle O \rangle-COOCH_3 + (2 + x) \ HOCH_2CH_2OH \xrightarrow{420-470 \ K}$$

$$HOCH_2CH_2OOC-\langle O \rangle-COOCH_2CH_2OH + 2CH_3OH + xHOCH_2CH_2OH$$

or more commonly nowadays by direct esterification of terephthalic acid with an excess of ethylene glycol

$$HOOC-\langle O \rangle-COOH + (2 + x) \ HOCH_2CH_2OH \xrightarrow{500-530 \ K}$$

$$HOCH_2CH_2OOC-\langle O \rangle-COOCH_2CH_2OH + 2H_2O + x \ HOCH_2CH_2OH$$

On completion of the first stage, the reaction temperature is raised to about 550 K so that the excess ethylene glycol and the ethylene glycol produced by further ester interchange reactions can be removed, and so that the polymer is formed above its melting temperature (538 K)

$$nHOCH_2CH_2OOC-\langle O \rangle-COOCH_2CH_2OH \rightarrow$$

$$HOCH_2CH_2O+OC-\langle O \rangle-COOCH_2CH_2O+_n H + (n - 1) \ HOCH_2CH_2OH$$

Thus by using ester interchange reactions the need for strict stoichiometric control is eliminated.

The preferred method for preparing polyamides from diamines and diacids is *melt polymerization* of the corresponding nylon salt. For example, in the preparation of nylon 6.6, hexamethylene diamine and adipic acid are first reacted together at low temperature to form hexamethylene diammonium adipate (nylon 6.6 salt) which then is purified by recrystallization. The salt is heated gradually up to about 550 K to effect melt polymerization and maintained at this temperature whilst removing the water produced as steam

$$nH_2N(CH_2)_6NH_2 \quad + \quad nHOOC(CH_2)_4COOH$$
$$\text{Hexamethylene diamine} \qquad \text{Adipic acid}$$
$$\downarrow$$
$$nH_3\overset{+}{N}(CH_2)_6\overset{+}{N}H_3\overset{-}{O}OC(CH_2)_4COO^-$$
$$\downarrow$$
$$H+NH(CH_2)_6NHOC(CH_2)_4CO+_nOH + (2n - 1)H_2O$$

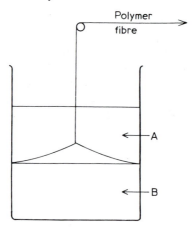

Fig. 2.3 *Schematic illustration of interfacial polymerization. Solution A is an aqueous solution of hexamethylene diamine and B a solution of sebacoyl chloride in carbon tetrachloride. The polymer can be drawn off in the form of a fibre. It is usual to include a base (e.g. NaOH) in solution A in order to neutralize the HCl formed by the reaction.*

A major advantage of melt polymerization by *salt dehydration* is that the use of a pure salt guarantees exact 1:1 stoichiometry.

A convenient method for preparation of polyesters and polyamides in the laboratory is the reaction of diacid chlorides with diols and diamines respectively (i.e. Schotten–Baumann reactions). These reactions proceed rapidly at low temperatures and often are performed as *interfacial polymerizations* in which the two reactants are dissolved separately in immiscible solvents which are then brought into contact. The best known example of this is the 'nylon rope trick' where a continuous film of nylon is drawn from the interface as illustrated in Fig. 2.3 for the preparation of nylon 6.10

$$nH_2N(CH_2)_6NH_2 \quad + \quad nClOC(CH_2)_8COCl$$
Hexamethylene diamine   Sebacoyl chloride
$$\downarrow$$
$$H\text{---}[NH(CH_2)_6NHOC(CH_2)_8CO]_n\text{---}Cl + (2n-1)\ HCl$$

The reaction takes place at the organic solvent side of the interface and because it usually is diffusion-controlled there is no need for strict control of stoichiometry.

2.3 Non-linear step polymerization

The inclusion of a monomer with a functionality greater than two has a dramatic effect upon the structure and molar mass of the polymer formed.

In the early stages of such reactions the polymer has a branched structure and, consequently, increases in molar mass much more rapidly with the extent of reaction than for a linear step polymerization. As the reaction proceeds, further branching reactions lead ultimately to the formation of complex network structures which have properties that are quite different from those of the corresponding linear polymer. For example, reaction of a dicarboxylic acid $R(COOH)_2$ with a triol $R'(OH)_3$ would lead to structures of the type

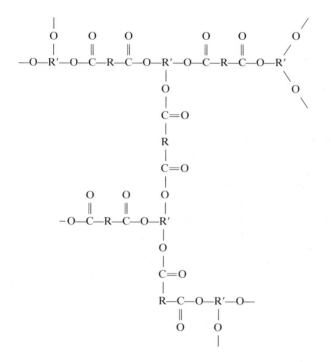

The point at which the first network molecule is formed is known as the *gel-point* because it is manifested by *gelation*, i.e. an abrupt change of the reacting mixture from a viscous liquid to a solid gel which shows no tendency to flow.

Before dealing with theoretical predictions of the gel-point, some important network-forming step polymerizations will be described.

2.3.1 *Network polymers*

Formaldehyde-based resins were the first network polymers prepared by step polymerization to be successfully commercialized. They are prepared in two stages. The first involves the formation of a prepolymer of low molar mass which may either be liquid or solid. In the second stage the prepolymer is forced to flow under pressure to fill a heated mould in which further reaction takes place to yield a highly crosslinked, rigid polymer in the shape of the mould. Since formaldehyde is difunctional, the co-reactants must have a functionality, f, greater than two and those most commonly employed are phenol ($f = 3$), urea ($f = 4$) and melamine ($f = 6$)

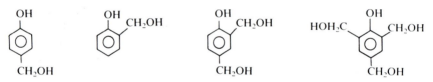

Formaldehyde   Phenol   Urea   Melamine

The hydroxyl group in phenol activates the benzene ring towards substitution in the 2-, 4- and 6-positions. Upon reaction of phenol with formaldehyde, methylol substituent groups are formed, e.g.

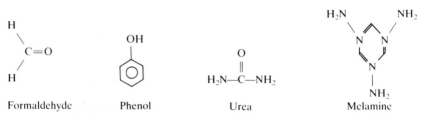

Further reaction leads principally to the formation of methylene bridges but also to dimethylene ether links

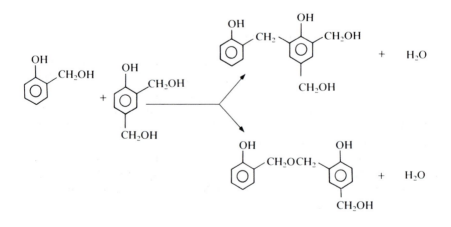

There are two types of *phenol–formaldehyde resin*. Those prepared using an excess of formaldehyde with base catalysis are known as *resoles*. The resole prepolymers possess many unreacted methylol groups that upon further heating react to produce the network structure.

Novolaks are prepared using an excess of phenol and acid catalysis which promotes condensation reactions of the methylol groups. Thus the prepolymers produced contain no methylol groups and are unable to crosslink, e.g.

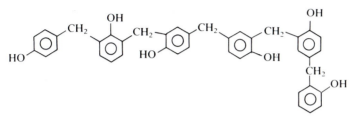

Normally they are dried and ground to a powder, mixed with fillers (e.g. mica, glass fibres, sawdust), colourants and hardeners, and then cured (i.e. crosslinked) in a hot mould. Hexamethylenetetramine, with magnesium or calcium oxide as a catalyst, usually is employed as the hardener to facilitate crosslinking reactions. Most of the crosslinks formed are methylene bridges, though some dimethylene amine links can be produced. The fillers are added to improve the electrical or mechanical properties of the resin.

The chemistry of *urea–* and *melamine–formaldehyde* resins involves the formation and condensation reactions of N-methylol groups, e.g. in general terms

$$—NH_2 + CH_2O \rightarrow —NHCH_2OH \xrightarrow{\;HN\diagdown\;} —NHCH_2N\diagup + H_2O$$

$$\Big\downarrow \; HOCH_2NH—$$

$$—NHCH_2OCH_2NH— + H_2O$$

The reactions usually are arrested at the prepolymer stage by adjusting the pH of the reacting mixture to slightly alkaline. After blending (e.g. with fillers, pigments, hardeners, etc.) the prepolymers are cured by heating in a mould. The hardeners are compounds (e.g. ammonium sulphamate) which decompose at mould temperatures to give acids that catalyse the condensation reactions.

Epoxy resins are formed from low molar mass prepolymers containing epoxide end-groups. The most important are the diglycidyl ether prepolymers prepared by reaction of excess epichlorohydrin with bisphenol-A in the presence of a base

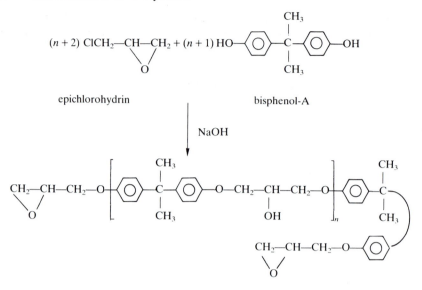

epichlorohydrin   bisphenol-A

NaOH

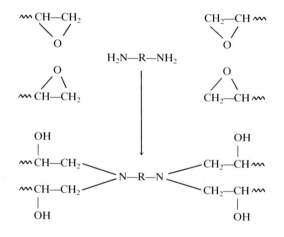

These prepolymers are either viscous liquids or solids depending upon the value of *n*. Usually they are cured by the use of multi-functional amines which undergo a polyaddition reaction with the terminal epoxide groups in the manner indicated below

Epoxy resins are characterized by low shrinkage on curing and find use as adhesives, electrical insulators, surface coatings and matrix materials for fibre-reinforced composites.

Polyurethane networks find a wide variety of uses (e.g. elastomers, flexible foams, and rigid foams) and usually are prepared by reaction of diisocyanates with branched polyester or polyether prepolymers that have hydroxyl end groups, e.g. reaction of

$$O=C=N-\!\!\!\!\bigcirc\!\!\!\!-CH_2-\!\!\!\!\bigcirc\!\!\!\!-N=C=O$$

with

$$
\begin{array}{c}
\qquad\qquad\qquad CH_3 \\
\qquad\qquad\qquad | \\
CH_2\!\!-\!\!\!\!+\!\!O\!-\!CH_2\!-\!CH\!\!+_n\!OH \\
| \\
CH_2\!\!-\!\!\!\!+\!\!O\!-\!CH_2\!-\!CH\!\!+_n\!OH \\
| \qquad\qquad\qquad | \\
(CH_2)_3 \qquad\qquad CH_3 \\
| \\
CH_2\!\!-\!\!\!\!+\!\!O\!-\!CH_2\!-\!CH\!\!+_n\!OH \\
\qquad\qquad\qquad | \\
\qquad\qquad\qquad CH_3
\end{array}
$$

via the chemistry described in Section 2.2.2. The molar mass and functionality of the prepolymer determine the crosslink density and hence the flexibility of the network formed.

2.3.2 *Carothers theory of gelation*

A simple theory for prediction of gel-points can be derived using the principles employed in the linear step polymerization theory of Carothers. When there is a stoichiometric balance in the numbers of mutually-reactive functional groups, the number-average functionality, f_{av}, is used and is defined by

$$f_{av} = \sum_i N_i f_i \Big/ \sum_i N_i$$

where N_i is the initial number of molecules of monomer i which has functionality f_i. Thus if there are N_o molecules present initially, the total number of functional groups present is $N_o f_{av}$. If at time t there are N molecules present, then the number of functional groups that have reacted is $2(N_o - N)$ since the number of molecules decreases by one for each link produced by reaction together of *two* functional groups. Therefore the extent of reaction p, which is the probability that a functional group present initially has reacted, is given by

$$p = \frac{2(N_o - N)}{N_o f_{av}}$$

TABLE 2.4 *Values of $\bar{x}_n$ for different values of p and f_{av} calculated using Equation (2.23)*

p	0.5	0.7	0.9	0.95	0.99
$f_{av} = 2.0$	2	3.33	10	20	100
$f_{av} = 2.1$	2.10	3.77	18.18	400	Gelled
$f_{av} = 2.2$	2.22	4.35	100	Gelled	Gelled
$f_{av} = 2.3$	2.35	5.13	Gelled	Gelled	Gelled

Simplifying and substituting for N/N_o using Equation (2.1) gives

$$p = \frac{2}{f_{av}}\left(1 - \frac{1}{\bar{x}_n}\right) \tag{2.22}$$

which can be rearranged to

$$\bar{x}_n = \frac{2}{2 - pf_{av}} \tag{2.23}$$

This equation reduces to the simple Carothers Equation (2.3) when $f_{av} = 2.0$. Slight increases in f_{av} above 2.0 give rise to substantial increases in the values of $\bar{x}_n$ attained at specific extents of reaction, as is demonstrated in Table 2.4.

If it is postulated that gelation occurs when $\bar{x}_n$ goes to infinity then it follows from Equation (2.22) that the critical extent of reaction, p_c, for gelation is given by

$$p_c = 2/f_{av} \tag{2.24}$$

It is possible to extend the theory to prediction of p_c when there is an imbalance of stoichiometry. However, this simple method of analysis will not be pursued further here because it is rather inelegant and always yields overestimates of p_c. The approach is flawed because it is based upon $\bar{x}_n$, an average quantity, tending to infinity. Molecules with degrees of polymerization both larger and smaller than $\bar{x}_n$ are present, and it is the largest molecules that undergo gelation first.

2.3.3 *Statistical theory of gelation*

The basic statistical theory of gelation was first derived by Flory who considered the reaction of an f-functional monomer RA_f ($f > 2$) with the difunctional monomers RA_2 and RB_2. It is necessary to define a parameter called the *branching coefficient*, α, which is the probability that an f-functional unit is connected via a chain of difunctional units to another *f*-functional unit. In other words α is the probability that the general sequence of linkages shown below exists

$$_{(f-1)}AR—AB—R—B[A—R—AB—R—B]_iA—R—A_{(f-1)}$$

In order to derive an expression for α from statistical considerations it is necessary to introduce another term, γ, which is defined as the initial ratio of A groups from RA_f molecules to the total number of A groups. Using γ it is possible to calculate the probabilities for the existence of each of the linkages in the general sequence given above. If the extent of reaction of the A groups is p_A and of the B groups is p_B then

probability of $_{(f-1)}AR—\overrightarrow{AB}—R—B$ = p_A
probability of $B—R—\overrightarrow{BA}—R—A$ = $p_B(1 - \gamma)$
probability of $A—R—\overrightarrow{AB}—R—B$ = p_A
probability of $B—R—\overrightarrow{BA}—RA_{(f-1)}$ = $p_B\gamma$

where the arrows indicate the linkages under consideration and the direction in which the chain is extending. Thus the probability that the general sequence of linkages has formed is

$$p_A[p_B(1 - \gamma)p_A]_i^i p_B\gamma$$

The branching coefficient is the probability that sequences with all values of i have formed and so is given by

$$\alpha = p_A p_B \gamma \sum_{i=0}^{\infty} \left[p_A p_B(1 - \gamma) \right]^i$$

Using the mathematical relation

$$\sum_{i=0}^{\infty} x^i = 1/(1 - x) \text{ for } x < 1$$

then

$$\alpha = p_A p_B \gamma [1 - p_A p_B(1 - \gamma)]^{-1} \tag{2.25}$$

Network molecules can form when n chains are expected to lead to more than n chains through branching of some of them. The maximum number of chains that can emanate from the end of a single chain, such as that analysed above, is $(f - 1)$ and so the probable number of chains emanating from the chain end is $\alpha(f - 1)$. Network molecules can form if this probability is not less than one, i.e. $\alpha(f - 1) \geq 1$. Thus the *critical branching coefficient*, α_c, for gelation is given by

$$\alpha_c = \frac{1}{(f - 1)} \tag{2.26}$$

Substitution of this equation into Equation (2.25) followed by rearrangement gives an expression for the product of the critical extents of reaction at gelation.

$$(p_A p_B)_c = \frac{1}{1 + \gamma(f - 2)} \tag{2.27}$$

If the *reactant ratio*, r, is defined as the initial ratio of A groups to B groups then $p_B = r p_A$ and

$$(p_A)_c = [r + r\gamma(f - 2)]^{-1/2}$$
$$(p_B)_c = r^{1/2}[1 + \gamma(f - 2)]^{-1/2}$$

When RA_2 molecules are absent $\gamma = 1$ and $(p_A p_B)_c = 1/(f - 1)$, i.e. $(p_A p_B)_c = a_c$.

The predictions of the statistical theory and of Carothers theory can now be compared with each other and with experimental gelation data. Fig. 2.4 shows the variation of p, $\bar{x}_n$ and viscosity, η, with time for a reaction between diethylene glycol (a diol), succinic acid (a diacid) and 1,2,3-propane tricarboxylic acid (a triacid) in which $r = 1.000$ and $\gamma = 0.293$. Gelation occurred after about 230 min and was manifested by η becoming infinite. The observed extent of reaction at gelation was 0.911 when $\bar{x}_n$ was only about 25. The Carothers Equation (2.24) predicts a p_c of 0.951 which as expected is high, whereas Equation (2.27) gives a value of 0.879. The simple statistical theory underestimates p_c because it does not take into account the effect of intramolecular reactions between end-groups. These

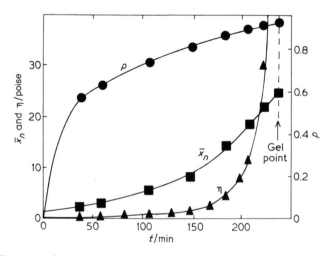

Fig. 2.4 *Variation of p, $\bar{x}_n$ and η with time for the polymerization of ethylene glycol with succinic acid and 1,2,3-propane tricarboxylic acid (after Flory).*

give rise to loops in the polymer structure and the polymerization must proceed to higher extents of reaction in order to overcome this wastage of functional groups. When the effects of intramolecular reaction are eliminated (e.g. extrapolation of p_c values to infinite reactant concentrations, see Section 2.2.7) the simple statistical theory gives accurate predictions of the gel-point. More advanced statistical theories in which intramolecular reactions are included have been derived but are beyond the scope of this book.

The theories of gelation presented here can only be applied when it is clear that the assumption of equal reactivity of functional groups is satisfactory. In terms of the polymerizations described earlier, the theories generally are applicable to the formation of polyester and polyurethane networks, but not to the formation of formaldehyde-based resins and epoxy resins. Failure of the principle of equal reactivity for the latter systems results from modification of the reactivity of a particular functional group by reaction of another functional group in the same molecule of monomer.

Chain polymerization

2.4 Free-radical polymerization

Free radicals are independently-existing species which possess an unpaired electron and normally are highly reactive with short lifetimes. Free-radical polymerizations are chain polymerizations in which each polymer molecule grows by addition of monomer to a terminal free-radical reactive site known as an *active centre*. Consequent upon every addition of monomer, the active centre is transferred to the newly-created chain end.

Free-radical polymerization is the most widely practised method of chain polymerization and is used almost exclusively for the preparation of polymers from monomers of the general structure $CH_2{=}CR_1R_2$ (see examples 1 and 3–7 of Table 1.1, Section 1.2.2). In common with other types of chain polymerization the reaction can be divided into three distinct stages: *initiation, propagation* and *termination*. The general chemistry associated with each stage is described in the following three sections by considering for simplicity the polymerization of a general vinyl monomer, $CH_2{=}CHX$.

2.4.1 *Initiation*

This stage involves creation of the free-radical active centre and usually takes place in two steps. The first is the formation of free radicals from an *initiator* and the second is the addition of one of these free radicals to a molecule of monomer.

There are two principal ways in which free radicals can be formed: (i) homolytic scission (i.e. homolysis) of a single bond, and (ii) single electron transfer to or from an ion or molecule (e.g. redox reactions).

Homolysis can be effected by the application of heat (Δ) and there are many compounds, in particular those containing peroxide (—O—O—) or azo (—N=N—) linkages, which undergo *thermolysis* in the convenient temperature range of 50–100°C

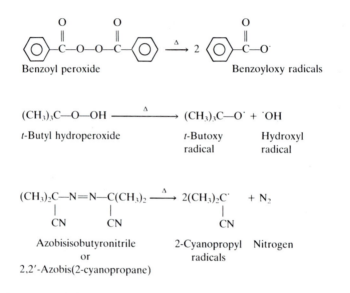

Benzoyl peroxide   Benzoyloxy radicals

$$(CH_3)_3C—O—OH \xrightarrow{\Delta} (CH_3)_3C—O^{\cdot} + {^{\cdot}}OH$$

t-Butyl hydroperoxide   *t*-Butoxy   Hydroxyl
                                     radical   radical

$$(CH_3)_2C—N=N—C(CH_3)_2 \xrightarrow{\Delta} 2(CH_3)_2C^{\cdot} \ + N_2$$
$$\quad\ \ |\qquad\qquad |\qquad\qquad\qquad\ \ |$$
$$\quad\ \ CN\qquad\quad CN\qquad\qquad\quad\ CN$$

Azobisisobutyronitrile   2-Cyanopropyl   Nitrogen
           or   radicals
2,2′-Azobis(2-cyanopropane)

Many of the radicals produced undergo further breakdown before reaction with monomer, for example β-scissions such as

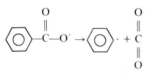

Phenyl   Carbon
radical   dioxide

$$(CH_3)_2C—O^{\cdot} \rightarrow {^{\cdot}}CH_3 + (CH_3)_2C=O$$
$$\quad\ \ |$$
$$\quad\ \ CH_3$$

Methyl   Acetone
radical

In each of these examples the dot indicates the site of the unpaired electron.

Additionally, homolysis can be brought about by *photolysis*, i.e. the action of radiation (usually ultraviolet). Examples are the dissociation of azobisisobutyronitrile and the formation of free radicals from benzophenone and benzoin. An advantage of photolysis is that the formation of free radicals begins at the instant of exposure and ceases as soon as the light source is removed.

Redox reactions often are used when it is necessary to perform polymerizations at low temperatures. Two examples are given below

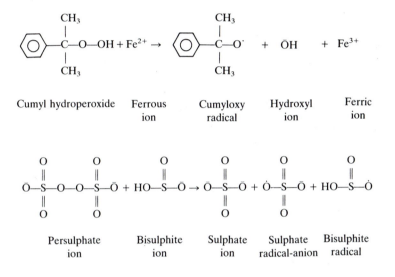

Cumyl hydroperoxide	Ferrous ion	Cumyloxy radical	Hydroxyl ion	Ferric ion

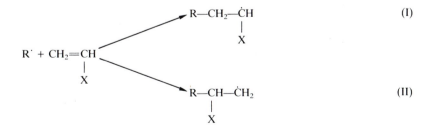

Persulphate ion	Bisulphite ion	Sulphate ion	Sulphate radical-anion	Bisulphite radical

An active centre is created when a free radical ($R\dot{}$) generated from an initiator attacks the π-bond of a molecule of monomer. There are two possible modes of addition

R—CH$_2$—ĊH   (I)
|
X

R˙ + CH$_2$=CH
|
X

R—CH—ĊH$_2$ (II)
|
X

Mode (I) predominates because attack at the methylene carbon is less sterically-hindered and yields a product free-radical that is more stable because of the effects of the adjacent X group (usually both steric and mesomeric stabilization).

Not all of the free radicals formed from the initiator are destined to react with monomer. Some are lost in side reactions such as those shown below for benzoyl peroxide

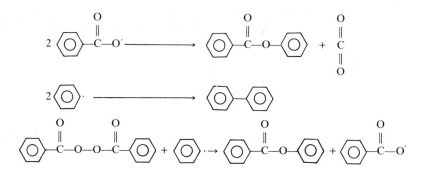

The latter type of reaction is known as induced decomposition and makes a significant contribution to wastage of peroxide initiators.

2.4.2 Propagation

This involves growth of the polymer chain by rapid sequential addition of monomer to the active centre. As with the second step of initiation, there are two possible modes of propagation

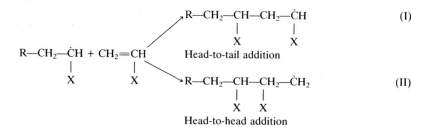

Mode (I) again predominates for the same reasons as described in the previous section. Thus the polymer chains are principally of the structure

$$-CH_2-CH-CH_2-CH-CH_2-CH-CH_2-CH-$$
$$\qquad | \qquad | \qquad | \qquad |$$
$$\qquad X \qquad X \qquad X \qquad X$$

though an occasional head-to-head linkage can be expected.

The time required for each monomer addition typically is of the order of a millisecond. Thus several thousand additions can take place within a few seconds.

2.4.3 *Termination*

In this stage, growth of the polymer chain is terminated. The two most common mechanisms of termination involve bimolecular reaction of growing polymer chains. *Combination* involves the coupling together of two growing chains to form a single polymer molecule

$$\text{ⱮⱮ}CH_2\text{—}\dot{C}H + \dot{C}H\text{—}CH_2\text{ⱮⱮ} \rightarrow \text{ⱮⱮ}CH_2\text{—}CH\text{—}CH\text{—}CH_2\text{ⱮⱮ}$$
$$\qquad\quad\underset{X}{|}\quad\underset{X}{|}\qquad\qquad\qquad\qquad\underset{X}{|}\quad\underset{X}{|}$$

Note that this results in a 'head-to-head' linkage. Alternatively, a hydrogen atom can be abstracted from one growing chain by another in a reaction known as *disproportionation*

$$\text{ⱮⱮ}CH_2\text{—}\dot{C}H + \dot{C}H\text{—}CH_2\text{ⱮⱮ} \rightarrow \text{ⱮⱮ}CH_2\text{—}CH_2 + CH{=}CH\text{ⱮⱮ}$$
$$\qquad\quad\underset{X}{|}\quad\underset{X}{|}\qquad\qquad\qquad\qquad\underset{X}{|}\quad\underset{X}{|}$$

Thus two polymer molecules are formed, one with a saturated end-group and the other with an unsaturated end-group. Also the chains have initiator fragments at only one end whereas combination yields polymer molecules with initiator fragments at both ends.

In general, both types of termination reaction take place but to different extents depending upon the monomer and the polymerization conditions. For instance it is found that polystyrene chain radicals terminate principally by combination whereas poly(methyl methacrylate) chain radicals terminate predominantly by disproportionation, especially at temperatures above 60°C.

2.4.4 *Rate of polymerization*

The general reaction scheme indicated in the previous sections can be analysed to obtain equations that describe the kinetics of free-radical polymerization.

Initiation may be represented by the two steps

$$I \xrightarrow{\text{slow}} nR\dot{}$$
$$R\dot{} + M \xrightarrow{\text{fast}} RM_1\dot{}$$

where n is the number of free radicals $R\dot{}$ formed upon breakdown of one molecule of the initiator I (usually $n = 1$ or 2), and M represents either a molecule of monomer or a monomer unit in a polymer chain. Since the formation of R˙ from the initiator proceeds much more slowly than the reaction of R˙ with monomer, the first step is rate-determining and controls the rate, R_i, of formation of active centres which then is defined as

$$R_i = \frac{d[R˙]}{dt} \tag{2.28}$$

If it is assumed that the rate constant, k_p, for propagation is independent of the length of the growing chain then *propagation* can be represented by a single general reaction

$$M_i˙ + M \xrightarrow{k_p} M_{i+1}˙$$

in which the initiator fragment R has been omitted for clarity. The amount of monomer consumed in the initiation stage is negligible compared to that consumed by the growing chains and so the rate of consumption of monomer is given by

$$-\frac{d[M]}{dt} = k_p[M_1˙][M] + k_p[M_2˙][M] + \ldots + k_p[M_i˙][M] + \ldots$$

or

$$-\frac{d[M]}{dt} = k_p[M] \left([M_1˙] + [M_2˙] + \ldots + [M_i˙] + \ldots\right)$$

If $[M˙]$ is the total concentration of all radical species (i.e. $[M˙] = \sum_{i=1}^{\infty} [M_i˙]$) then

$$-\frac{d[M]}{dt} = k_p[M][M˙] \tag{2.29}$$

Termination can be represented by

$$M_i˙ + M_j˙ \begin{array}{c} \xrightarrow{k_{tc}} M_{i+j} \\ \xrightarrow{k_{td}} M_i + M_j \end{array}$$

where k_{tc} and k_{td} are the rate constants for combination and disproportionation respectively. Thus the overall rate at which radicals are consumed is given by

$$-\frac{d[M˙]}{dt} = 2k_{tc}[M˙][M˙] + 2k_{td}[M˙][M˙]$$

The factors of 2 and use of the total concentration, $[M^{\cdot}]$, of radical species arise because *two* growing chains of *any length* are consumed by each termination reaction. The equation can be simplified to

$$-\frac{d[M^{\cdot}]}{dt} = 2k_t[M^{\cdot}]^2 \tag{2.30}$$

where k_t is the overall rate constant for termination and is given by

$$k_t = k_{tc} + k_{td}$$

At the start of the polymerization the rate of formation of radicals greatly exceeds the rate at which they are lost by termination. However, $[M^{\cdot}]$ increases rapidly and so the rate of loss of radicals by termination increases. A value of $[M^{\cdot}]$ is soon attained at which the latter rate exactly equals the rate of radical formation. The net rate of change in $[M^{\cdot}]$ is then zero and the reaction is said to be under *steady-state conditions*. In practice, most free-radical polymerizations operate under steady-state conditions for all but the first few seconds. If this were not so and $[M^{\cdot}]$ continuously increased, then the reaction would go out of control and could lead to an explosion. Steady-state conditions are defined by

$$\frac{d[R^{\cdot}]}{dt} = -\frac{d[M^{\cdot}]}{dt}$$

and from Equations (2.28) and (2.30)

$$R_i = 2k_t[M^{\cdot}]^2$$

so that the steady-state total concentration of all radical species is given by

$$[M^{\cdot}] = \left(\frac{R_i}{2k_t}\right)^{1/2} \tag{2.31}$$

This equation can be substituted into Equation (2.29) to give a general expression for the rate of monomer consumption which more commonly is called the *rate of polymerization* and signified by R_p

$$R_p = -\frac{d[M]}{dt} = k_p[M]\left(\frac{R_i}{2k_t}\right)^{1/2}$$

or

$$R_p = \left(\frac{k_p}{2^{1/2}k_t^{1/2}}\right)R_i^{1/2}[M] \tag{2.32}$$

It must be emphasized that Equation (2.32) and any other equation derived from it, applies only under steady-state conditions.

The initiation stage can now be considered in more detail so that equations for R_i can be derived for substitution into Equation (2.32). The most common method of initiation is thermolysis for which

$$I \xrightarrow{k_d} 2R^{\cdot}$$

and the rate of formation of active centres is given by

$$R_i = 2fk_d[I] \tag{2.33}$$

where k_d is the rate constant for initiator dissociation, f, the *initiator efficiency*, is the fraction of primary free radicals $R^{\cdot}$ that successfully initiate polymerization, and the factor of 2 enters because two radicals are formed from one molecule of initiator. Normally, the initiator efficiency is in the range 0.3–0.8 due to side reactions (see Section 2.4.1). Substitution of Equation (2.33) into Equation (2.32) gives for *initiation by thermolysis*

$$R_p = k_p \left(\frac{fk_d}{k_t} \right)^{1/2} [M][I]^{1/2} \tag{2.34}$$

When initiation is brought about by *photolysis* of an initiator (i.e. $I \xrightarrow{hv} 2R^{\cdot}$)

$$R_i = 2\Phi \, \varepsilon \, I_o[I] \tag{2.35}$$

and

$$R_p = k_p \left(\frac{\Phi \, \varepsilon \, I_o}{k_t} \right)^{1/2} [M][I]^{1/2} \tag{2.36}$$

where I_o is the intensity of the incident light, ε is the molar absorptivity of the initiator and Φ is the quantum yield (i.e. the photochemical equivalent of initiator efficiency).

2.4.5 Degree of polymerization

The number-average degree of polymerization, $\bar{x}_n$, of the polymer produced is given by

$$\bar{x}_n = \frac{\text{Moles of monomer consumed in unit time}}{\text{Moles of polymer formed in unit time}} \tag{2.37}$$

On the basis of the simple kinetic scheme analysed in the previous section

$$\bar{x}_n = \frac{k_p[M][M^{\cdot}]}{k_{tc}[M^{\cdot}]^2 + 2k_{td}[M^{\cdot}]^2}$$

The denominator takes into account the fact that combination produces one polymer molecule whereas disproportionation produces two. Under steady-state conditions [M˙] can be substituted using Equation (2.31) to give

$$\bar{x}_n = \frac{k_p[M]}{(1+q)k_t^{1/2}(R_i/2)^{1/2}} \tag{2.38}$$

where $q = k_{td}/k_t$ and is the fraction of termination reactions that proceed by disproportionation $(0 \leqslant q \leqslant 1)$. When termination occurs only by combination $q = 0$, whereas when it occurs only by disproportionation $q = 1$.

Substitution of Equation (2.33) into Equation (2.38) gives for *initiation by thermolysis*

$$\bar{x}_n = \frac{k_p[M]}{(1+q)(fk_dk_t)^{1/2}[I]^{1/2}} \tag{2.39}$$

Similarly, from Equation (2.35), when initiation is by *photolysis*

$$\bar{x}_n = \frac{k_p[M]}{(1+q)(\Phi \varepsilon I_o k_t)^{1/2}[I]^{1/2}} \tag{2.40}$$

Thus for both types of initiation $\bar{x}_n \propto [M][I]^{-1/2}$ which may be contrasted with $R_p \propto [M][I]^{1/2}$ from Equations (2.34) and (2.36). By increasing [M] it is possible to increase both $\bar{x}_n$ and R_p. However, increasing [I] gives rise to a reduction in $\bar{x}_n$ whilst causing an increase in R_p. As a particular reaction proceeds both [M] and [I] decrease, though [M] decreases more rapidly because a very large number of monomer molecules are consumed by each active centre created. Thus the simple expressions predict that both R_p and $\bar{x}_n$ for the formation of polymer at a *specific instant in time* (i.e. when the monomer and initiator concentrations are [M] and [I]) should decrease with reaction time (ultimately becoming zero).

The validity of the expressions for R_p and $\bar{x}_n$, i.e. of the simple kinetics scheme, will be examined in the following two sections. Experimentally this is done by measuring the initial values of R_p and $\bar{x}_n$, and correlating them with the initial concentrations $[M]_0$ and $[I]_0$.

2.4.6 Autoacceleration

There is ample experimental evidence to support the relationship $(R_p)_{\text{initial}} \propto [M]_0[I]_0^{1/2}$ for most systems. However, when $[M]_0$ is high there often is observed a sharp increase in R_p as the conversion of monomer increases. This phenomenon is known as *autoacceleration* or the *Trommsdorff–Norrish effect* or the *gel effect* and is demonstrated in Fig. 2.5 by some experimental conversion curves for polymerization of methyl methacrylate (MMA) at different $[MMA]_0$. Since R_p is proportional to the

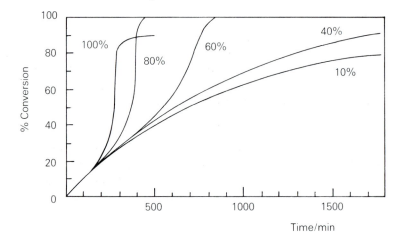

Fig. 2.5 *Effect of initial monomer concentration ([MMA]₀ given as a percentage) upon the % conversion-time curves for polymerization of methyl methacrylate at 323 K using benzoyl peroxide as initiator and benzene as diluent (after Schultz and Harborth, (1947) Makromol. Chem.* **1**, *106).*

slope of the conversion curve, autoacceleration is observed to occur earlier in the reaction and to be more pronounced as $[MMA]_0$ increases. Accompanying the increase in R_p is an increase in the molar mass of the polymer formed.

Autoacceleration arises as a consequence of the increase in viscosity of the reaction medium caused by the formation of polymer molecules. As the viscosity increases the mobility of the long-chain radical species is greatly reduced and they begin to have difficulty in moving into close proximity for termination to occur. When the viscosity attains a critical value termination becomes controlled by the rate of translational diffusion of the growing chains, which is of course very low. The result is a dramatic decrease in the rate constant, k_t, for termination and, from Equations (2.32) and (2.38), a large increase in both R_p and $\bar{x}_n$. The initiation and propagation reactions are not affected because the monomer molecules are small and have relatively high mobility even in media of high viscosity.

Free-radical polymerizations are exothermic and so energy is evolved at an increasing rate as autoacceleration begins. If the dissipation of energy is poor there may well be an explosion. The simplest ways of avoiding autoacceleration are to stop the reaction before chain diffusion becomes difficult or to use dilute solutions of monomer, though the latter is complicated by the occurrence of chain transfer reactions.

2.4.7 *Chain transfer*

The values of $\bar{x}_n$ found experimentally often are very much lower than those calculated from Equation (2.38) and its analogues, indicating the existence of additional reactions that terminate the growth of a chain radical. The reactions which are responsible for the reduction in $\bar{x}_n$ are known collectively as *chain transfer* reactions and may be represented by

$$M_i^{\cdot} + T-A \rightarrow M_i-T + A^{\cdot}$$

where T and A are fragments linked by a single bond in a hypothetical molecule TA. The chain radical abstracts T$^{\cdot}$ (often a hydrogen or halogen atom) from TA to yield a 'dead' polymer molecule and the radical A$^{\cdot}$, which can then react with a molecule of monomer to initiate the growth of a new chain

$$A^{\cdot} + M \rightarrow AM_i^{\cdot}$$

If re-initiation is rapid the rate of polymerization is not affected, but since growth of the chain radicals is terminated prematurely $\bar{x}_n$ is reduced.

All molecular species present in a free-radical polymerization are potential sources of chain transfer. Thus when applying Equation (2.37), a series of terms are required in the denominator in order to take account of the polymer molecules formed by chain transfer to molecules of *monomer*, *initiator* and *solvent* as well as those formed by combination and disproportionation. Hence

$$\bar{x}_n = \frac{k_p[M][M^{\cdot}]}{k_{tc}[M^{\cdot}]^2 + 2k_{td}[M^{\cdot}]^2 + k_{trM}[M^{\cdot}][M] + k_{trI}[M^{\cdot}][I] + k_{trS}[M^{\cdot}][S]}$$

where k_{trM}, k_{trI} and k_{trS} are respectively the rate constants for chain transfer to monomer, initiator and solvent (S). Under steady-state conditions $[M^{\cdot}]$ is given by Equation (2.31) and the above equation can be reduced to

$$\frac{1}{\bar{x}_n} = \frac{(1+q)k_t^{1/2}(R_i/2)^{1/2}}{k_p[M]} + \frac{k_{trM}}{k_p} + \frac{k_{trI}[I]}{k_p[M]} + \frac{k_{trS}[S]}{k_p[M]}$$

which is known as the *Mayo–Walling Equation* and normally is written in the form

$$\frac{1}{\bar{x}_n} = \frac{1}{(\bar{x}_n)_0} + C_M + C_I\frac{[I]}{[M]} + C_S\frac{[S]}{[M]} \tag{2.41}$$

where $(\bar{x}_n)_0$ is the number-average degree of polymerization obtained in the absence of chain transfer, and the *transfer constants* C_M, C_I and C_S are the ratios of k_{tr}/k_p for each type of chain transfer reaction.

Table 2.5 lists for various compounds the transfer constants observed in the polymerization of styrene at 333 K. The transfer constant for styrene shows that one chain transfer to monomer occurs for approximately every 10^4 propagation steps. Chain transfer to the initiator benzoyl peroxide is significant and is an example of induced decomposition (see Section 2.4.1). In contrast, chain transfer to azobisisobutyronitrile is negligible ($C_I \approx 0$) and so it is a better initiator for studies of reaction kinetics. The wide variations in transfer constant for the other compounds listed highlight the need for careful choice of the solvent to be used in a free radical polymerization. Compounds with high transfer constants (e.g. carbon tetrabromide and dodecyl mercaptan) cannot be used as solvents but instead are employed at low concentrations to control (reduce) molar mass and are known as *chain transfer agents* or *regulators* or *modifiers*. In general, transfer constants increase as the strength of the bond cleaved decreases and as the stability of the product radical increases.

Chain transfer to polymer has no effect upon $\bar{x}_n$ but it does result in the formation of branched polymer molecules. *Intramolecular* or *back-biting* reactions give rise to short-chain branches, e.g. in the polymerization of ethylene

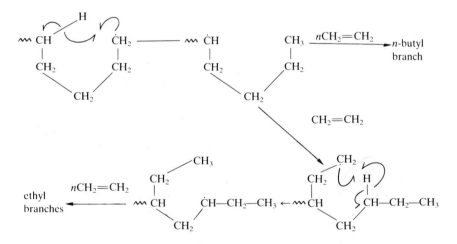

TABLE 2.5 *Transfer constants observed in the polymerization of styrene at 333 K*

Compound	Bond cleaved (T-A)	Transfer constant (k_{tr}/k_p)
Styrene	H—C(Ph):CH$_2$	7×10^{-5}
Benzoyl peroxide	PhCOO—OOCPh	5×10^{-2}
Benzene	H—Ph	1.8×10^{-6}
Toluene	H—CH$_2$Ph	1.2×10^{-5}
Chloroform	H—CCl$_3$	5×10^{-5}
Carbon tetrabromide	Br—CBr$_3$	2.2
Dodecyl mercaptan	H—SC$_{12}$H$_{25}$	14.8

Long-chain branches result from *intermolecular* chain transfer reactions, e.g. in the polymerization of methyl acrylate

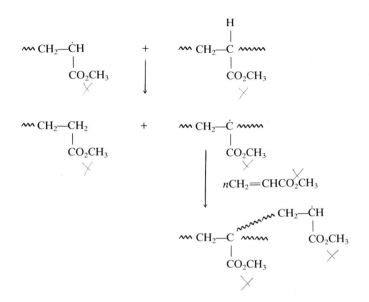

2.4.8 *Inhibition and retardation*

Certain substances react with the free-radical active centres to produce species (either radical or non-radical) which are incapable of re-initiating polymerization. If the reaction is highly efficient polymerization is prevented and the substance is said to be an *inhibitor*. When the reaction is less efficient or yields species that slowly re-initiate polymerization, the rate of polymerization is reduced and the substance is known as a *retarder*. The effect of addition of an inhibitor or a retarder to a free-radical polymerization is shown schematically in Fig. 2.6.

Nitrobenzene is a retarder for polymerization of styrene and acts via chain transfer reactions. The product radicals are of relatively low reactivity and only slowly add to molecules of styrene. In this case both the rate and degree of polymerization are reduced.

Quinones inhibit the polymerization of most monomers by scavenging every active centre formed. An *induction period* is observed whilst the molecules of inhibitor are consumed (and hence deactivated). After this period, polymerization proceeds in the normal way. Oxygen can act as a retarder or an inhibitor and must be excluded from all polymerizations in

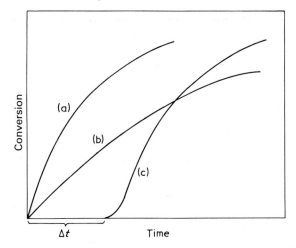

Fig. 2.6 *Effect of different additivies upon free-radical polymerization. (a) reaction without additives, (b) reaction in the presence of a retarder, (c) reaction in the presence of an inhibitor (Δt is the induction period).*

order to obtain reproducible results. Thus it is usual to perform free-radical polymerizations under an inert atmosphere.

Inhibitors are added to monomers at low levels to prevent premature polymerization during transportation and storage. Therefore, monomers should be purified prior to use. On an industrial scale this often is not practical and either the level of initiator is increased to overcome the inhibitor or the monomer is supplied with an inhibitor (e.g. hydroquinone) which is inactive in the absence of oxygen.

Autoinhibition is observed in the polymerization of certain monomers due to degradative chain transfer to monomer, e.g. for allylic monomers

$$R^{\cdot} + CH_2{=}CH{-}CH_2X \rightarrow R{-}H + CH_2{=}CH{-}\dot{C}HX$$

where the product radical is stabilized by resonance. Typically the rates of polymerization are very low and $\bar{x}_n < 20$.

2.4.9 *Molar mass distribution*

The statistical analysis of linear step polymerization (Section 2.2.5) can be extended to the prediction of molar mass distributions for polymers prepared by free-radical polymerization. However, the situation is more complex because the growth of an active centre can be terminated by various different reactions and because the molar mass of the polymer formed decreases with time due to the different rates of decrease in [M] and [I]. In order to simplify the analysis, [M] and [I] will be taken as constants, i.e. the equations derived will be applicable only at *low conversions*.

It is necessary to define a new parameter, β, as the probability that an active centre will propagate rather than terminate. Thus β is given by the ratio of the rate of propagation to the sum of the rates of all reactions that an active centre can undergo (i.e. propagation, combination, disproportionation and chain transfer). It is analogous to the parameter p (the extent of reaction) in step polymerization. However, for free-radical polymerization N is the total number of polymer molecules formed and N_o is the total number of monomer molecules polymerized. (cf. Section 2.2.4).

The simplest case to consider is when the growth of an active centre is terminated by disproportionation and/or chain transfer (though not to polymer). In each case the length of the 'dead' polymer chain is the same as that of the chain radical immediately prior to its termination. A chain of i repeat units is formed after initiation (i.e. from RM_i) by a sequence of $(i - 1)$ propagation reactions. The probability that one of these reactions has taken place is β and the probability that $(i - 1)$ successive monomer additions have occurred is the product of the individual probabilities, i.e. $\beta^{(i-1)}$. The probability that the next reaction is termination rather than propagation is $(1 - \beta)$ and so the probability, $P(i)$, that a chain with degree of polymerization i is formed is given by

$$P(i) = (1 - \beta)\beta^{(i-1)} \tag{2.42}$$

Furthermore, since one termination event has taken place for each polymer molecule formed

$$\beta = \frac{N_o - N}{N_o} \text{ or } N = N_o(1 - \beta) \tag{2.43}$$

It can be seen that Equations (2.42) and (2.43) are exactly analogous to the linear step polymerization Equations (2.8) and (2.2) respectively (i replaces x and β replaces p). Thus under these conditions of termination, polymers prepared by free-radical polymerization also have the most probable distribution of molar mass (Section 2.2.5) and

$$\overline{M}_w/\overline{M}_n = 1 + \beta$$

For the formation of long molecules $\beta \to 1$ and $\overline{M}_w/\overline{M}_n \to 2$ as for linear polymers prepared by step polymerization.

The analysis of termination by *combination* is slightly more complex since two growing chains are terminated by their coupling together. A 'dead' polymer molecule of degree of polymerization i is formed when a growing chain of length j terminates by combination with a growing chain of length $(i - j)$ where j can be any integer from 1 to $(i - 1)$. Thus in this case the probability $P(i)$ is given by

$$P(i) = \sum_{j=1}^{i-1} P(j)P(i - j)$$

where from Equation (2.42)

$$P(j) = (1 - \beta)\beta^{(j-1)}$$
$$P(i - j) = (1 - \beta)\beta^{(i-j-1)}$$

Substituting these expressions into the equation for $P(i)$ gives

$$P(i) = \sum_{j=1}^{i-1} (1 - \beta)^2 \beta^{(i-2)}$$

which easily simplifies to the following relation defining $P(i)$ for termination by combination

$$P(i) = (i - 1)(1 - \beta)^2 \beta^{(i-2)} \tag{2.44}$$

Equation (2.44) describes the mole fraction distribution for the polymer formed and may be used to derive the number-average molar mass from

$$\overline{M}_n = \sum P(i)M_i$$

Since $M_i = iM_0$, where M_0 is the molar mass of the repeat unit, then applying Equation (2.44) gives

$$\overline{M}_n = M_0(1 - \beta)^2 \sum i(i - 1)\beta^{(i-2)}$$

Using the mathematical relation

$$\sum_{i=1}^{\infty} i(i - 1)\beta^{(i-2)} = 2(1 - \beta)^{-3} \text{ for } \beta < 1$$

the equation for $\overline{M}_n$ reduces to

$$\overline{M}_n = \frac{2M_0}{(1 - \beta)} \tag{2.45}$$

In order to evaluate the weight-average molar mass, $\overline{M}_w$, it is necessary to obtain an expression for the weight fraction, w_i, of i-mers. Equation (2.11) from Section 2.2.5 is applicable but requires N_i, the total number of i-mers. This is given by

$$N_i = NP(i) \tag{2.46}$$

Since two propagating chains are terminated by each combination event

$$\beta = \frac{N_0 - 2N}{N_0} \text{ or } N = \frac{N_0(1 - \beta)}{2} \tag{2.47}$$

and so from Equations (2.11), (2.44), (2.46) and (2.47)

$$w_i = (1/2)i(i - 1)(1 - \beta)^3 \beta^{(i-2)} \tag{2.48}$$

and from

$$\overline{M}_w = \sum w_i M_i \qquad \text{with } M_i = iM_0$$

is obtained

$$\overline{M}_w = (1/2)M_0(1 - \beta)^3 \sum i^2(i - 1)\beta^{(i-2)}$$

Another mathematical relation

$$\sum_{i=1}^{\infty} i^2(i - 1)\beta^{(i-2)} = 2(2 + \beta)(1 - \beta)^{-4} \text{ for } \beta < 1$$

leads to

$$\overline{M}_w = \frac{M_0(2 + \beta)}{(1 - \beta)} \tag{2.49}$$

Thus the polydispersity index for termination by combination is given by

$$\overline{M}_w/\overline{M}_n = (2 + \beta)/2 \tag{2.50}$$

and for the formation of long molecules $\beta \to 1$ and so $\overline{M}_w/\overline{M}_n \to 1.5$. Therefore, the molar mass distribution of the polymer formed is narrower when chain growth is terminated by combination than when it occurs by disproportionation and/or chain transfer to molecules other than polymer molecules.

Chain transfer to polymer, which leads to branching, has no effect upon $\overline{M}_n$ because there is no change in the number of polymer molecules formed. The molar mass distribution is not affected when the reaction is intramolecular, but is broadened if it proceeds intermolecularly.

2.4.10 *Determination of individual rate constants*

In order to use the equations for the rate and degree of polymerization predictively, it is necessary to know the values of the individual rate constants for the particular polymerizations of interest. For initiation by thermolysis fk_d, k_p, k_t and k_{tr} are required. The quantity fk_d (typically about 10^{-6}s^{-1}) is most easily determined by measuring the time required for complete consumption of a known quantity of an efficient inhibitor which reacts stoichiometrically with primary free radicals. The stable diphenylpicrylhydrazyl radical often is used for this purpose since it reacts by 1:1 coupling with other free radicals and changes from an intense purple colour to colourless upon reaction. Measurements of the initial rate of polymerization (e.g. by dilatometry) at known $[M]_0$ and $[I]_0$ then enables the ratio $k_p/k_t^{1/2}$ to be evaluated from Equation (2.34). Whilst this ratio could also be determined from measurements of $\bar{x}_n$ using Equation (2.39), the values obtained often are unreliable due to the effects of chain transfer and uncertainty in the value of the termination parameter q. In order to separate k_p and k_t it is necessary to obtain the average lifetime, τ, of an active centre, i.e. the average time elapsing between creation of an active centre and its termination

$$\tau = \frac{\text{concentration of active centres}}{\text{rate of loss of active centres}}$$

and so

$$\tau = \frac{[M^\cdot]}{2k_t[M^\cdot]^2}$$

Since $R_p = k_p[M][M^\cdot]$ the equation may be written in the form

$$\tau = \frac{k_p[M]}{2k_t R_p}$$

Thus simultaneous measurements of τ and R_p at known $[M]$ enable the ratio k_p/k_t to be evaluated. Normally these measurements are made under non-steady-state conditions using a technique known as the *rotating sector method*. Knowledge of $k_p/k_t^{1/2}$ and k_p/k_t allows the value of the individual rate constants to be determined (see Table 2.6). Once k_p is known, values of k_{tr} can be evaluated by application of the Mayo–Walling Equation (2.41) to experimental $\bar{x}_n$ data. This also enables the determination of $(\bar{x}_n)_o$ from which q can be calculated.

TABLE 2.6 *Some typical values for the rate constants for propagation and termination in free-radical polymerization*

Monomer	$k_p/\text{dm}^3\,\text{mol}^{-1}\,\text{s}^{-1}$	$k_t/\text{dm}^3\,\text{mol}^{-1}\,\text{s}^{-1}$
Styrene	176	3.6×10^5
Methyl methacrylate	367	9.35×10^6
Methyl acrylate	2090	4.75×10^6
Vinyl acetate	3700	7.4×10^5

2.4.11 *Effects of temperature*

The temperature dependences of the rate of polymerization and the degree of polymerization arise from the dependences upon temperature, T, of the individual rate constants. The latter are described in terms of the appropriate Arrhenius equations

$$k_d = A_d \exp(-E_d/RT)$$
$$k_p = A_p \exp(-E_p/RT)$$
$$k_t = A_t \exp(-E_t/RT)$$
$$k_{tr} = A_{tr} \exp(-E_{tr}/RT)$$

where A_d, A_p, A_t and A_{tr} are the (nominally temperature-independent) *collision factors* and E_d, E_p, E_t and E_{tr} are the *activation energies* for the individual reactions, and R is the gas constant. Hence it is the activation energies that determine the sensitivity of the value of a rate constant to changes in temperature. Typical values of activation energy for the above processes are given in Table 2.7.

From Equations (2.34) and (2.39) for initiation by thermolysis $R_p \propto k_p k_d^{1/2}/k_t^{1/2}$ and $\bar{x}_n \propto k_p/k_d^{1/2}k_t^{1/2}$. Applying the Arrhenius equations, taking logarithms and differentiating with respect to temperature gives

$$\frac{d\ln(R_p)}{dT} = \frac{(2E_p + E_d) - E_t}{2RT^2} \tag{2.51}$$

and

$$\frac{d\ln(\bar{x}_n)}{dT} = \frac{2E_p - (E_d + E_t)}{2RT^2} \tag{2.52}$$

Since $E_d \gg E_p > E_t$ then $d\ln(R_p)/dT$ is positive and the rate of polymerization increases with temperature but $d\ln(\bar{x}_n)/dT$ is negative and the degree of polymerization decreases with temperature. In qualitative terms, increasing the temperature increases $[M^\cdot]$, i.e. more chains are growing at any given time, which increases the rate of monomer consumption but also increases the probability of termination thereby reducing $\bar{x}_n$.

TABLE 2.7 *Typical ranges for the activation energies of the individual processes in free-radical polymerization*

Process	Activation energy/kJ mol^{-1}
Initiator dissociation	$110 < E_d < 160$
Propagation	$15 < E_p < 40$
Termination	$2 < E_t < 20$
Chain transfer	$40 < E_{tr} < 80$

The rate constant dependences of the kinetics Equations (2.36) and (2.40) for polymerization initiated by photolysis differ from those for thermolysis only in that they have no term in k_d, and so analysis of them yields equations that are simply Equations (2.51) and (2.52) with $E_d = 0$. Thus both $d\ln(R_p)/dT$ and $d\ln(\bar{x}_n)/dT$ are positive, showing that both R_p and $\bar{x}_n$ increase with temperature (because the propagation rate constant increases more rapidly with temperature than does the termination rate constant).

The variation of $\bar{x}_n$ with temperature also depends upon the temperature dependence of the transfer constants, C, in the Mayo–Walling Equation (2.41). In general terms $C = k_{tr}/k_p$ and a similar analysis to that for R_p and $\bar{x}_n$ leads to

$$\frac{d\ln(C)}{dT} = \frac{E_{tr} - E_p}{RT^2}$$

Normally $E_{tr} > E_p$ and so transfer constants increase with temperature. Hence chain transfer becomes more significant as temperature increases and contributes increasingly to the reduction in $\bar{x}_n$.

The predicted continuous increase in R_p with temperature is based upon an assumption that the propagation reaction is irreversible. This assumption is not correct and the reverse reaction, *depropagation*, also can occur, i.e. the propagation step is more accurately represented by

$$M_i^{\cdot} + M \rightleftharpoons M_{i+1}^{\cdot}$$

Under equilibrium conditions

$$\Delta G_p^0 = -RT\ln(K_p) \tag{2.53}$$

where ΔG_p^0 is the standard Gibbs free energy for propagation and K_p is the equilibrium constant which is given by

$$K_p = \frac{[M^{\cdot}]_{eq}}{[M^{\cdot}]_{eq}[M]_{eq}} = [M]_{eq}^{-1} \tag{2.54}$$

where $[M]_{eq}$ is the equilibrium monomer concentration. Substitution of $\Delta G_p^0 = \Delta H_p^0 - T\Delta S_p^0$ and Equation (2.54) into Equation (2.53) leads to

$$\ln[M]_{eq} = \frac{\Delta H_p^0}{RT} - \frac{\Delta S_p^0}{R} \tag{2.55}$$

where ΔH_p^0 and ΔS_p^0 are respectively the standard enthalpy and entropy changes for propagation. Since propagation involves the formation of a σ-bond from a less stable π-bond, ΔH_p^0 is negative (typically -50 to $-100\,\text{kJ}\,\text{mol}^{-1}$), i.e. *polymerization is exothermic*. However, because propagation increases the degree of order in the system, ΔS_p^0 also is

TABLE 2.8 *Ceiling temperatures for different monomers*

Monomer	Ceiling temperature T_c
α-Methyl styrene	334 K
Styrene	583 K
Methyl methacrylate	493 K

negative (typically -100 to $-120 \, \text{J K}^{-1} \, \text{mol}^{-1}$) and opposes growth of the active centre. Thus depropagation is promoted by an increase in temperature and from Equation (2.55) $[\text{M}]_{eq}$ increases with increasing temperature. At the *ceiling temperature*, T_c, $[\text{M}]_{eq}$ becomes equal to the concentration of monomer in the pure monomer, $[\text{M}]_{\text{pure}}$, and propagation does not occur. Hence a monomer will not polymerize above its T_c. Rearrangement of Equation (2.55) gives

$$T_c = \frac{\Delta H_p^0}{\Delta S_p^0 + \mathbf{R} \ln [\text{M}]_{\text{pure}}} \tag{2.56}$$

Some values of T_c are given in Table 2.8. For most monomers T_c is well above the temperatures (0–80 °C) normally employed for free-radical polymerization and therefore is not restrictive. However, certain monomers have low values of T_c and can only be polymerized at low temperatures. This generally is the case for 1,1-disubstituted monomers with bulky substituents and results from the reduction in magnitude of ΔH_p^0 caused by steric interactions between the substituent groups in adjacent repeat units of the polymer chain (e.g. ΔH_p^0 for styrene is approximately $-70 \, \text{kJ mol}^{-1}$ but for α-methylstyrene is about $-35 \, \text{kJ mol}^{-1}$).

2.4.12 *Bulk polymerization*

The four commonly used methods for performing free-radical polymerization are described in this and the subsequent three sections. *Bulk polymerization* is the simplest and involves only the monomer and a monomer-soluble initiator. The high concentration of monomer gives rise to high rates of polymerization and high degrees of polymerization. However, the viscosity of the reaction medium increases rapidly with conversion (i.e. as polymer forms), making it difficult to remove the heat evolved upon polymerization because of the inefficient stirring, and leading to autoacceleration. These problems can be avoided by restricting the reaction to low conversions, though on an industrial scale the process economics necessitate recovery and recycling of unreacted monomer. A different complication arises when the polymer is insoluble in its monomer, (e.g. acrylonitrile, vinyl chloride) since the polymer precipitates as it forms and the usual kinetics do not apply.

The principal advantage of bulk polymerization is that it produces high molar mass polymer of high purity. For example, it is used to prepare transparent sheets of poly(methyl methacrylate) in a two-stage process. The monomer is first partially polymerized to yield a viscous solution which then is poured into a sheet mould where polymerization is completed. This method reduces the problems of heat transfer and contraction in volume upon polymerization.

2.4.13 *Solution polymerization*

Many of the difficulties associated with bulk systems can be overcome if the monomer is polymerized in solution. The solvent lowers the viscosity of the reaction medium, thus assisting heat transfer and reducing the likelihood of autoacceleration. However, the presence of solvent leads to other complications. The reduced monomer concentration gives rise to proportionate decreases in the rate and degree of polymerization. Furthermore, if the solvent is not chosen with care chain transfer to solvent may be appreciable and can result in a major reduction in the degree of polymerization. Finally, isolation of the polymer requires either evaporation of the solvent or precipitation of the polymer by adding the solution to an excess of a non-solvent. For these reasons, commercial use of solution polymerization tends to be restricted to the preparation of polymers for applications which require the polymer to be used in solution.

2.4.14 *Suspension polymerization*

A better way of avoiding the problems of heat transfer associated with bulk polymerization on an industrial scale is to use *suspension polymerization*. This is essentially a bulk polymerization in which the reaction mixture is suspended as droplets in an inert medium. The initiator, monomer and polymer must be insoluble in the suspension medium which usually is water. A solution of initiator in monomer is prepared and then added to the pre-heated aqueous suspension medium. Droplets of the organic phase are formed and maintained in suspension by the use of (i) vigorous agitation throughout the reaction and (ii) dispersion stabilizers dissolved in the aqueous phase (e.g. surfactants and/or low molar mass polymers such as poly(vinyl alcohol) or hydroxymethylcellulose). The low viscosity of the aqueous continuous phase and the high surface area of the dispersed droplets provide for good heat transfer. Each droplet acts as a small bulk polymerization reactor for which the normal kinetics apply and polymer is produced in the form of beads (typically 0.1–2 mm diameter) which are easily isolated by filtration provided that they are rigid and not tacky. Thus the reaction must be taken to complete conversion and normally is not used to prepare polymers that have low glass transition temperatures. At high conversions autoacceleration can occur but is better controlled than in bulk polymerization due to the greatly improved heat dissipation.

Suspension polymerization is widely used on an industrial scale (e.g. for styrene, methyl methacrylate and vinyl chloride), though care has to be taken to remove the dispersion stabilizers by thorough washing of the beads.

2.4.15 *Emulsion polymerization*

Another heterogeneous process of great industrial importance is *emulsion polymerization*. The reaction components differ from those used in suspension polymerization only in that the initiator must not be soluble in monomer but soluble only in the aqueous dispersion medium. Nevertheless, this difference has far-reaching consequences for the mechanism and kinetics of polymerization, and also for the form of the reaction product which is a colloidally-stable dispersion of particulate polymer in water known as a *latex*. The polymer particles generally have diameters in the range 0.05–1 μm, i.e. considerably smaller than for suspension polymerization.

Anionic surfactants most commonly are used as dispersion stabilizers. They consist of molecules with hydrophobic hydrocarbon chains at one end of which is a hydrophilic anionic head group and its associated counter-ion (e.g. sodium lauryl sulphate, $CH_3(CH_2)_{11}SO_4^- Na^+$). Due to their hydrophobic tails they have a low molecular-solubility in water and above a certain characteristic concentration, the *critical micelle concentration* (CMC), the surfactant molecules form into spherical aggregates known as micelles which contain of the order of 100 molecules and typically are about 5 nm in diameter. The surfactant molecules in the micelles have their hydrophilic head groups in contact with the water molecules and their hydrocarbon chains pointing inwards to form a hydrophobic core. Micelles have the ability to absorb considerable quantities of water-insoluble substances into their interior, and so when a water-insoluble monomer is added to an aqueous solution containing a surfactant well above its CMC three phases are established: (i) the aqueous phase in which small quantities of surfactant and monomer are molecularly dissolved, (ii) large (about 1 μm) droplets of monomer maintained in suspension by adsorbed surfactant molecules and agitation, and (iii) small (about 10 nm) monomer-swollen micelles which are far greater in number than the monomer droplets but contain a relatively small amount of the total monomer. This is the situation at the beginning of an emulsion polymerization, but in addition, the aqueous phase also contains the initiator which usually is either a redox system (Section 2.4.1) or a persulphate ($S_2O_8^{2-} \xrightarrow{\Delta} 2\,SO_4^{-}$).

Primary free radicals formed from the initiator react with monomer in the aqueous phase to produce oligomeric radical species which then diffuse into monomer-swollen micelles to initiate polymerization. Their entry into monomer droplets is very unlikely because the smaller micelles have a much higher total surface area and also are more efficient at capturing radicals. Propagation within the micelles is supported by absorption of monomer from the aqueous phase, there being concurrent diffusion of monomer from droplets into the aqueous phase to maintain equilibrium. The monomer-swollen polymer-particle nuclei thus formed soon exceed the size of the original micelles; the micellar surfactant becomes adsorbed surfactant and is supplemented by additional surfactant in order to maintain the colloidal stability of the particles (which is due to electrostatic charge repulsion between adsorbed surfactant layers). This additional surfactant derives principally from disruption of monomer-swollen micelles which have not undergone initiation, the monomer released being re-distributed throughout the system. As the reaction progresses there comes a point, usually well below 10% conversion, at which micelles are consumed completely, thus signifying the end of particle nucleation (i.e. *interval I*). The number, N_p, of latex particles per unit volume of latex then remains constant. Polymerization within these particles continues, supported by diffusion of monomer as described above. Thus the monomer droplets have only one function which is to serve as reservoirs of monomer. The rate of monomer diffusion exceeds the rate of polymerization so that the concentration, $[M]_p$, of monomer within a particle remains constant. Since N_p is constant, the rate of polymerization also is constant. Eventually monomer droplets are exhausted, marking the end of the period of constant rate (i.e. *interval II*). Thereafter $[M]_p$ and the rate of polymerization decrease continuously as the remaining monomer present in the particles is polymerized (*interval III*). The three intervals of emulsion polymerization are shown by the conversion curve in Fig. 2.7.

If $\bar{n}$ is the average number of radicals per latex particle, then the corresponding number of moles of radicals is $\bar{n}/N_A$, where N_A is the Avogadro constant. Thus the rate of polymerization in an 'average' particle is $k_p[M]_p(\bar{n}/N_A)$ and so the *rate of polymerization per unit volume of latex* is given by

$$R_p = k_p[M]_p(\bar{n}/N_A)N_p \tag{2.57}$$

where R_p and k_p have their usual meaning and $[M]_p$ and N_p are as defined above. The value of $\bar{n}$ can vary widely dependent upon the reaction formulation and the conditions used. Only the simplest situation will be

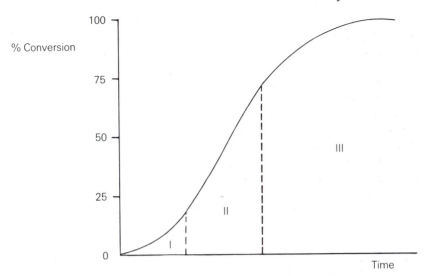

Fig. 2.7 *Variation of percentage conversion with time for emulsion polymerization.*

considered here and involves the assumption that, because the particles are very small, two radical species can only exist independently within a particle for very short periods of time before reacting together. Thus termination can be considered to occur immediately upon entry of a second radical species into a particle containing a single propagating chain radical. The particle then remains dormant until entry of another radical initiates the propagation of a new chain radical. Therefore, on average, each particle contains one propagating chain radical for half the time of its existence and none for the remaining half. Under these conditions $\bar{n} = \frac{1}{2}$ and

$$R_p = \frac{k_p[\text{M}]_p N_p}{2\text{N}_A} \tag{2.58}$$

Furthermore, if the molar rate of formation of radical species from the initiator is ρ_i then the average time interval between successive entries of radicals into a particle is $(N_p/\rho_i\text{N}_A)$. Since each propagating chain radical adds molecules of monomer at a rate $k_p[M]_p$, the number-average degree of polymerization of the polymer formed is given by

$$\bar{x}_n = k_p[\text{M}]_p(N_p/\rho_i\text{N}_A) \tag{2.59}$$

Equations (2.58) and (2.59) are known as the *Smith–Ewart Equations* and are best applied to interval II when $[\text{M}]_p$ and N_p are constant. They show that both R_p and $\bar{x}_n$ can be increased by increasing N_p (e.g. by using a higher concentration of surfactant). Thus because simultaneously-propagating chain radicals are segregated into separate particles, high molar mass polymer is formed at high rates of polymerization.

Further important features of emulsion polymerization are the good heat transfer, the relatively low viscosity of the product latexes at high polymer concentrations and the ability to control particle morphology (e.g. formation of core-shell particle structures by successive additions of different monomers). In addition, it is worth noting that the reaction mechanism usually is more complex than indicated here. Chain transfer agents are often used to reduce $\bar{x}_n$, desorption of small radical species from the particles takes place, and other modes of particle nucleation can operate.

Polymers prepared by emulsion polymerization are used either directly in the latex form (e.g. emulsion paints, adhesives, foamed carpet-backings) or after isolation by coagulation or spray-drying of the latex (e.g. synthetic rubbers and thermoplastics). In this respect, contamination by the inorganic salts and dispersion stabilizers is often the most significant problem.

2.5 Ionic polymerization

Chain polymerization of olefinic monomers can also be effected via active centres which possess an ionic charge. There are two types of *ionic polymerization*: those in which the active centre is positively charged are termed *cationic polymerizations* and those in which it is negatively charged are termed *anionic polymerizations*. Because the active centre has an ionic charge, these polymerizations are more monomer-specific than free-radical polymerization and will proceed only with monomers that have substituent groups which can stabilize the active centre (e.g. by inductive and/or mesomeric effects). For cationic active centres of the type $\sim\!\!CH_2\!\!-\!\!\overset{+}{C}HX$, polymerization will proceed if the substituent group X is able to donate electrons and/or delocalize the positive charge. However, for polymerization via anionic active centres (e.g. $\sim\!\!CH_2\!\!-\!\!\overset{-}{C}HX$), the substituent group must be able to withdraw electrons and/or delocalize the negative charge. Thus, most monomers cannot be polymerized both by cationic and by anionic polymerization (see Table 2.9). Only when the substituent group has a weak inductive effect and is capable of delocalizing both positive and negative charges will a monomer undergo both cationic and anionic polymerization (e.g. styrene and 1,3-dienes).

There are further important distinctions between free-radical and ionic polymerizations. For example, many ionic polymerizations proceed at very much higher rates than is usual for free-radical polymerization, largely because the concentration of actively propagating chains is very much

TABLE 2.9 *Susceptibility of various types of monomers to free-radical, cationic and anionic polymerization**

Monomer	Structure[†]	Free-radical	Cationic	Anionic
Ethylene	$CH_2 = CH_2$	✓	✓	×
1-Alkyl olefins +	$CH_2 = CHR_1$	(×)	(×)	×
1,1-Dialkyl olefins	$CH_2 = CR_1R_2$	(×)	✓	×
1,3-Dienes	$CH_2 = CH - CR = CH_2$	✓	✓	✓
Styrene, α-methyl styrene	$CH_2 = CRPh$	✓	✓	✓
Vinyl halides	$CH_2 = CHX$	✓	×	(×)
Vinyl esters	$CH_2 = CHOCOR_1$	✓	×	(×)
Vinyl ethers	$CH_2 = CHOR_1$	(×)	✓	×
Acrylates, methacrylates	$CH_2 = CRCOOR_1$	✓	×	✓

* ✓ means susceptible, × means not susceptible, and (×) indicates that whilst the monomer might be expected to polymerize (e.g. on the basis of inductive and/or mesomeric effects) it does not do so successfully because of side-reactions (e.g. chain transfer).

[†] R_1 and R_2 = alkyl, R = H or CH_3, Ph = phenyl, and X = halogen.

+ Also called α-olefins.

higher (typically by a factor of 10^4-10^6). A further difference is that a propagating ionic active centre is accompanied by a *counter-ion* of opposite charge. Both the rate and stereochemistry of propagation are influenced by the counter-ion and its degree of association with the active centre. Thus the polarity of a polymerization solvent and its ability to solvate the counter-ion can have very significant effects upon ionic polymerizations. Furthermore, termination cannot occur by reaction between two ionic active centres because they are of similar charge and hence repel each other.

2.6 Cationic polymerization

Since cationic polymerization is a chain reaction it also can be considered in terms of initiation, propagation and termination stages.

2.6.1 *Initiation*

Cationic active centres are created by reaction of monomer with electrophiles (e.g. R^+). Protonic acids such as sulphuric acid (H_2SO_4) and perchloric acid ($HClO_4$) are of use as initiators and involve addition of a proton (H^+) to monomer. However, hydrogen halide acids (e.g. HCl) are not suitable as initiators because the halide counter-ion rapidly combines with the carbocationic active centre to form a stable covalent bond. Lewis acids such as boron trifluoride (BF_3), aluminium chloride ($AlCl_3$) and tin tetrachloride ($SnCl_4$) are the most important initiators but must be used in conjunction with a so-called 'co-catalyst' which very often is water but also can be an acid or an organic halide

$$BF_3 + H_2O \rightleftharpoons H^+(BF_3OH)^-$$
$$AlCl_3 + RCl \rightleftharpoons R^+(AlCl_4)^-$$

For the BF_3/H_2O system, the second step of initiation is

$$H^+(BF_3OH)^- + CH_2{=}CR_1R_2 \rightarrow CH_3{-}\overset{+}{C}R_1R_2(BF_3OH)^-$$

In general the second step can be represented by

$$R^+A^- + CH_2{=}CR_1R_2 \rightarrow R{-}CH_2{-}\overset{+}{C}R_1R_2A^-$$

where R^+ is the electrophile and A^- is the counter-ion.

2.6.2 Propagation

For the same reasons as described for free-radical polymerization, propagation proceeds predominantly via successive head-to-tail additions of monomer to the active centre.

$$\text{\Large m} CH_2{-}\overset{+}{C}R_1R_2A^- + CH_2{=}CR_1R_2 \rightarrow \text{\Large m}CH_2{-}CR_1R_2{-}CH_2{-}\overset{+}{C}R_1R_2A^-$$

2.6.3 Termination

Growth of individual chains is terminated most commonly either by unimolecular rearrangement of the ion pair, e.g.

$$\text{\Large m} CH_2{-}\overset{+}{C}R_1R_1A^- \rightarrow \text{\Large m}CH{=}CR_1R_2 + H^+A^-$$

or through chain transfer. Chain transfer to monomer often makes a significant contribution

$$\text{\Large m} CH_2{-}\overset{+}{C}R_1R_2A^- + CH_2{=}CR_1R_2 \rightarrow \text{\Large m}CH{=}CR_1R_2 + CH_3{-}\overset{+}{C}R_1R_2A^-$$

Additionally, chain transfer to solvent, reactive impurities (e.g. H_2O) and polymer can occur, the latter resulting in the formation of branched polymers.

2.6.4 Kinetics

The exact mechanism of cationic polymerization often depends upon the type of initiator, the structure of the monomer and the nature of the solvent. Often the reaction is heterogeneous because the initiator is only partially soluble in the reaction medium. These features make the formulation of a general kinetics scheme somewhat difficult. Nevertheless a kinetics scheme based upon the chemistry given in the previous three sections will be analysed, though of course the equations obtained must be applied with discretion.

The following general kinetics scheme will be assumed

Initiation: $\quad R^+A^- + M \xrightarrow{k_i} RM_1^+ A^-$

Propagation: $\quad RM_n^+ A^- + M \xrightarrow{k_p} RM_{n+1}^+ A^-$

Termination: $\quad RM_n^+ A^- \xrightarrow{k_t} RM_n + H^+ A^-$

Chain transfer: $RM_n^+ A^- + M \xrightarrow{k_{tr.M}} RM_n + HM_1^+ A^-$

The rate of polymerization is given by

$$R_p = -\frac{d[M]}{dt} = k_p[M][M^+] \tag{2.60}$$

where $[M^+] = \sum_{n=1}^{\infty} [RM_n^+]$. Applying the steady-state condition

$$d[M^+]/dt = -d[M^+]/dt$$

gives

$$k_i[R^+A^-][M] = k_t[M^+]$$

which can be rearranged to

$$[M^+] = \left(\frac{k_i}{k_t}\right)[R^+A^-][M]$$

Substitution of this relation into Equation (2.60) gives an expression for the steady-state rate of polymerization

$$R_p = \left(\frac{k_i k_p}{k_t}\right)[R^+A^-][M]^2 \tag{2.61}$$

which has a second order dependence upon [M] resulting directly from the usually satisfactory assumption that the second step of initiation is rate-determining (cf. free-radical polymerization, Section 2.4.4). However, application of the steady-state condition when the formation of R^+A^- is rate-determining for initiation leads to a first order dependence of R_p upon [M].

If the ionic product of termination by ion-pair rearrangement or chain transfer to monomer is capable of rapidly initiating polymerization, then $[M^+] = [R^+A^-]$ and $R_p = k_p[M][R^+A^-]$.

For each of the above kinetics possibilities the equilibrium constant for formation of R^+A^- must be taken into account when evaluating $[R^+A^-]$.

Analysis for the number-average degree of polymerization, $\bar{x}_n$, is somewhat simpler. Equation (2.37) of Section 2.4.5 again is valid and leads to

$$\bar{x}_n = \frac{k_p[M][M^+]}{k_t[M^+] + k_{tr,M}[M^+][M]}$$

for the above kinetics scheme where consumption of monomer by processes other than propagation is assumed to be negligible. Inversion and simplification of the equation gives

$$\frac{1}{\bar{x}_n} = \frac{k_t}{k_p[M]} + \frac{k_{tr,M}}{k_p} \qquad (2.62)$$

This is the cationic polymerization equivalent of the Mayo–Walling Equation (2.41). Additional terms can be added to the right-hand side to take account of other chain transfer reactions, e.g. for chain transfer to solvent the additional term is $k_{tr,S}[S]/k_p[M]$. In each case $\bar{x}_n$ is independent of initiator concentration and in the absence of chain transfer is given by $(\bar{x}_n)_o = (k_p/k_t)[M]$.

2.6.5 *Effects of temperature*

Following the method of analysis used in Section 2.4.11, appropriate Arrhenius expressions can be substituted into Equation (2.61). Taking logarithms and then differentiating with respect to temperature gives

$$\frac{d \ln (R_p)}{dT} = \frac{E_i + E_p - E_t}{RT^2} \qquad (2.63)$$

The substituent group effects which give rise to stabilization of the cationic active centre also cause polarization of the $C{=}C$ bond in the monomer, activating it towards attack by electrophiles. Thus the activation energies E_i and E_p are relatively small and individually must be less than E_t (and also $E_{tr,M}$) if polymer is formed. In view of this, the overall activation energy $E_i + E_p - E_t$ can be positive or negative depending upon the particular system. When the overall activation energy is negative, R_p is observed to increase as the temperature is reduced (e.g. polymerization of styrene in 1,2-dichloroethane using $TiCl_4/H_2O$ as the initiator).

A similar analysis for $(\bar{x}_n)_o$ gives

$$\frac{d \ln (\bar{x}_n)_o}{dT} = \frac{E_p - E_t}{RT^2} \qquad (2.64)$$

Since $E_p < E_t$ this shows that $(\bar{x}_n)_o$ increases as temperature decreases. Analysis of the transfer constants $(C = k_{tr}/k_p)$ gives the same equation as for free-radical polymerization

$$\frac{d \ln (C)}{dT} = \frac{E_{tr} - E_p}{RT^2}$$

and since $E_{tr} > E_p$ the effect of chain transfer becomes more significant as temperature increases. Thus $\bar{x}_n$ always decreases if the reaction temperature is increased.

2.6.6 *Solvent and counter-ion effects*

The various states of association between an active centre and its counter-ion can be represented by the following equilibria

$$M_n\text{--}A \rightleftharpoons M_n^+ A^- \rightleftharpoons M_n^+ \| A^- \rightleftharpoons M_n^+ + A^-$$

Covalent	Contact	Solvent-separated	Free ions
bond	ion-pair	ion-pair	

Free carbocationic active centres propagate faster than the contact ion-pairs (typically by an order of magnitude) and so as the equilibria shift to the right the overall rate constant, k_p, for propagation increases. Thus polar solvents which favour ion-pair separation (e.g. dichloromethane) and larger counter-ions which are less strongly associated (e.g. $SbCl_6^-$) give rise to higher values of k_p. Furthermore, an increase in the degree of dissociation of the ions can result in a reduction in the overall rate constant, k_t, for ion-pair rearrangement.

An interesting feature of certain so-called *pseudocationic polymerizations* is that they involve propagation of polarized covalently-bonded species (e.g. $\sim CH_2CR_1R_2\text{---}OClO_3$ species when initiation is by perchloric acid).

The kinetics scheme analysed in Section 2.6.4 naturally leads to formation of polymers with a most probable distribution of molar mass (cf. Section 2.4.9). However, polymers formed by cationic polymerization sometimes have complex (e.g. bimodal) molar mass distributions and in certain cases (e.g. pseudocationic polymerizations) this is thought to be due to simultaneous growth of different types of active species with very different but characteristic values of the rate constants for propagation and ion-pair rearrangement.

2.6.7 *Practical considerations*

The propensity for reactions which terminate the growth of propagating chains greatly restricts the usefulness of cationic polymerization. Usually it is necessary to perform the reactions at sub-zero temperatures in order to produce polymers with suitably high molar masses. If water is employed as a co-catalyst then it should only be used in stoichiometric quantities with the initiator, otherwise chain transfer to water will result in a substantial decrease in molar mass. Thus all reactants and solvents should be rigorously dried and purified to remove impurities which could take part in chain transfer reactions. Many cationic polymerizations are extremely rapid and complete conversion of the monomer can be achieved in a matter of seconds. This creates additional problems with respect to heat transfer and non-steady-state conditions.

The only cationic polymerization of major technological importance is the synthesis of butyl rubber by copolymerization of isobutylene with small quantities of isoprene at low temperature (e.g. $-90°C$) using Lewis acid initiators (e.g. $AlCl_3$) and chlorinated solvents (e.g. chloromethane).

2.7 Anionic polymerization

An important feature of anionic polymerization is the absence of inherent termination processes. Termination by ion-pair rearrangement does not occur because it requires the highly unfavourable elimination of a hydride ion. Furthermore, the alkali metal (or alkaline earth metal) counter-ions used have no tendency to combine with the carbanionic active centres to form unreactive covalent bonds. Thus in the absence of chain transfer reactions the propagating polymer chains retain their active carbanionic end-groups. If more monomer is added after complete conversion of the initial quantity, the chains will grow further by polymerization of the additional monomer and will again remain active. Such polymer molecules which permanently retain their active centres (of whatever type) and continue to grow so long as monomer is available are termed *living polymers*.

The polymerization of styrene in liquid ammonia, initiated by potassium amide, was one of the first anionic polymerizations to be studied in detail. In this polymerization chain transfer to ammonia terminates the growth of polymer chains and living polystyrene is not formed. The reaction is of minor importance nowadays but will be briefly considered since it provides a further example of kinetics analysis.

Interest in anionic polymerization grew enormously following the work of Michael Szwarc in the mid-1950s. He demonstrated that under carefully controlled conditions carbanionic living polymers could be formed using electron transfer initiation.

2.7.1 *Polymerization of styrene in liquid NH_3 initiated by KNH_2*

Initiation involves dissociation of potassium amide followed by addition of the amide ion to styrene. Termination occurs by proton abstraction from ammonia (i.e. chain transfer to solvent). Hence the kinetics scheme to be analysed is

Initiation:

$$KNH_2 \rightleftharpoons K^+ + \bar{N}H_2$$

$$\bar{N}H_2 + CH_2=CHPh \xrightarrow{k_i} H_2NCH_2-\bar{C}HPh$$

Propagation:

$$H_2N+CH_2-CHPh)_{n-1}CH_2-\bar{C}HPh + CH_2=CHPh$$

$$\downarrow k_p$$

$$H_2N+CH_2-CHPh)_nCH_2-\bar{C}HPh$$

Chain transfer:

$$H_2N+CH_2-CHPh)_nCH_2-\bar{C}HPh + NH_3$$

$$\downarrow k_{tr}$$

$$H_2N+CH_2-CHPh)_nCH_2-CH_2Ph + \bar{N}H_2$$

The second step of initiation usually is rate-determining and so the amide ion produced upon chain transfer to ammonia can initiate polymerization, but at a rate controlled by the rate constant, k_i, for initiation. Therefore, it is normal to consider this chain transfer reaction as a true kinetic-chain termination step so that application of the steady-state condition gives

$$k_i[\bar{N}H_2][CH_2=CHPh] = k_{tr}[\sim\!\!CH_2-\bar{C}HPh][NH_3]$$

which upon rearrangement yields

$$[\sim\!\!CH_2-\bar{C}HPh] = \left(\frac{k_i}{k_{tr}}\right)\frac{[\bar{N}H_2][CH_2=CHPh]}{[NH_3]}$$

Thus the steady-state rate of polymerization is given by

$$R_p = \left(\frac{k_i k_p}{k_{tr}}\right)\frac{[\bar{N}H_2][CH_2=CHPh]^2}{[NH_3]} \tag{2.65}$$

Analysis for the number-average degree of polymerization using Equation (2.37) leads to

$$\bar{x}_n = \frac{k_p[CH_2=CHPh]}{k_{tr}[NH_3]} \tag{2.66}$$

Examination of Equations (2.65) and (2.66) for the effects of temperature yields equations which are identical in form to Equations (2.63) and (2.64) of Section 2.6.5, but in which E_t is replaced by the activation energy, E_{tr}, for chain transfer to ammonia. Reducing the reaction temperature increases $\bar{x}_n$ (because $E_p - E_{tr} \approx -17\,\text{kJ mol}^{-1}$) but decreases R_p (because $E_i + E_p - E_{tr} \approx +38\,\text{kJ mol}^{-1}$). Due to the small magnitude of $E_p - E_{tr}$ and the high concentration of ammonia, chain transfer to ammonia is highly competitive with propagation and only low molar mass polystyrene is formed even at low temperatures.

2.7.2 *Polymerization without termination: organometallic initiators*

Organolithium compounds (e.g. butyllithium) are the most widely used organometallic initiators. They are soluble in non-polar hydrocarbons but tend to aggregate in such media, i.e.

$$(RLi)_n \rightleftharpoons nRLi$$

where $n = 6$ for ^nBuLi and $n = 4$ for ^sBuLi and ^tBuLi. The reactivity of the aggregated species is very much lower than that of the free species and it is the latter which principally are responsible for initiation, e.g.

$$^sBu^- \; Li^+ \; + \; CH_2{=}CR_1R_2 \rightarrow {}^sBu{-}CH_2{-}CR_1R_2Li^+$$

Disaggregation can be brought about by the addition of small quantities of polar solvents which are able to solvate the lithium ions, e.g. tetra-hydrofuran (THF). If the reaction is performed in polar solvents such as THF, then initiation is rapid because the organolithium compounds exist as free species with more strongly developed ionic character than in non-polar media.

Organometallic compounds of other alkali metals (e.g. benzylsodium and cumylpotassium) are insoluble in non-polar hydrocarbons and so are used only for reactions performed in polar media.

Propagation proceeds in the usual way with predominantly head-to-tail additions of monomer due to steric and electronic effects

$$\sim\!\!\!\sim CH_2{-}\bar{C}R_1R_2M_I^+ \; + \; CH_2{=}CR_1R_2 \rightarrow \sim\!\!\!\sim CH_2{-}CR_1R_2{-}CH_2{-}\bar{C}R_1R_2M_I^+$$

where M_I^+ is the metallic counter-ion.

2.7.3 *Polymerization without termination: electron transfer initiation*

Electron transfer initiation involves donation of single electrons to molecules of monomer to form monomeric radical-anion species which then couple together to give dicarbanion species that initiate polymerization of the remaining monomer. Whilst alkali metals can be used as electron donors, they are insoluble in most organic solvents so that the reaction is heterogeneous and initiation is slow and difficult to reproduce. Homogeneous initiation by electron transfer can be achieved in ether solvents such as THF using soluble electron transfer complexes formed by reaction of alkali metals with polycyclic aromatic compounds (e.g. naphthalene, biphenyl). Sodium naphthalide was one of the first to be used and is formed by the addition of sodium to an ethereal solution of naphthalene

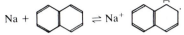

The naphthalide radical-anion is green in colour and stabilized by resonance, though only one canonical form is shown above. The position of the equilibrium depends upon the solvent; in THF it lies completely to the right. The complex initiates polymerization by donation of an electron to monomer

$$Na^+ \quad + CH_2{=}CR_1R_2 \rightleftharpoons \dot{C}H_2{-}\bar{C}R_1R_2 \; Na^+ \; + $$

and this equilibrium is shifted to the right by rapid coupling of the monomeric radical-anions

$$2\,\dot{C}H_2{-}\bar{C}R_1R_2\overset{+}{Na} \rightarrow \overset{+}{Na}\bar{R}_2R_1\bar{C}{-}CH_2{-}CH_2{-}\bar{C}R_1R_2\overset{+}{Na}$$

The green colour rapidly disappears and is replaced by the colour of the carbanionic active centres (e.g. for polymerization of styrene, $R_1 = H$ and $R_2 = Ph$, the colour is red).

Propagation takes place by sequential addition of monomer to both ends of the dicarbanionic species.

2.7.4 *Polymerization without termination: kinetics*

Regardless of whether the initiator is an organometallic compound or an electron transfer complex, each initiator species generates a single carbanionic active centre. It is usual to assume that the initiator reacts completely before any of the active centres begin to propagate, i.e. that all of the initiator species exist in active form free in solution and that the rate constant, k_i, for initiation is very much greater than the rate constant, k_p, for propagation. Under these conditions the total concentration of propagating carbanionic active centres is equal to the concentration, $[I]$, of initiator used. Thus in the absence of termination reactions

$$R_p = -\frac{d[M]}{dt} = k_p[I][M] \qquad (2.67)$$

This shows the reaction to be pseudo-first-order since for each particular reaction $k_p[I]$ is a constant. Equation (2.67) is usually satisfactory for homogeneous polymerizations performed in polar solvents. However, for reactions carried out in non-polar solvents the effects of slow initiation and of aggregation of both the initiator species and the carbanionic active centres must be taken into account when evaluating the concentration of propagating carbanionic active centres.

The number-average degree of polymerization is given by a slight modification of Equation (2.37) in recognition of the living nature of the polymerization

$$\bar{x}_n = \frac{\text{moles of monomer consumed}}{\text{moles of polymer chains produced}} \qquad (2.68)$$

For complete conversion of both monomer and initiator, irrespective of the relative rates of initiation and propagation, application of Equation (2.68) gives

$$\bar{x}_n = \frac{K[M]_0}{[I]_0} \qquad (2.69)$$

where $K = 1$ for initiation by organometallic compounds, $K = 2$ for electron transfer initiation, and $[M]_0$ and $[I]_0$ are respectively the initial concentrations of monomer and initiator.

For monomers with low ceiling temperatures (e.g. α-methylstyrene; see Section 2.4.11) the propagation–depropagation equilibrium must be considered when applying Equation (2.68) since an appreciable monomer concentration may exist at equilibrium.

2.7.5 *Polymerization without termination: molar mass distribution*

Analysis of chain polymerization without termination for the molar mass distribution of the polymer formed requires a different approach to that used in Sections 2.2.5 and 2.4.9 for the analyses of step polymerization and of chain polymerization with termination. It is necessary to define a rate parameter Φ as the average rate at which a molecule of monomer is added to a single active centre that can be either an initiator species or a propagating chain. The precise form of Φ is immaterial, though it can be expected to contain a rate constant and a term in monomer concentration. By using Φ for both the initiation and propagation steps it is implicitly assumed that $k_i = k_p$. However, provided that long polymer chains are formed there is no tangible effect upon the predictions of the analysis if $k_i > k_p$.

At the beginning of the polymerization there are N initiator species, each of which is equally active. After a time t, N_0 of these initiator species have added no molecules of monomer, N_1 have added one molecule of monomer, N_2 have added two molecules of monomer and in general N_x have added x molecules of monomer. Since N_0 decreases by one for each reaction between an initiator species and monomer

$$-\frac{dN_o}{dt} = \Phi N_o$$

Rearranging this equation and applying integration limits

$$\int_N^{N_o} \frac{dN_o}{N_o} = -\int_0^t \Phi dt$$

which by letting $v = \int_0^t \Phi dt$ may be solved to give

$$N_0 = Ne^{-v} \qquad (2.70)$$

The rate of change in the numbers of all other species (i.e. N_x for $x = 1$, $2, 3, \ldots, \infty$) is given by the following general equation

$$\frac{dN_x}{dt} = \Phi N_{x-1} - \Phi N_x \qquad (2.71)$$

since an x-mer is formed by addition of monomer to an $(x - 1)$-mer and is lost by addition of monomer. Recognizing that $dv = \Phi dt$ then Equation (2.71) can be rearranged to give the following first-order linear differential equation

$$\frac{dN_x}{dv} + N_x = N_{x-1}$$

the solution of which gives

$$N_x = e^{-v} \int_0^v e^v N_{x-1} dv \qquad (2.72)$$

Substitution of Equation (2.70) into Equation (2.72) yields $N_1 = Nve^{-v}$ which can then be substituted back into Equation (2.72) to give $N_2 = N(v^2/2)e^{-v}$ which also can be substituted into Equation (2.72) to give $N_3 = N(v^3/6)e^{-v}$. This process can be continued for all values of x, but it soon becomes evident that the solutions have the general form

$$N_x = Nv^x e^{-v}/x! \qquad (2.73)$$

or

$$P(x) = v^x e^{-v}/x! \qquad (2.74)$$

where $P(x)$ is the mole fraction of x-mers. Equation (2.74) is of identical form to the frequency function of the Poisson distribution and so *the polymer formed has a Poisson distribution of molar mass.*

Neglecting initiator fragments the number-average molar mass is given by

$$\overline{M}_n = \sum_{x=1}^{\infty} P(x)M_x$$

where the molar mass of an x-mer $M_x = xM_0$ and M_0 is the molar mass of the monomer. Thus

$$\bar{M}_n = M_0 e^{-v} \sum_{x=1}^{\infty} x v^x / x!$$

i.e.

$$\bar{M}_n = M_0 v e^{-v} \sum_{x=1}^{\infty} v^{x-1} / (x-1)!$$

Using the standard mathematical relation $\sum_{r=0}^{\infty} v^r / r! = e^v$ the equation for $\bar{M}_n$ reduces to

$$\bar{M}_n = M_0 v \qquad (2.75)$$

and so v is *the number-average degree of polymerization*.

Since N living polymer molecules are formed, the total mass of polymer is NM_0v. Thus the weight-fraction, w_x, of x-mers is given by

$$w_x = \frac{N_x M_x}{NM_0 v}$$

neglecting initiator fragments. Substituting $M_x = xM_0$ and Equation (2.73) gives

$$w_x = e^{-v} v^{x-1} / (x-1)! \qquad (2.76)$$

The weight-average molar mass is given by

$$\bar{M}_w = \sum_{x=1}^{\infty} w_x M_x$$

which upon substitution of Equation (2.76) and $M_x = xM_0$ leads to

$$\bar{M}_w = M_0 e^{-v} \sum_{x=1}^{\infty} x v^{x-1} / (x-1)!$$

Using the mathematical relation $\sum_{r=1}^{\infty} r v^{r-1} / (r-1)! = (v+1)e^v$ the equation for $\bar{M}_w$ becomes

$$\bar{M}_w = M_0(v+1) \qquad (2.77)$$

Thus $\bar{M}_w - \bar{M}_n = M_0$ and the molar mass distribution is extremely narrow, as is revealed by the expression for the polydispersity index

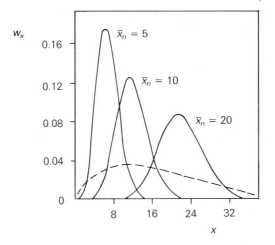

Fig. 2.8 *Weight-fraction Poisson distribution of chain lengths for various values of $\bar{x}_n$ in a polymerization without termination. For comparison, the broken line represents the most probable distribution of molar mass when $\bar{x}_n = 11$ (after Flory).*

$$\frac{\bar{M}_w}{\bar{M}_n} = 1 + \frac{1}{v} \tag{2.78}$$

from which it can be seen that $\bar{M}_w/\bar{M}_n \rightarrow 1$ as $v(= \bar{x}_n) \rightarrow \infty$. This result arises from the difference (M_0) between $\bar{M}_w$ and $\bar{M}_n$ becoming negligible as molar mass increases and is not due to a further narrowing of the molar mass distribution. In fact the distribution becomes broader as $\bar{x}_n$ increases, as is shown by the plots of Equation (2.76) given in Fig. 2.8, thus highlighting the deficiences in using $\bar{M}_w/\bar{M}_n$ as an absolute measure of the breadth of a molar mass distribution. Nevertheless, provided that $k_i \geqslant k_p$, polymerization without termination always yields polymers with narrow molar mass distributions and anionic polymerization is widely used for the preparation of such polymer standards.

2.7.6 *Deactivation of carbanionic living polymers*

Most commonly the active end-groups of carbanionic living polymers are deactivated by reaction with proton donors (usually alcohols) added at the end of the polymerization to give saturated end-groups in the final polymer

$$\text{ⱽⱽCH}_2\text{—}\bar{\text{C}}\text{R}_1\text{R}_2\text{M}_t^+ + \text{ROH} \rightarrow \text{ⱽⱽCH}_2\text{—CHR}_1\text{R}_2 + \text{R}\bar{\text{O}}\text{M}_t^+$$

However, it also is possible to introduce terminal functional groups by making use of reactions that are well established in the field of organic chemistry. For example, carboxylic acid and hydroxyl end-groups can be formed by reactions analogous to those of Grignard reagents

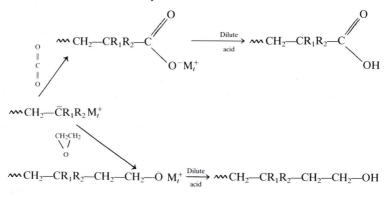

Thus polymer molecules which possess either one or two functionalized end-groups can be prepared by controlled deactivation of living polymers formed using either organometallic initiators or electron transfer initiation respectively. Furthermore, these molecules can be prepared with pre-defined molar masses (by application of Equation (2.69)) and narrow molar mass distributions.

2.7.7 *Solvent and counter-ion effects*

A carbanionic active centre and its counter-ion exist in a series of states of association similar to those described for cationic polymerization in Section 2.6.6. However, in anionic polymerization the ion-pairs are much tighter because the counter-ions are considerably smaller than those employed in cationic polymerization. Consequently, the ion-pairs propagate at rates which typically are two to four orders of magnitude lower than those for propagation of free carbanionic active centres. The reactivity of ion-pairs is greatly enhanced by the use of a more polar solvent especially if it is able to solvate the counter-ion. When solvation is absent or weak (e.g. benzene or 1,4-dioxan) the rate constant for ion-pair propagation increases as the size of the counter-ion increases (i.e. $K^+ > Na^+ > Li^+$) due to the consequent increase in the separation of the ions. However, in polar solvating solvents (e.g. tetrahydrofuran or 1,2-dimethoxyethane) the opposite trend is often observed because the smaller counter-ions are more strongly solvated.

Generally, in non-polar solvents free ions and solvent-separated ion-pairs are not present and the contact ion-pairs are aggregated. Furthermore, initiation can be slow relative to propagation, usually due to stronger aggregation of the initiator species than the propagating ion-pairs, and leads to the formation of polymers with molar mass distributions which are broader than the Poisson distribution. Since the activation energy for propagation is positive, the rates of polymerization increase as the reaction temperature is increased.

In polar solvating solvents the formation of solvent-separated ion-pairs from contact ion-pairs is exothermic whereas the formation of free ions from solvent-separated ion-pairs is essentially athermal. Therefore, the effect of reducing the reaction temperature is to increase the concentration of solvent-separated ion-pairs relative to the concentration of the less reactive contact ion-pairs and the rate of polymerization increases. In contrast to the ion-pairs, the reactivity of free carbanionic active centres is not greatly enhanced by an increase in the strength of solvation of the counter-ions.

The existence of several kinds of propagating species (e.g. free ions and ion-pairs) with vastly different rates of propagation could lead to the formation of polymers with broad or complex molar mass distributions. However, the rates of interconversion between the different species normally are greater than their rates of propagation and so there is no significant effect upon the molar mass distribution. Thus polymers prepared by anionic polymerization using fast initiation have narrow molar mass distributions typically with $\overline{M}_w/\overline{M}_n \leq 1.05$.

The solvent and counter-ion also can influence the stereochemistry of propagation and these effects will be considered in Section 2.8.

2.7.8 *Practical considerations*

The high reactivity and low concentration of carbanionic active centres makes anionic polymerization very susceptible to inhibition by trace quantities of reactive impurities (e.g. H_2O, CO_2, O_2). Thus all reactants and solvents must be rigorously purified and the reactions must be carried out under inert conditions in scrupulously-clean sealed apparatus. High vacuum techniques are often used for this purpose.

Anionic polymerization of polar monomers is complicated by side reactions that involve the polar groups. For example in the polymerization of acrylates and methacrylates, the initiator species and propagating active centres can react with the C=O groups in the monomer (or polymer)

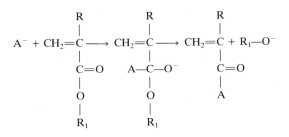

The side reactions lead to loss of initiator, termination of chain growth and formation of polymers with broad molar mass distributions. For methyl methacrylate ($R = R_1 = CH_3$) the side reactions essentially can be eliminated by using polar solvents, low temperatures, bulky initiators (for which reaction with the $C{=}O$ group is sterically-hindered) and large counter-ions, e.g. by polymerization in tetrahydrofuran at $-75°C$ using cumylcaesium as initiator.

The ability to produce polymers of well-defined structure using anionic polymerization is of great importance and despite the above difficulties it is widely used for this purpose. Thus polymers with narrow molar mass distributions, terminally-functionalized polymers, and perhaps most important of all, well-defined block copolymers (Section 2.16.9) can be prepared using anionic polymerization.

2.8 Stereochemistry of polymerization

In addition to the effects of skeletal structure and of the chemical composition of the repeat units, the properties of a polymer are strongly influenced by its molecular microstructure. Variations in the geometric and configurational arrangements of the atoms in the repeat unit, and the distribution of these different spatial arrangements for the repeat units along the chain, are of particular importance.

Different molecular microstructures arise from there being several possible modes of propagation. The possibility of head-to-tail and head-to-head placements of the repeat units has been encountered already, with the observation that for both steric and energetics reasons the placement is almost exclusively head-to-tail for most polymers. Therefore in the subsequent sections dealing with the stereochemistry of propagation only head-to-tail placements will be considered.

2.8.1 *Tacticity*

For polymers prepared from monomers of the general structure $CH_2 = CXY$, where X and Y are two different substituent groups, there are two distinct configurational arrangements of the repeat unit.

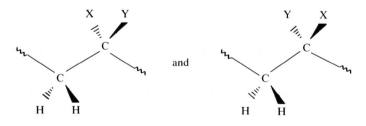

where ▼ and ⫤ indicate bonds which are extending above and below the plane of the paper respectively. These two *stereoisomers* of the repeat unit cannot be interchanged by bond rotation and exist because the substituted carbon atom is attached to four different groups (i.e. it is asymmetric). Unlike simple organic compounds with asymmetric carbon atoms, the stereoisomers indicated above show no significant optical activity because the two polymer chain residues attached to the asymmetric carbon atom are almost identical. Nevertheless, the existence of two isomeric forms of the repeat unit, and in particular their distribution along the polymer chain, are of great significance. In *isotactic* polymers all the repeat units have the same configuration, whereas in *syndiotactic* polymers the configuration alternates from one repeat unit to the next. *Atactic* polymers have a random placement of the two configurations. These three stereochemical forms are shown for short segments of polymer chains in Fig. 2.9.

Polypropylene (X=H, Y=CH₃) provides a good example of the importance of tacticity. The commercial material is essentially isotactic and

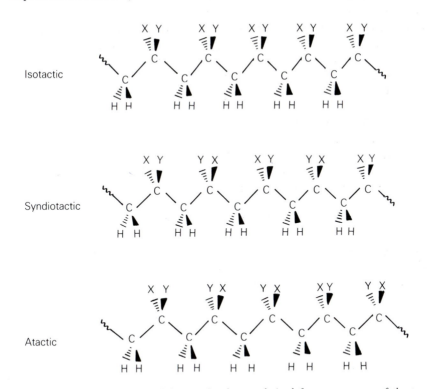

Fig. 2.9 *Different stereochemical forms of polymers derived from monomers of the type CH₂ = CXY.*

due to its regular structure is crystalline (*c.* 65%). It is the crystalline regions that give rise to the good mechanical properties of commercial polypropylene. In contrast, atactic polypropylene is unable to crystallize because of its irregular structure and is a soft, wax-like amorphous material which has no useful mechanical properties.

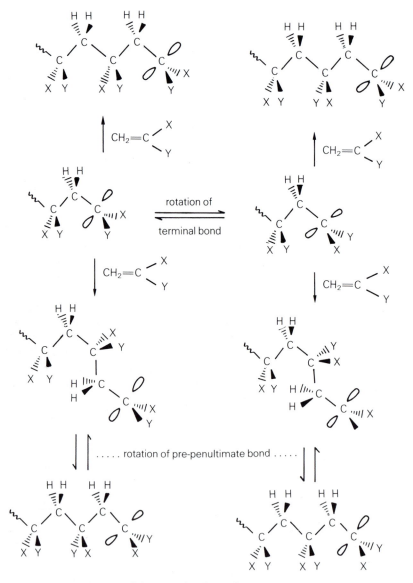

Fig. 2.10 *Elementary features of the stereochemistry of propagation.*

The tacticity of a polymer is controlled by the stereochemistry of propagation, some elementary aspects of which are illustrated in Fig. 2.10. The terminal active centres of propagating chains in free-radical, cationic and anionic polymerizations can be considered to be sp^2 hybridized, the remaining p-orbital containing one, none and two electrons respectively. This form of hybridization is normal for free radicals and carbocations, but for carbanionic active centres is a consequence of resonance with the substituent group(s) (a requirement for which is coplanarity and a change from the normal sp^3 to sp^2 hybridization). Thus in each case there is a planar arrangement of the groups about the terminal active carbon atom and so its configuration in the resulting polymer molecule is determined by the way in which monomer adds to it in a propagation step. As indicated in Fig. 2.10, the orientation of the substituent groups on the terminal active carbon atom relative to the orientation of those on the asymmetric carbon atom of the penultimate repeat unit, and the face of the planar active centre to which the molecule of monomer adds, are of great importance. Usually, steric and/or electronic repulsion between similar substituent groups results in a slight preference for syndiotactic rather than isotactic placements. This preference is accentuated by reducing the reaction temperature and highly syndiotactic polymers can be formed by ionic polymerization in a polar solvent at low temperature (e.g. anionic polymerization of methyl methacrylate initiated by 9-fluorenyllithium at $-78°C$ in tetrahydrofuran). In contrast, the relatively high temperatures normally employed for free-radical polymerizations result in the formation of essentially atactic polymers.

Highly isotactic polymers can be prepared by ionic polymerization if there is strong coordination of the counter-ion with the terminal units in the polymer chain and with the incoming molecule of monomer. However, this is difficult to achieve with non-polar monomers and usually requires the monomer to have polar substituent groups which can act as sites for strong coordination (e.g. cationic polymerization: vinyl ethers; anionic polymerization: methacrylate esters). In order to prepare highly isotactic polymers from such polar monomers the reaction must be carried out at low temperature in a non-polar solvent using an initiator which yields a small counter-ion so that ion-pair association is promoted (e.g. solvent: toluene at $-78°C$; cationic initiator: boron trifluoride etherate; anionic initiator: 1,1-diphenylhexyllithium). The coordination is easily disrupted (e.g. by addition of a small quantity of a polar solvent) resulting in loss of the stereochemical control and formation of predominantly syndiotactic polymer. Isotactic polymers are more easily prepared from non-polar monomers by polymerizations involving coordination to transition metals (Section 2.9).

Polymers with more complex tacticities are formed from monomers of the general structure $XCH=CHY$ since each backbone carbon atom is asymmetric. However, since these monomers do not readily form homopolymers they will not be considered here.

The complications of tacticity are absent in polymers prepared from monomers of the type $CH_2=CX_2$ because they contain no asymmetric backbone carbon atoms and therefore must be stereoregular.

2.8.2 *Polymerization of conjugated dienes*

The most important conjugated dienes are the following 1,3-dienes

$$CH_2=CH-CH=CH_2 \qquad CH_2=\overset{\displaystyle CH_3}{\underset{\displaystyle |}{C}}-CH=CH_2 \qquad CH_2=\overset{\displaystyle Cl}{\underset{\displaystyle |}{C}}-CH=CH_2$$

Butadiene   Isoprene   Chloroprene

which have the general structure $CH_2=CR-CH=CH_2$. There are four basic modes for addition of such 1,3-dienes to a growing polymer chain and these are shown in Table 2.10 (for butadiene there are only three modes because the 1,2- and 3,4-additions are identical since R=H).

The importance of repeat unit isomerism in poly(1,3-dienes) is very clearly demonstrated by the naturally-occurring polyisoprenes. *Gutta*

TABLE 2.10 *Basic modes for addition of 1,3-dienes* $(CH_2=CR-CH=CH_2)$ *to a growing polymer chain* $(M_n^*)^{\dagger}$

Mode of addition	Product of addition	Repeat unit structure
1,2-addition	$M_n-CH_2-\overset{\displaystyle R}{\underset{\displaystyle |}{\underset{\displaystyle CH=CH_2}{C^*}}}$	$+CH_2-\overset{\displaystyle R}{\underset{\displaystyle |}{\underset{\displaystyle CH=CH_2}{C}}}+$
3,4-addition	$M_n-CH_2-\overset{*}{\underset{\displaystyle CR=CH_2}{\underset{\displaystyle |}{CH}}}$	$+CH_2-\underset{\displaystyle CR=CH_2}{\underset{\displaystyle |}{CH}}+$
cis-1,4-addition	$M_n-CH_2\diagdown \diagup\overset{*}{CH_2}$ $C=C$ $H\diagup \diagdown R$	$+CH_2\diagdown \diagup CH_2+$ $C=C$ $H\diagup \diagdown R$
trans-1,4-addition	$M_n-CH_2\diagdown \diagup R$ $C=C$ $H\diagup \diagdown \overset{*}{CH_2}$	$+CH_2\diagdown \diagup R$ $C=C$ $H\diagup \diagdown CH_2+$

† For each mode there is the possibility of head-to-head or head-to-tail placement, and for 1,2- and 3,4-addition the additional complication of isotactic or syndiotactic placement.

percha and *balata* are predominantly *trans*-1,4-polyisoprene, and due to their regular structure are able to crystallize which causes them to be hard rigid materials. However, *natural rubber* is *cis*-1,4-polyisoprene, which has a less symmetrical structure that does not allow easy crystallization under normal conditions and so is an amorphous rubbery material. The difference in regularity between these structures is shown schematically for chain segments containing four head-to-tail repeat units.

cis-1,4-polyisoprene

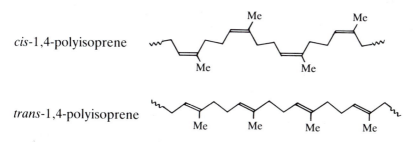

trans-1,4-polyisoprene

Table 2.11 shows the proportions of the different repeat units in homopolymers of butadiene and isoprene prepared using various polymerization conditions. The factors which are of importance in determining these proportions are:

(i) The conformation of the 1,3-diene molecule when it adds to the growing chain, since this at least initially is retained in the new active unit formed by its addition. In the absence of specific effects, the molecules exist mainly in the transoid conformation ( ) which is more stable than the cisoid ( ) and leads to a preponderance of initially *trans*-active units.

(ii) The relative stabilities of the various structures for the active unit.

(iii) For 1,4-addition, the rate of isomerisation between the *cis*- and *trans*- forms of an active unit relative to their individual rates of propagation. Transformation of one form into the other results from the combined effects of resonance and bond rotation, e.g.

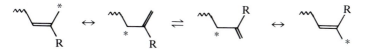

where * represents a single electron or a positive or negative charge.

In free-radical polymerization there are no special effects and the polymers obtained have a high proportion of *trans*-1,4 repeat units which increase in number at the expense of *cis*-1,4 repeat units as the reaction temperature is reduced. The preference for *trans*-1,4 addition is more pronounced for isoprene due to the presence of the methyl substituent group.

TABLE 2.11 *Molecular microstructures of butadiene and isoprene homopolymers prepared using various polymerization conditions*

Monomer	Polymerization conditions	Microstructure (mole fractions)			
		cis-1,4	*trans*-1,4	1,2	3,4
Butadiene	Free radical at −20°C	0.06	0.77	0.17	—
Butadiene	Free radical at 100°C	0.28	0.51	0.21	—
Butadiene	Anionic in hexane with Li$^+$ counter-ion at 20°C	0.68	0.28	0.04	—
Butadiene	Anionic in diethyl ether with Li$^+$ counter-ion at 0°C	0.08	0.17	0.75	—
Isoprene	Free radical at −20°C	0.01	0.90	0.05	0.04
Isoprene	Free radical at 100°C	0.23	0.66	0.05	0.06
Isoprene	Anionic in cyclohexane with Li$^+$ counter-ion at 30°C	0.94	0.01	0.00	0.05
Isoprene	Anionic in diethyl ether with Li$^+$ counter-ion at 20°C	← 0.35 →		0.13	0.52

Anionic polymerization in a non-polar solvent using Li$^+$ as the counter-ions leads to the formation of polymers with high proportions of *cis*-1,4 repeat units. Under these conditions the monomer is held in the cisoid conformation by strong coordination to the small Li$^+$ counter-ion as it adds to the growing chain

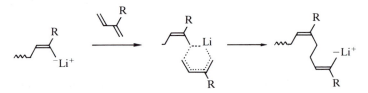

and so the active units initially are in the *cis*- form. Also, the electron density in the active unit is greatest at the terminal carbon atom, thus favouring 1,4 propagation. As long as the monomer concentration is sufficiently high, the rate of propagation of the *cis*- form of the active unit exceeds its rate of isomerization and *cis*-1,4 propagation predominates. The rate of *cis*- to *trans*- isomerization of the active unit in the polymerization of isoprene is much lower than for butadiene and gives rise to the very high *cis*-1,4 content of polyisoprene prepared in this way.

If anionic polymerization is performed in non-polar solvents using counter-ions other than Li$^+$ or in polar solvents (regardless of the counter-ion) stereochemical control is lost and the proportion of 1,4-repeat units is reduced considerably. High proportions of 1,2- and 3,4-addition occur for butadiene and isoprene respectively, partly because coordination effects are much weaker but also because in polar solvents the electron density in the active unit is greatest at the carbon atom in the γ-position relative to the terminal carbon atom.

Cationic polymerization is of little use for the preparation of homopolymers from conjugated dienes because side reactions lead to cyclic structures in the polymer chain and loss of a significant proportion of the expected residual unsaturation.

2.9 Ziegler–Natta coordination polymerization

The use of ionic polymerizations for the preparation of highly stereoregular polymers is restricted to specific monomers, in particular polar monomers. Generally this method is not appropriate for non-polar monomers because they require stronger coordination than can be achieved with the counter-ions used in ionic polymerizations.

In 1953 Karl Ziegler reported the preparation of linear polyethylene by polymerization of ethylene using catalysts prepared from aluminium alkyl compounds and transition metal halides (cf. free-radical polymerization of ethylene yields polyethylene with a large number of both short- and long-chain branches; Section 2.4.7). Giulio Natta quickly recognized, and pursued, the potential of this new type of polymerization for the preparation of stereoregular polymers. By slightly modifying the catalysts used in Ziegler's work, he was able to prepare highly isotactic linear crystalline polymers from non-polar α-olefins (e.g. propylene). The enormous academic and industrial importance of these discoveries was recognized in 1963 by the joint award to Ziegler and Natta of the Nobel Prize for Chemistry.

2.9.1 *Ziegler–Natta catalysts*

Usually *Ziegler–Natta catalysts* are broadly defined in terms of their preparation which involves reacting compounds (commonly halides) of groups IV–VIII transition metals (e.g. Ti, V, Cr, Zr) with organometallic compounds (e.g. alkyls, aryls or hydrides) of groups I–III metals (e.g. Al, Mg, Li). This definition is in fact too broad since not all such reactions yield catalysts suitable for preparing stereoregular polymers. Nevertheless, for each monomer there is a wide range of catalysts that are suitable.

The catalysts which are useful for the preparation of isotactic polymers are *heterogeneous*, i.e. they are insoluble in the solvent, or diluent, in which they are prepared. Their activity and stereoregulating ability are greatly affected by the components, and method, used for their preparation. For example, the α-form of $TiCl_3$ can be used to prepare catalysts suitable for the synthesis of isotactic polypropylene, whereas the β-form yields catalysts which give no stereochemical control. If α-$TiCl_3$ is reacted with $AlEt_2Cl$ it gives a catalyst of lower activity but much higher stereospecificity than that obtained from its reaction with $AlEt_3$. The inclusion of electron donors such as Lewis bases (e.g. ethers, ketones and esters) during preparation of the catalyst also can improve stereospecificity, often but not always with a loss of activity. Ball milling of the catalyst usually improves its activity, not only by increasing the surface area available but also by inducing crystal–crystal transformations.

In the search for higher efficiency, *supported Ziegler–Natta catalysts* have been developed in which the transition metal is either bonded to or occupies lattice sites in the support material. Magnesium compounds are widely used as supports (e.g. $Mg(OH)_2$, $Mg(OEt)_2$, $MgCl_2$) and catalysts with both high activity and high stereospecificity can be obtained from $TiCl_4$ supported on $MgCl_2$ which has been ball-milled in the presence of aromatic esters (e.g. ethyl benzoate).

Ziegler–Natta catalysts that are soluble in the solvent in which they are prepared (i.e. *homogeneous*) are of limited use because in general they do not provide stereochemical control. Nevertheless there are some notable exceptions. For example, syndiotactic polypropylene can be prepared at low temperatures (e.g. $-78°C$) using soluble catalysts based upon vanadium compounds (e.g. $VCl_4 + AlEt_3$). In addition, homogeneous catalysts prepared from benzyl derivatives of Ti and Zr have yielded isotactic polypropylene but are of low activity.

The factors which control catalyst activity and stereospecificity will not be considered here since they are complex and not completely understood. Furthermore, although there is strong experimental evidence that propagation occurs by monomer insertion at a metal–carbon bond, there still is no single definitive mechanism for propagation in Ziegler–Natta polymerizations. In the following sections two mechanisms which are representative of those that have been postulated will be described for heterogeneous catalysts prepared by reaction of α-$TiCl_3$ with trialkylaluminiums (AlR_3).

2.9.2 *Propagation: monomer insertion at group I–III metal–carbon bonds*

A number of mechanisms have been proposed for propagation by insertion of monomer at groups I–III metal–carbon bonds after initial polarization of the monomer by coordination to the transition metal. Since both metals are involved, these are often termed *bimetallic mechanisms*. An example is the mechanism proposed by Natta in which the active site is an electron-deficient bridge complex formed by reaction between a surface Ti atom and AlR_3. Propagation may be represented by

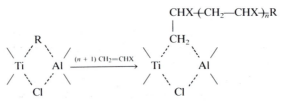

where peripheral ligands (i.e. Cl for the Ti atom and R for the Al atom) are omitted for clarity. The proposed mechanism is shown below and involves initial coordination of monomer to the Ti atom. This is followed by cleavage of the Ti—C bridging bond and polarization of the monomer in a six-membered cyclic transition state. The molecule of monomer then inserts into the Al—C bond and the bridge reforms.

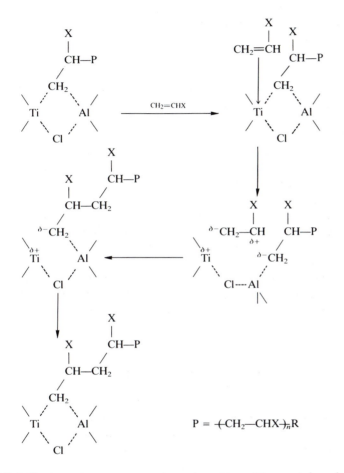

2.9.3 *Propagation: monomer insertion at transition metal–carbon bonds*

A *monometallic mechanism* proposed by Cossee and Arlman is the most widely accepted mechanism in which propagation occurs by insertion of monomer at transition metal–carbon bonds. They recognized that for electrical neutrality in α-TiCl$_3$ crystals the octahedrally-coordinated surface Ti atoms must have Cl vacancies (i.e. empty d-orbitals) and proposed that the active sites are surface Ti atoms which have been alkylated by reaction with AlR$_3$. The overall propagation reaction is represented by

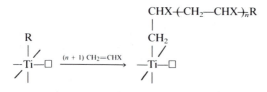

where the peripheral Cl ligands have been omitted for clarity and —☐ indicates an empty d-orbital. Details of the mechanism are presented below. After initial coordination of the monomer at the vacant d-orbital, it is inserted into the Ti-C bond via a cyclic transition state. The polymer chain then migrates back to its original position thus preserving the stereochemical control associated with the specific nature of the catalyst surface.

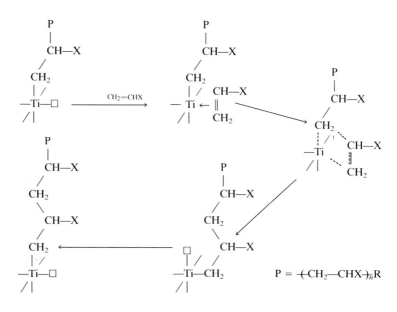

2.9.4 *Propagation: mechanistic overview*

Whilst the precise mechanism of propagation is not known, there are certain mechanistic features which now are widely accepted on the basis of experimental evidence:

(i) Monomer initially is coordinated at vacant d-orbitals of transition metal atoms at the catalyst surface.

(ii) The orientation of a coordinated molecule of monomer is determined by its steric and electronic interactions with the ligands around the transition metal atom. One particular orientation is of lowest energy.

(iii) The propagation step is completed by insertion of the coordinated molecule of monomer into a metal–carbon bond.

(iv) The orientation of the molecule of monomer as it inserts into a metal–carbon bond determines the configuration of the asymmetric carbon atom in the newly-formed terminal repeat unit.

(v) Isotactic polymer is formed when the preferred orientation for coordination of monomer is of much lower energy than other possible orientations; each successive molecule of monomer then adopts the same preferred orientation as it undergoes coordination and then insertion.

(vi) The mechanism of monomer insertion always leads to the formation of linear polymer chains, irrespective of the detailed stereochemistry.

The weight of current experimental evidence is in favour of mechanisms involving successive insertions of monomer into a transition metal–carbon bond. In addition, it is found that the methylene carbon atom from the monomer is always bonded to the transition metal atom (i.e. Cat—CH_2—CHX$+$(CH_2—CHX$)_n$R and not Cat—CHX—$CH_2$$+$(CHX—$CH_2$)$_n$R where Cat represents the catalyst surface).

2.9.5 *Termination*

There are several reactions which can cause termination of the growth of a propagating chain and some of the more common ones are summarized below:

(i) Internal hydride transfer

$$Cat\text{—}CH_2\text{—}CH\text{+}CH_2\text{—}CH\text{+}_nR \rightarrow Cat\text{—}H + CH_2\text{=}C\text{+}CH_2\text{—}CH\text{+}_nR$$
$$\qquad\qquad |\qquad\quad |\qquad\qquad\qquad\qquad\quad |\qquad\quad |$$
$$\qquad\qquad X\qquad\quad X\qquad\qquad\qquad\qquad\quad X\qquad\quad X$$

(ii) Chain transfer to monomer

$$Cat\text{—}CH_2\text{—}CH\text{+}CH_2\text{—}CH\text{+}_nR + CH_2\text{=}CH \rightarrow$$
$$\qquad\qquad |\qquad\quad |\qquad\qquad\qquad\qquad |$$
$$\qquad\qquad X\qquad\quad X\qquad\qquad\qquad\qquad X$$

$$Cat\text{—}CH_2\text{—}CH_2 + CH_2\text{=}C\text{+}CH_2\text{—}CH\text{+}_nR$$
$$\qquad\qquad\qquad |\qquad\qquad\quad |\qquad\quad |$$
$$\qquad\qquad\qquad X\qquad\qquad\quad X\qquad\quad X$$

(iii) Chain transfer to the organometallic compound (M_tR_m)

$$Cat\text{—}CH_2\text{—}CH\text{+}CH_2\text{—}CH\text{+}_nR + M_tR_m \rightarrow Cat\text{—}R + R_{m-1}M_t\text{+}CH_2\text{—}CH\text{+}R$$
$$\qquad\qquad |\qquad\quad |\qquad\qquad\qquad\qquad\qquad\qquad\qquad\qquad\qquad\qquad\qquad\qquad |_{n+1}$$
$$\qquad\qquad X\qquad\quad X\qquad\qquad\qquad\qquad\qquad\qquad\qquad\qquad\qquad\qquad\qquad X$$

(iv) Chain transfer to compounds (H-T) with active hydrogen(s)

$$Cat—CH_2—CH\!\!+\!\!CH_2—CH\!\!+_n\!\!R + H—T \rightarrow Cat—T + CH_3—CH\!\!+\!\!CH_2—CH\!\!+_n\!\!R$$

$$\quad\quad\quad | \quad\quad\quad | \quad\quad\quad\quad\quad\quad\quad\quad\quad\quad | \quad\quad\quad |$$

$$\quad\quad\quad X \quad\quad\quad X \quad\quad\quad\quad\quad\quad\quad\quad\quad\quad X \quad\quad\quad X$$

where Cat and R have their usual meaning, M_t is a group I–III metal of oxidation number m (e.g. $M_t = Al$, $m = +3$) and T is a molecular fragment bonded to an active hydrogen atom.

Under normal conditions of polymerization internal hydride transfer is negligible and termination of propagating chains is dominated by chain transfer processes. Polymer molar mass is often controlled by using hydrogen as a chain transfer agent (i.e. via process (iv) with $T = H$).

2.9.6 *Kinetics*

The kinetics of Ziegler–Natta polymerization are complicated by the heterogeneous nature of the reaction and so will only be considered in simple outline here. The rate of polymerization is given by

$$R_p = k_p C_p^* \theta_M \tag{2.79}$$

where k_p is the rate constant for propagation, C_p^* is the concentration of active catalyst sites and θ_M is the fraction of these sites at which monomer is adsorbed. Usually θ_M is expressed in terms of a standard adsorption isotherm (e.g. Langmuir) and is assumed to have an equilibrium value which depends upon competition between the monomer, the organometallic compound and other species (e.g. hydrogen) for adsorption at the active catalyst sites.

A general equation for the number-average degree of polymerization, $\bar{x}_n$, can be obtained by application of Equation (2.37) of Section 2.4.5

$$\bar{x}_n = \frac{k_p C_p^* \theta_M}{k_{ht} C_p^* + k_{tr,M} C_p^* \theta_M + k_{tr,A} C_p^* \theta_A + k_{tr,H_2} C_p^* \theta_{H_2}}$$

where k_{ht}, $k_{tr,M}$, $k_{tr,A}$ and k_{tr,H_2} are the rate constants for internal hydride transfer and for chain transfer to monomer, organometallic compound and hydrogen respectively, and θ_A and θ_{H_2} are the respective fractions of the active catalyst sites at which the organometallic compound and hydrogen are adsorbed. This equation can be inverted and simplified to yield a general Mayo–Walling equation for Ziegler–Natta polymerization

$$\frac{1}{\bar{x}_n} = \frac{k_{ht}}{k_p \theta_M} + \frac{k_{tr,M}}{k_p} + \frac{k_{tr,A} \theta_A}{k_p \theta_M} + \frac{k_{tr,H_2} \theta_{H_2}}{k_p \theta_M} \tag{2.80}$$

At high concentrations of monomer (i.e. high θ_M) in the absence of hydrogen, Equation (2.80) takes the limiting form $\bar{x}_n = k_p/k_{tr,M}$.

It is well established that there are differences in activity between individual active sites on the same catalyst surface. Furthermore, termination processes can alter the nature of a given active site and thereby modify its activity (Section 2.9.5). Thus the rate constants in Equations (2.79) and (2.80) must be regarded as average quantities. A consequence of the differences in activity between the catalyst sites is that the polymer formed has a broad distribution of molar mass (typically $5 < \bar{M}_w/\bar{M}_n < 30$).

2.9.7 *Practical considerations*

In general organometallic compounds are highly reactive and many ignite spontaneously upon exposure to the atmosphere. For this reason Ziegler–Natta catalysts are prepared and used under inert, dry conditions typically employing hydrocarbons (e.g. cyclohexane, heptane) as solvents and diluents. Normally, polymerization is carried out at temperatures in the range 50–150°C with the general observation that rates of polymerization increase but stereospecificity decreases as temperature increases. Most catalysts have some active sites which do not yield stereoregular polymer. Thus when preparing crystalline isotactic poly(α-olefins) it is often necessary to remove amorphous atactic polymer from the product by solvent extraction.

The three general types of process used for polymerization of ethylene and α-olefins employ heterogeneous catalysts and are the solution, slurry and gas-phase processes. *Solution processes* operate at high temperatures (>130°C) so that as the polymer forms it dissolves in the hydrocarbon solvent used. At the lower temperatures (50–100°C) used in *slurry processes* the polymer is insoluble in the hydrocarbon diluent and precipitates as it forms to give a dispersion (or slurry) of polymer in the diluent. Advances in catalyst technology have led to a major increase in the use of *gas-phase processes* which have the distinct advantage of not requiring a solvent or diluent. These processes involve dispersion of the particulate catalyst in gaseous monomer and operate at low temperatures and pressures. Each of the processes is used for commercial production of high density (linear) polyethylene, isotactic polypropylene and copolymers of ethylene with α-olefins.

Ziegler–Natta catalysts can also be used for preparation of stereoregular polymers from 1,3-dienes. For example, polyisoprene with 96–97 per cent *cis*-1,4 content (i.e. synthetic 'natural rubber') can be prepared using catalysts obtained from $TiCl_4 + Al^iBu_3$.

Attempts have been made to prepare stereoregular polymers from polar monomers (e.g. vinyl chloride, methyl methacrylate) using modified Ziegler–Natta catalysts, but without success. When polymerization does occur it yields non-stereospecific polymer and is thought to proceed by free-radical mechanisms.

2.9.8 *Other catalysts for coordination polymerization*

Chromium trioxide based catalysts supported on silica were developed by Phillips Petroleum at the same time as the original work of Ziegler and Natta. These catalysts polymerize non-polar olefins by mechanisms which are similar to those involved in Ziegler–Natta polymerization but do not give such good stereochemical control and are used principally for the preparation of linear polyethylene. More recently, supported catalysts of very high activity for the polymerization of ethylene have been prepared from chromates and also from chromacene.

2.10 **Ring-opening polymerization**

Polymers with the general structure

$$\left[R{-}Z\right]_n$$

where $-Z-$ is a linking group (e.g. $-O-$, $-O\overset{\overset{\displaystyle O}{\|}}{C}-$, $-NH\overset{\overset{\displaystyle O}{\|}}{C}-$) can be prepared either by step polymerization (Sections 2.2.1 and 2.2.2) or by *ring-opening polymerization* of the corresponding cyclic monomer, i.e.

$$n \ \overset{R}{\underset{Z}{\bigcirc}} \longrightarrow \left[R{-}Z\right]_n$$

Additionally, ring-opening polymerization can be used to prepare polymers which cannot easily be prepared by other methods, e.g. poly(phosphazene)s. Some important ring-opening polymerizations are listed in Table 2.12.

The driving force for ring-opening of cyclic monomers is the relief of bond-angle strain and/or steric repulsions between atoms crowded into the centre of the ring (cf. Section 2.2.7). Therefore, as for other types of polymerization, the enthalpy change for ring-opening is negative. Relief of bond-angle strain is most important for 3- and 4-membered rings, whereas for 8- to 11-membered rings it is the relief of steric crowding that matters. These enthalpic effects are much smaller for 5-, 6- and 7-membered rings (especially 6-membered) and such monomers are more difficult to

TABLE 2.12 *Some important ring-opening polymerizations*

Monomer	Polymer

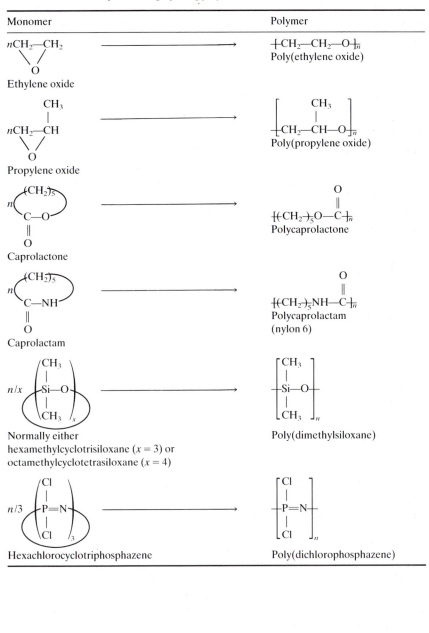

Ethylene oxide → Poly(ethylene oxide)

Propylene oxide → Poly(propylene oxide)

Caprolactone → Polycaprolactone

Caprolactam → Polycaprolactam (nylon 6)

Normally either hexamethylcyclotrisiloxane ($x = 3$) or octamethylcyclotetrasiloxane ($x = 4$) → Poly(dimethylsiloxane)

Hexachlorocyclotriphosphazene → Poly(dichlorophosphazene)

polymerize. Usually ring-opening is initiated by acids or bases and polymer molecules are formed by chain polymerization mechanisms which most commonly involve sequential additions of monomer to cationic or anionic active centres respectively. However, the precise mechanism of polymerization depends greatly upon the initiator, monomer and polymerization conditions. For this reason it is not possible to treat generally the ring-opening polymerization of cyclic monomers. In the subsequent sections some specific cationic and anionic ring-opening polymerizations are considered in order to illustrate the different types of mechanism which can operate.

2.10.1 *Cationic ring-opening polymerization*

The *initiators* used in cationic ring-opening polymerizations are of the same type as those used for cationic polymerization of ethylenic monomers (Section 2.6.1), e.g. strong protonic acids (e.g. H_2SO_4, CF_3SO_3H, CF_3CO_2H), and Lewis acids used in conjunction with co-catalysts (e.g. $Ph_3C^+PF_6^-$, $CH_3\overset{+}{C}O\ SbF_6^-$). For simplicity the initiator will be generally represented as R^+A^-.

In the *polymerization of ethylene oxide*, initiation takes place by addition of R^+ to the epoxide oxygen atom to yield a cyclic oxonium ion (I) which is in equilibrium with the corresponding open-chain carbocation (II)

$$R^+A^- + CH_2{-}\overset{+}{C}H_2 \rightarrow R{-}\overset{+}{O}\underset{CH_2}{\overset{CH_2}{\diagup\;|\;\diagdown}}\ A^- \rightleftharpoons R{-}O{-}CH_2{-}\overset{+}{C}H_2A^-$$

$$\text{(I)} \qquad\qquad\qquad\qquad \text{(II)}$$

Both species can propagate: (I) via ring-opening of the cyclic oxonium ion upon nucleophilic attack at a ring carbon atom by the epoxide oxygen atom in another molecule of monomer, and (II) via its addition to monomer in a reaction similar to the initiation step. In each case the initial product of propagation has a terminal cyclic oxonium ion formed from the newly added molecule of monomer

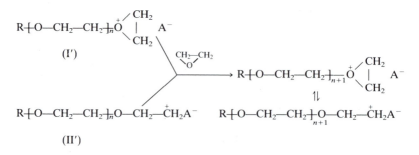

Termination can occur via ion-pair rearrangement of I' or II' to give

$$R \{ O-CH_2-CH_2 \}_n O-CH=CH_2 + H^+A^-$$

Also, when less stable counter-ions such as $AlCl_4^-$ are employed, both I′ and II′ can rearrange to give

$$R \{ O-CH_2-CH_2 \}_n O-CH_2-CH_2Cl + AlCl_3$$

Intramolecular and intermolecular chain transfer to polymer are significant. The former leads to the formation of rings

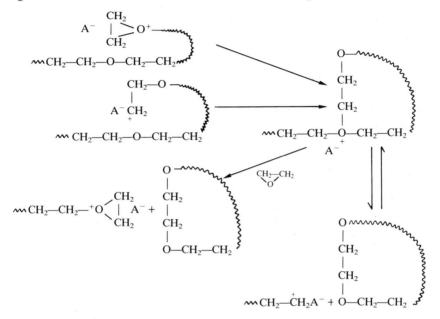

whereas the latter can be considered as interchange reactions and can take place with both linear-chain and ring molecules

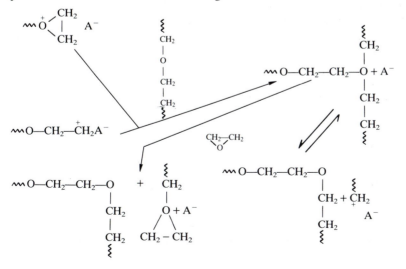

These reactions give rise to *ring-chain equilibria* which are a characteristic feature of many ring-opening polymerizations (i.e. linear-chain molecules are in equilibrium with ring molecules).

This polymerization is further complicated by other modes of propagation (e.g. via a rearranged form of II′, i.e. $\sim\!\!O\!\!-\!\!\overset{+}{C}H\!\!-\!\!CH_3$, or via terminal —OH groups if an alcohol is present or if H^+ is the initiating species). Thus even for ethylene oxide, cationic ring-opening polymerization is quite complex. For epoxide monomers with a substituent group (e.g. propylene oxide) the complexity is greatly increased because it is possible to have head-to-tail or head-to-head placements of repeat units each of which possesses an asymmetric carbon atom that has two possible configurations. The head-to-tail stereoregular forms of such polyethers may be compared to those for ethylenic polymers (Section 2.8.1). When each asymmetric carbon atom has the same configuration (i.e. in the *isotactic* form) the substituent groups alternate from one side to the other of the planar fully-extended backbone

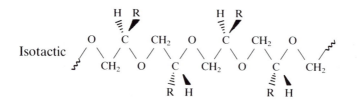

whereas in the *syndiotactic* form the configurations alternate and all the substituent groups are on the same side

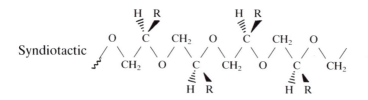

Thus the positions of the substituent groups relative to the plane of the backbone are opposite to those in isotactic and syndiotactic ethylenic polymers (cf. Fig. 2.9). Furthermore, the polyethers are optically active because each repeat unit has a truly asymmetric carbon atom. However, cationic ring-opening polymerization of an optically pure monomer (e.g.) does not result in the formation of isotactic polymer since species of type II′ have planar active centres which usually are stabilized by the substituent group and make a more significant contribution to propagation than for ethylene oxide

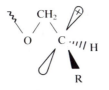

The specific configuration of the asymmetric carbon atom in the monomer is lost when the planar carbocationic active centre is formed.

Lactones (i.e. cyclic esters) of general structure

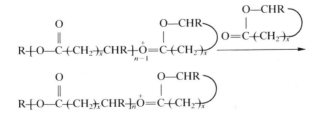

undergo cationic ring-opening polymerization via two principal mechanisms. The dominant mechanism for unsubstituted lactones (i.e. $R = H$) involves addition of R^+ to the carbonyl oxygen atom to form an oxonium ion, followed by scission of the O—CHR bond upon nucleophilic attack by another molecule of monomer, i.e.

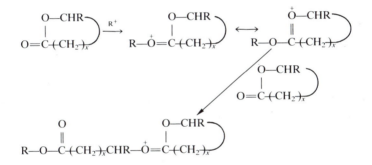

in which the counter-ion has been omitted. This mechanism of propagation may be represented by

For substituted lactones (e.g. $R =$ alkyl) this mechanism results in an inversion of the configuration of the asymmetric carbon atom, —CHR—

(i.e. the configuration in the polyester is opposite to that in the monomer). However, for such substituted lactones a second mechanism of propagation is more probable due to the inductive and steric effects of the substituent. This mechanism involves initial addition of R^+ to the ring oxygen atom, followed by scission of the CO—O bond upon nucleophilic attack by another molecule of monomer, i.e.

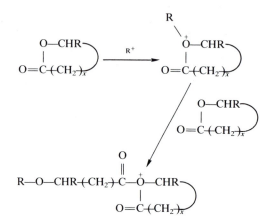

in which the counter-ion again has been omitted. Propagation via this mechanism may be represented by

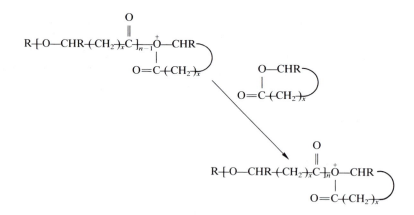

and results in retention of the configuration of the asymmetric carbon atom.

Intramolecular and intermolecular chain transfer to polymer takes place in the polymerization of lactones in much the same way as for epoxides.

Thus ring-chain equilibria are established and are especially important at high conversions of monomer.

2.10.2 *Anionic ring-opening polymerization*

A very wide range of *initiators* have been used to effect anionic ring-opening polymerization and these inclu le alkali metals (e.g. Na, K), inorganic bases (e.g. NaOH, KOH, NaNH$_2$), metal alkoxides (e.g. LiOCH$_3$, NaOEt), metal alkyls and hydrides (e.g. Li^{n}Bu, NaH), electron transfer complexes (e.g. sodium naphthalide) and many other compounds including coordination catalysts. Only the use of the first three types of initiator will be considered here.

Polymerization of *epoxides* by inorganic bases or metal alkoxides (both represented by M$_i^+$B$^-$) proceeds as follows

Initiation: M$_i^+$B$^-$ + CH$_2$—CHR —→ B—CH$_2$—CHR—O$^-$M$_i^+$

Propagation: B$\left[\text{CH}_2\text{—CHR—O}\right]CH_2$—CHR—O$^-M_i^+$ + CH$_2$—CHR

—————→ B$\left[\text{CH}_2\text{—CHR—O}\right]_nCH_2$—CHR—O$^-M_i^+$

For substituted epoxides (i.e. R ≠ H) nucleophilic attack of monomer by the active anion takes place at the least sterically-hindered CH$_2$ carbon atom to give the head-to-tail structures shown above. Also, the configuration of the asymmetric carbon atom (—CHR—) in the monomer is retained in the polymer. Chain transfer to monomer can be a significant side reaction if the substituent group possesses hydrogen atoms on the α-carbon atom, e.g. for propylene oxide

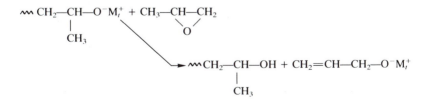

and produces allylic alkoxide ions which initiate polymerization to produce molecules with C=C end groups.

Intermolecular (interchange) and intramolecular (back-biting) chain transfer to polymer can occur but usually are less significant than in cationic ring-opening polymerization.

Lactones polymerize via nucleophilic attack at the ring carbonyl group followed by scission of the CO—O bond

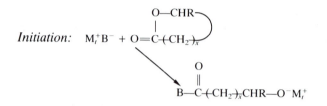

Initiation:

Propagation:

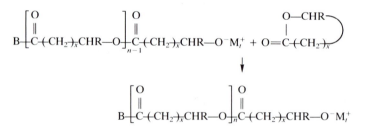

The configuration of the asymmetric carbon atom $-\!(CHR)\!-$ in a substituted monomer is retained in the polymer. An exception to this general mechanism is the polymerization of β-propiolactone ($x=1$, $R=H$) by weak bases (e.g. $CH_3COO^-Na^+$) in which nucleophilic attack takes place at the O—CH_2 carbon atom causing scission of this bond and formation of carboxylate ion active species (i.e. $\sim\!\!\sim CHR\!-\!(CH_2)_x COO^-$).

Polymerization of *lactams* (i.e. cyclic amides) usually is initiated with alkali metals or strong bases and proceeds via a mechanism which initially involves formation of a lactamate ion, e.g.

The lactamate ion is stabilized by resonance

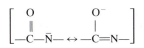

and relatively slowly attacks the carbonyl carbon atom in a molecule of monomer causing scission of the CO—NH bond to produce a highly reactive terminal −N̄H ion which rapidly abstracts H⁺ from another molecule of monomer

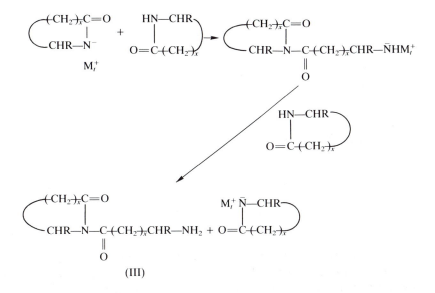

(III)

The ring carbonyl carbon atom in the product (III) is much more strongly activated towards nucleophilic attack than that in the monomer because of the second carbonyl group bonded to the ring nitrogen atom (i.e. (III) is an N-acyllactam). Propagation can be represented generally by

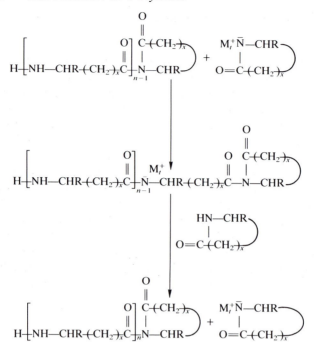

where monomeric lactamate ion attacks the ring carbonyl carbon atom of the terminal N-acyllactam causing scission of the CO—N bond to produce a —$\bar{\text{N}}$— ion which abstracts H^+ from a molecule of monomer to form another lactamate ion. It also is possible for the —$\bar{\text{N}}$— ion to attack a N-acyllactam end-group on another propagating chain and this leads to the formation of branches. For each propagation step, the configuration of the asymmetric carbon atom on a substituted monomer is retained in the polymer.

   Induction periods are often observed in the polymerization of lactams whilst the concentration of (III) increases. These can be eliminated by inclusion of acylating agents (e.g. acid halides and anhydrides) that react rapidly with the lactam to form N-acyllactams which then propagate

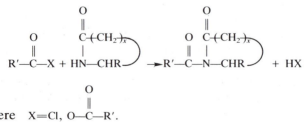

where $X = Cl, O—C—R'$.

Specialized Methods of Polymerization

2.11 Solid-state polymerization

There are many monomers which are capable of undergoing polymerization in the solid (most commonly crystalline) state, though in general the mechanisms of such polymerizations are not clearly understood. In many cases the lattice structure of the crystalline monomer is considerably disrupted upon polymerization and therefore is not maintained in the polymer produced. Nevertheless in favourable cases it is possible to prepare highly oriented polymers by solid-state polymerization.

Certain polyamides and polyesters can be prepared by solid-state polycondensation step polymerization, usually by heating the appropriate monomer or monomer salt. For example, nylon 11 can be prepared by heating crystals of 11-aminoundecanoic acid (melting point = 188°C) at 160°C under vacuum. Polyaddition step polymerization of conjugated dialkene monomers can be induced in the solid state by exposure of the monomer to ultraviolet radiation. For example, irradiation of crystalline 2,5-distyrylpyrazine yields quantitatively a highly crystalline linear cyclobutane polymer

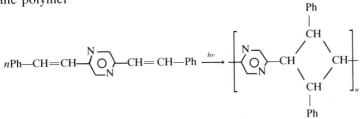

A number of ethylenic monomers and cyclic monomers undergo chain polymerization in the solid state when exposed to high-energy radiation (e.g. γ-radiation). Examples include acrylamide, acrylic acid, acrylonitrile, vinyl acetate, styrene, 1,3,5-trioxane, propriolactone and hexamethylcyclotrisiloxane. These polymerizations may involve free-radical and/or ionic species, but in most cases the precise mechanisms operating are not known.

The period of greatest research activity in solid-state polymerization was the 1960s. Currently, there is considerable interest in the solid-state polymerization of diacetylene (i.e. diyne) single crystals which yield macroscopic polymer single crystals virtually free of defects. These polymerizations can be induced thermally or by irradiation (ultraviolet or high-energy radiation) and are said to be *topochemical* since the direction of chain growth is defined by the geometry and symmetry of the monomer's crystal lattice structure. The polymerization of a symmetrical disubstituted diacetylene monomer is illustrated schematically in Fig. 2.11. The reaction is believed to proceed via carbene species, though free-radical

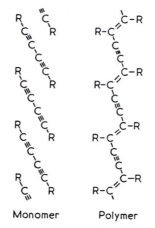

Monomer   Polymer

Fig. 2.11 *Schematic representation of the solid-state polymerization of a symmetrical disubstituted diacetylene monomer. Typical R groups are* $-CH_2-O-SO_2-C_6H_4-CH_3$ *and* $-CH_2-O-CO-NH-C_6H_5$.

species may also be involved. The result is a direct transition from monomer molecules to polymer without major atomic movements and so the three-dimensional order of the monomer's crystal lattice is, to a large extent, maintained. The polymers produced are completely crystalline and the polymer molecules are linear, stereoregular, of very high degree of polymerization and lie in the polymer crystals in a chain-extended conformation. Photographs of some polydiacetylene single crystals are shown in Fig. 4.12 (Section 4.1.6).

2.12 Metathesis polymerization

Olefin (or *alkene*) *metathesis* is a disproportionation reaction which may be represented generally by

$$2R_1-CH=CH-R_2 \rightleftharpoons R_1-CH=CH-R_1 + R_2-CH=CH-R_2$$

and has been applied to the preparation of polymers by *ring-opening metathesis polymerization* of cycloalkenes and bicycloalkenes. A simple example is the metathesis polymerization of cyclopentene

$$n\,\bigcirc \rightarrow \leftarrow CH_2-CH_2-CH_2-CH=CH\rightarrow_n$$

Metathesis polymerization is similar to Ziegler–Natta polymerization in that it is catalysed by the products of reactions between transition metal compounds and either metal alkyls or Lewis acids. However, in contrast to

Ziegler–Natta systems, metathesis reactions normally are performed using soluble (i.e. homogeneous) catalysts. Tungsten-based catalysts are used most widely, especially those obtained by reaction of WCl_6 with aluminium, tin or zinc alkyls (e.g. $AlEtCl_2$, $SnMe_4$, $ZnMe_2$). In addition, catalysts based on molybdenum, ruthenium and rhenium can be used.

The active species in propagation is a transition metal carbene with a vacant d-orbital. For polymerization of cyclopentene by tungsten-based catalysts the overall propagation reaction can be represented by

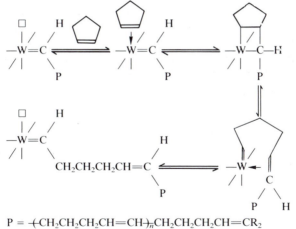

where peripheral tungsten ligands are omitted and —☐ is an empty d-orbital. A detailed mechanism is shown below and involves initial coordination of the C=C bond in the cycloalkene at the vacant d-orbital. An unstable metallocyclobutane intermediate then forms and breaks down to give a new transition metal carbene and a new C=C bond which decoordinates upon migration of the carbene to the original site of the metallocarbene.

$$P = \,\, -(CH_2CH_2CH_2CH=CH)_nCH_2CH_2CH_2CH=CR_2$$

Thus the ring-opening reaction proceeds by cleavage of the C=C bond in the monomer.

Whilst metathesis polymerization has been known since the mid-1950s it is only during the last decade that it has been extensively studied. It has been used in a convenient route for the synthesis of inherently conductive polyacetylene and is used commercially to prepare polymers from norbornene

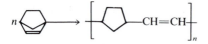

and dicyclopentadiene

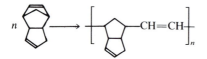

The latter polymerization results in the formation of a crosslinked polymer due to participation of the pendant C=C bonds in the metathesis polymerization.

It should be borne in mind that the polymer structures shown above are simplistic since there are often many stereochemical forms of the repeat units. In certain cases highly stereoregular polymers can be formed, e.g. polymerization of racemic 1-methylnorbornene by rhenium pentachloride gives a polymer with a head-to-tail *cis*-syndiotactic microstructure

2.13 Group transfer polymerization

Unlike other methods of polymerization, *group transfer polymerization* was discovered only recently. It was first reported in the early 1980s by research workers at Du Pont and is suitable for polymerization of acrylic and methacrylic monomers, particularly methacrylates. Propagation involves reaction of a terminal silyl ketene acetal with monomer by Michael addition during which the silyl group transfers to the added monomer thus creating a new terminal silyl ketene acetal group. Polymerization is initiated by monomeric silyl ketene acetals and is catalysed by anions (e.g. HF_2^-, CN^-, N_3^-) or Lewis acids (e.g. $ZnBr_2$, Al^iBu_2Cl). The following is an example of group transfer polymerization of a methacrylate monomer initiated by dimethylketene methyl trimethylsilyl acetal and catalysed by tris(dimethylamino)sulphonium bifluoride

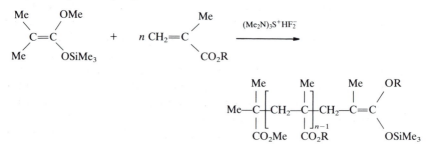

Propagation is believed to proceed via the following mechanism in which the catalyst (Cat) activates transfer of the trimethylsilyl group by association with the silicon atom

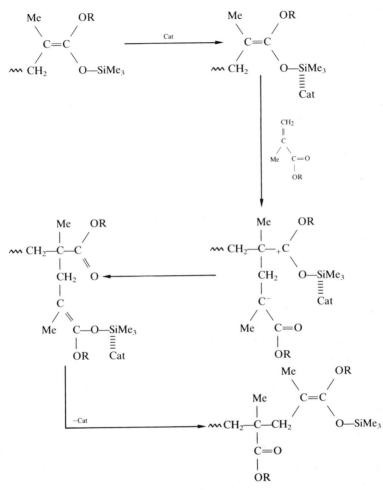

The polymerization is rapidly terminated by compounds containing active hydrogen(s) and so must be performed under dry conditions using reactants and solvents which have been rigorously dried and purified. If these precautions are taken, then living polymers are formed and in favourable cases the Poisson distribution of molar mass is obtained (Section 2.7.5). Polymerization is terminated by addition of a proton source (e.g. water or dilute acid) or by coupling of two active species (e.g. with a dihalide). Together with the use of initiators containing protected functional groups, this facilitates the preparation of terminally-functional polymers, e.g. poly(methyl methacrylate) with terminal carboxylic acid groups may be prepared as follows

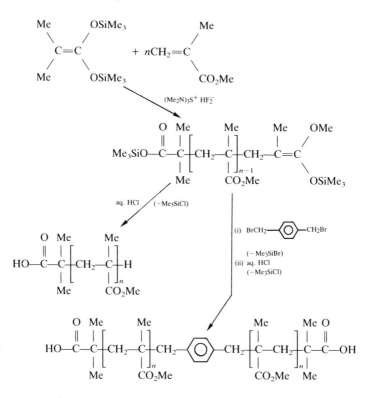

Whilst polymers have been prepared from acrylates, acrylonitrile and other similar monomers, group transfer polymerization is most suited to the preparation of low molar mass ($< 50\,\text{kg mol}^{-1}$) functionalized methacrylate homopolymers and copolymers.

More recently the workers at Du Pont have developed *aldol group transfer polymerization*, in which a silyl vinyl ether is polymerized using an aldehyde as the initiator to give a living silylated poly(vinyl alcohol), e.g.

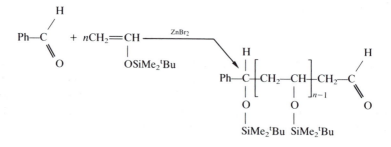

Thus the active species is an aldehyde group and propagation is believed to proceed via a mechanism of the type

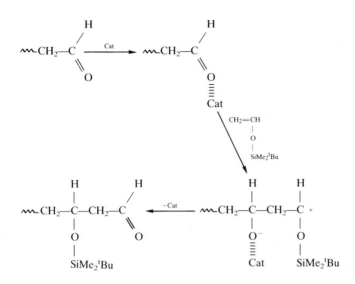

in which the silyl group is transferred from monomer to the previously terminal aldehyde oxygen atom thus generating a new terminal aldehyde group.

2.14 Other specialized methods of polymerization

The previous three sections give an outline of specialized polymerizations which currently are of considerable research interest. For more details concerning these and other specialized methods of polymerization (e.g. template polymerization, polymerization in clathrates, cyclopolymerization, plasma polymerization) an advanced up-to-date text should be consulted.

Copolymerization

2.15 Step copolymerization

Copolymers are formed by simultaneous polymerization of two or more different monomers. Thus the simplest step copolymerizations are of the general type $ARB + AR'B$ or $RA_2 + R'B_2 + R''B_2$. For example, reaction of hexamethylene diamine with a mixture of adipic and sebacic acids yields a copolyamide containing both nylon 6.6 and nylon 6.10 repeat units.

In the following two sections some of the more simple aspects and uses of linear step copolymerization will be described. The principles which will be introduced are equally applicable to non-linear step copolymerization which, therefore, will not be considered separately.

2.15.1 *Statistical copolymers*

Most step copolymerizations are taken to high extents of reaction in order to produce copolymers with suitably high molar masses (Sections 2.2.4 and 2.2.5). A consequence of this is that the overall compositions of the copolymers obtained correspond to those of the comonomer mixtures used to prepare them. However, it must be borne in mind that the sequence distribution of the different repeat units along the copolymer chains is an important factor controlling the properties of a copolymer and that the distribution is affected by differences in monomer reactivity.

When the mutually-reactive functional groups (i.e. A and B) have reactivities which are the same for each monomer, the probability of reaction of a particular monomer depends only upon the mole fraction of functional groups that it provides. Most commonly this situation obtains when the different monomers containing the same functional groups are of similar structure, as for adipic acid and sebacic acid in the example given above. Under these conditions *random copolymers* which have the most probable distribution of molar mass are formed and the reaction kinetics described in Section 2.2.6 apply.

Step copolymerizations which involve two (or more) monomers containing the same type of functional group, but of different reactivity, are more complex and do not yield random copolymers. The monomer containing the more reactive functional groups reacts preferentially and so is incorporated into the oligomeric chains formed early in the reaction in higher proportion than its initial mole fraction in the comonomer mixture. The less reactive monomer participates in the copolymerization to an increasing extent as the concentration of the more reactive monomer is depleted by preferential reaction. Thus the copolymer molecules formed contain significant sequences of the same repeat units, i.e. they have a 'blocky' structure, and the length of these homopolymer sequences increases as the difference in monomer reactivity increases. Such blocky copolymer structures are often formed when performing step copolymerizations of the type $RA_2 + R'B_2 + R''C_2$ where the functional groups of type B and C are unreactive towards each other and only react with the functional groups of type A. For example, reaction of a dicarboxylic acid with a mixture of a diol and a diamine yields a blocky poly(ester-*co*-amide) because the amino groups are more reactive than the hydroxyl groups.

2.15.2 *Block copolymers*

By using as comonomers low molar mass prepolymers with terminal functional groups, step copolymerization can be used to prepare alternating block copolymers. For example, the ester interchange reaction of dimethylterephthalate, CH_3OOC—⟨O⟩—$COOCH_3$, with poly(oxytetramethylene)diol, H—$[O$—$(CH_2)_4$—$]_m OH$, and ethylene glycol, HO—CH_2—CH_2—OH, yields poly[poly(oxytetramethylene)-*block*-poly (ethylene terephthalate)] a simplified structure of which is shown below

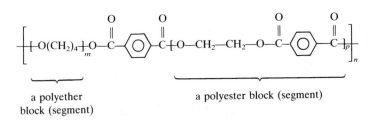

a polyether   a polyester block (segment)
block (segment)

For alternating block copolymers prepared by such step copolymerizations, it is usual to call the blocks *segments* and the copolymers *segmented copolymers*. In the above example the polyether blocks are called *soft segments* because they provide an amorphous rubbery phase, and the polyester blocks are called *hard segments* because they provide a rigid crystalline phase.

Some commercial polyurethane elastomers are in fact segmented copolymers prepared by reaction of a diisocyanate with a prepolymer polyol (hydroxyl group functionality $\geqslant 2$) and a short-chain diol (e.g. butan-1,4-diol)

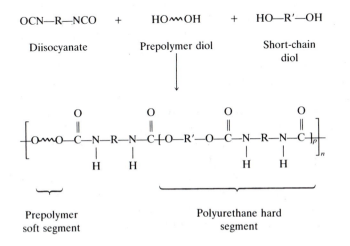

Prepolymer   Polyurethane hard
soft segment   segment

Most commonly, aliphatic polyester or poly(propylene oxide) prepolymers are used and often have functionalities >2, thus giving rise to the formation of segmented copolymer networks.

2.16 Chain copolymerization

As for step copolymerization, differences in monomer reactivity in chain copolymerization affect the sequence distribution of the different repeat units in the copolymer molecules formed. The most reactive monomer again is incorporated preferentially into the copolymer chains but, because of the different nature of chain polymerization, high molar mass copolymer molecules are formed early in the reaction. Thus, at low overall conversions of the comonomers, the high molar mass copolymer molecules formed can have compositions which differ significantly from the composition of the initial comonomer mixture. Also in contrast to step copolymerization, theoretical prediction of the relative rates at which the different monomers add to a growing chain is more firmly established. In the next section a general theoretical treatment of chain copolymerization of two monomers is presented and introduces an approach which can be applied to derive equations for more complex chain copolymerizations involving three or more monomers.

2.16.1 *Copolymer composition equation*

In order to predict the composition of the copolymer formed at a particular instant in time during a chain copolymerization, it is necessary to construct a kinetics model of the reaction. The simplest model will be analysed here and is the *terminal model* which assumes that the reactivity of an active centre depends only upon the terminal monomer unit on which it is located. It is further assumed that the amount of monomer consumed in reactions other than propagation is negligible and that copolymer molecules of high molar mass are formed. Thus for copolymerization of monomer A with monomer B, only two types of active centre need be considered

$$\text{\tiny\textasciitilde\textasciitilde\textasciitilde\textasciitilde}\,A^* \quad \text{and} \quad \text{\tiny\textasciitilde\textasciitilde\textasciitilde\textasciitilde}\,B^*$$

where the asterisk represents the active centre (e.g. an unpaired electron for free-radical copolymerizations). Both active centres propagate by addition of either an A or B monomer molecule and so there are four possible propagation reactions, each with its own rate constant

$$\text{\textasciitilde\textasciitilde}\,A^* + A \xrightarrow{k_{AA}} \text{\textasciitilde\textasciitilde}\,AA^*$$

$$\text{\textasciitilde\textasciitilde}\,A^* + B \xrightarrow{k_{AB}} \text{\textasciitilde\textasciitilde}\,AB^*$$

$$\text{\small$\sim\hspace{-4pt}\sim$} B^* + B \xrightarrow{k_{BB}} \text{\small$\sim\hspace{-4pt}\sim$} BB^*$$

$$\text{\small$\sim\hspace{-4pt}\sim$} B^* + A \xrightarrow{k_{BA}} \text{\small$\sim\hspace{-4pt}\sim$} BA^*$$

The reactions with rate constants k_{AA} and k_{BB} are known as *homopropagation* reactions and those with rate constants k_{AB} and k_{BA} are called *cross-propagation* reactions. Hence the rate of consumption of monomer A is given by

$$-\frac{d[A]}{dt} = k_{AA}[A^*][A] + k_{BA}[B^*][A] \tag{2.81}$$

where $[A^*]$ and $[B^*]$ are the total concentrations of propagating chains with terminal A-type and B-type active centres respectively. Similarly, the rate of consumption of monomer B is given by

$$-\frac{d[B]}{dt} = k_{BB}[B^*][B] + k_{AB}[A^*][B] \tag{2.82}$$

At any instant in time during the reaction, the ratio of the amount of monomer A to monomer B being incorporated into the copolymer chains is obtained by dividing Equation (2.82) into Equation (2.81)

$$\frac{d[A]}{d[B]} = \frac{[A]}{[B]} \left\{ \frac{k_{AA}\,[A^*]/[B^*] + k_{BA}}{k_{BB} + k_{AB}[A^*]/[B^*]} \right\} \tag{2.83}$$

An expression for the ratio $[A^*]/[B^*]$ is obtained by applying steady-state conditions to $[A^*]$ and $[B^*]$, i.e.

$$\frac{d[A^*]}{dt} = 0 \quad \text{and} \quad \frac{d[B^*]}{dt} = 0$$

In terms of the creation and loss of active centres of a particular type, the contribution of initiation and termination reactions is negligible compared to that of the cross-propagation reactions. Thus

$$\frac{d[A^*]}{dt} = k_{BA}[B^*][A] - k_{AB}[A^*][B]$$

and

$$\frac{d[B^*]}{dt} = k_{AB}[A^*][B] - k_{BA}[B^*][A]$$

Application of the steady-state condition to either of these equations leads to

$$\frac{[A^*]}{[B^*]} = \frac{k_{BA}[A]}{k_{AB}[B]} \tag{2.84}$$

Substituting Equation (2.84) into Equation (2.83) and simplifying yields one form of the *copolymer composition equation*

$$\frac{d[A]}{d[B]} = \frac{[A]}{[B]} \left(\frac{r_A[A] + [B]}{[A] + r_B[B]} \right) \tag{2.85}$$

where r_A and r_B are the respective *monomer reactivity ratios* defined by

$$r_A = \frac{k_{AA}}{k_{AB}} \text{ and } r_B = \frac{k_{BB}}{k_{BA}} \tag{2.86}$$

Equation (2.85) gives the *molar ratio* of A-type to B-type repeat units in the copolymer formed at any instant (i.e. a very small interval of time) during the copolymerization when the monomer concentrations are $[A]$ and $[B]$. Often it is more convenient to express compositions as mole fractions. The *mole fraction* f_A of monomer A in the comonomer mixture is $[A]/([A] + [B])$ and that of monomer B is $f_B = 1 - f_A$. The mole fraction F_A of A-type repeat units in the copolymer formed at a particular instant in time is $d[A]/(d[A] + d[B])$ and that of B-type repeat units is $F_B = 1 - F_A$. Addition of unity to both sides of Equation (2.85) allows it to be rearranged in terms of F_A (or F_B), f_A and f_B. The following *copolymer composition equations* are obtained

$$F_A = \frac{r_A f_A^2 + f_A f_B}{r_A f_A^2 + 2 f_A f_B + r_B f_B^2} \tag{2.87}$$

and

$$F_B = \frac{r_B f_B^2 + f_A f_B}{r_A f_A^2 + 2 f_A f_B + r_B f_B^2} \tag{2.88}$$

2.16.2 *Monomer reactivity ratios and copolymer composition/structure*

Monomer reactivity ratios are important quantities since for a given instantaneous comonomer composition, they control the overall composition of the copolymer formed at that instant and also the sequence distribution of the different repeat units in the copolymer. From Equation (2.86), they are the ratios of the homopropagation to the cross-propagation rate constants for the active centres derived from each respective monomer. Thus if $r_A > 1$ then $\sim\!\!\wedge\!\!A^*$ prefers to add monomer A (i.e. it prefers to homopolymerize), whereas if $r_A < 1$ $\sim\!\!\wedge\!\!A^*$ prefers to add monomer B and hence copolymerize. Similarly, r_B describes the behaviour

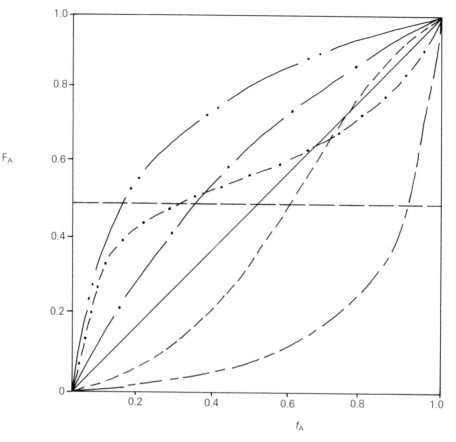

Fig. 2.12 *Plots of Equation (2.87) showing the variation of copolymer composition (F_A) with comonomer composition (f_A) for different pairs of r_A, r_B values:* —, $r_A = 1.0$, $r_B = 1.0$; —·—, $r_A = 2.0$, $r_B = 0.5$; —––—, $r_A = 0.1$, $r_B = 10.0$; —··—, $r_A = 4.0$, $r_B = 0.1$; ———, $r_A = 2.0$, $r_B = 4.0$; —··—··—, $r_A = 0.5$, $r_B = 0.1$; ———, $r_A = r_B = 0$.

of monomer B. Plots of Equation (2.87) for different r_A, r_B pairs are shown in Fig. 2.12 and are commented upon below.

Random copolymers with $F_A = f_A$ (for all values of f_A) are formed when $r_A = r_B = 1$, i.e. when the probability of adding monomer A is equal to the probability of adding monomer B for both types of active centre. There are very few copolymerizations which approximate to this condition and those that do involve copolymerization of monomers with very similar structures (e.g. free-radical copolymerization of tetrafluoroethylene with chloro-trifluoroethylene).

More commonly one monomer, assumed here to be monomer A, is

more reactive than the other, and both types of active centre prefer to add the more reactive monomer. In terms of reactivity ratios, this gives rise to $r_A > 1$ with $r_B < 1$. On this basis it is easy to appreciate why there are no simple copolymerizations for which $r_A > 1$ with $r_B > 1$.

Ideal copolymerization is a special case of $r_A > 1$, $r_B < 1$ (or $r_A < 1$, $r_B > 1$) copolymerization for which $r_A r_B = 1$. The name derives from the similarity of the F_A vs f_A curves to boiling point diagrams for ideal liquid–liquid mixtures. Under these conditions

$$r_A = \frac{1}{r_B}, \quad \text{i.e.} \quad \frac{k_{AA}}{k_{AB}} = \frac{k_{BA}}{k_{BB}}$$

which means that the *relative rates* at which the two monomers are incorporated into the copolymer chains are the same for both types of active centre (i.e. $\sim A^*$ and $\sim B^*$). Thus, even though $F_A \neq f_A$, the sequence distribution of the repeat units in the copolymer formed is random. Substitution of $r_B = r_A^{-1}$ into Equation (2.87) gives a simplified *copolymer composition equation for ideal copolymerization*

$$F_A = \frac{r_A f_A}{r_A f_A + f_B} \tag{2.89}$$

Ideal copolymerization occurs when there are no specific effects controlling either one or more of the four propagation reactions, since the relative rates of reaction of the two monomers then depend only upon their absolute relative reactivities. In many copolymerizations $r_A > 1$, $r_B < 1$ (or vice versa) with $r_A r_B \neq 1$ (usually $r_A r_B < 1$). Such copolymerizations give F_A vs f_A curves that are similar to those for ideal copolymerization but skewed towards copolymer compositions richer in the repeat units derived from the more reactive monomer.

The higher the ratio r_A / r_B for $r_A > 1$, $r_B < 1$ copolymerizations, the longer are the continuous sequences of A-type repeat units and the shorter are the continuous sequences of B-type repeat units in the copolymer molecules formed. When $r_A \gg 1$ with $r_B \ll 1$ there is a tendency towards consecutive homopolymerization of the two monomers. The molecules formed early in the reaction have very long sequences of A-type repeat units with the occasional B-type repeat unit (i.e. they are essentially molecules of homopolymer A). Later in the reaction, when monomer A has been consumed completely, molecules of homopolymer B are formed by polymerization of the residual monomer B.

Azeotropic copolymerization occurs when $r_A < 1$ with $r_B < 1$ and when $r_A > 1$ with $r_B > 1$, though the latter of these conditions is rarely observed in practice. The name again comes from analogy with boiling point diagrams of liquid–liquid mixtures, since the F_A vs f_A curves are characterized by their intersection of the $F_A = f_A$ line at one point which corresponds to the *azeotropic composition* $(f_A)_{azeo}$. Substituting $F_A = f_A = (f_A)_{azeo}$ into Equation (2.87) leads to

$$(f_A)_{azeo} = \frac{1 - r_B}{2 - r_A - r_B} \tag{2.90}$$

As the product $r_A r_B$ decreases there is an increasing tendency towards alternation in the additions of monomer molecules to the propagating chains. The extreme case of azeotropic copolymerization is $r_A = r_B = 0$ and always produces perfectly alternating copolymers, irrespective of the value of f_A (i.e. $F_A = 0.50$ for $0 < f_A < 1$), because the homopropagation reactions do not occur.

2.16.3 *Copolymer composition drift*

For a given pair of comonomers, the value of F_A for the copolymer formed early in the reaction is determined by the initial value of f_A via Equation (2.87). For most copolymerizations $F_A \neq f_A$ and one monomer is consumed preferentially causing f_A to change as the overall monomer conversion increases. Since Equation (2.87) is applicable to each increment of conversion, the change in f_A gives rise to a variation in F_A with conversion. This is known as *copolymer composition drift* and leads to copolymers which consist of copolymer molecules with significantly different compositions. This broadening of the *distribution of copolymer composition* beyond that arising from the normal statistical variation of copolymer composition about F_A at any specific value of f_A, clearly becomes more significant as the overall monomer conversion increases.

In $r_A > 1$, $r_B < 1$ copolymerizations f_A (and hence F_A) decreases with conversion as monomer A is consumed preferentially. Eventually monomer A is consumed completely leaving some unreacted monomer B (i.e. f_A becomes zero) and so thereafter the homopolymer of monomer B is formed.

In azeotropic copolymerizations f_A changes with conversion either until (i) for $r_A < 1$, $r_B < 1$ it becomes equal to either zero or unity and the corresponding homopolymer is formed from then onwards, or until (ii) for $r_A > 1$, $r_B > 1$ it is equal to $(f_A)_{azeo}$ and copolymer with $F_A = (f_A)_{azeo}$ is formed thereafter.

For many applications the tolerance for copolymer composition drift is small and its control is essential. The two strategies most commonly used for this purpose are:

(i) the overall monomer conversion is limited (usually to $\leqslant 5\%$) in order to reduce the drift in f_A;

(ii) additional quantities of the monomer which is consumed preferentially are fed to the reaction vessel at a controlled rate during the copolymerization in order to maintain f_A constant.

2.16.4 *Evaluation of monomer reactivity ratios*

In order to use the copolymer composition equations predictively, it is necessary to know the values of r_A and r_B. The most simple method for their evaluation involves determination of the compositions (i.e. F_A) of the copolymers formed at low conversions in a series of copolymerizations performed using known initial comonomer compositions (i.e. f_A). Equation (2.87) can be rearranged to give

$$\frac{f_A(1-2F_A)}{F_A(1-f_A)} = r_B + \left[\frac{f_A^2(F_A-1)}{F_A\,(1-f_A)^2} \right] r_A \qquad (2.91)$$

For each experimental f_A, F_A pair, the left-hand side of the equation is plotted against the coefficient of r_A. The data points are then fitted to a straight line, from which the slope gives r_A and the intercept r_B.

This method has been widely used in the past but there now are much more sophisticated and statistically-valid methods for evaluation of reactivity ratios. However, they are beyond the scope of this book and the reader is recommended to consult an up-to-date advanced review of copolymerization for details.

2.16.5 *Free-radical copolymerization and the Q–e scheme*

Many commercially-important copolymers are prepared by free-radical copolymerization of ethylenic monomers, e.g. styrene–butadiene rubber (SBR), acrylonitrile–butadiene–styrene copolymer (ABS), ethylene–vinyl acetate copolymer (EVA) and acrylonitrile–butadiene copolymer (nitrile rubber).

Most free-radical copolymerizations can be categorized into either one or the other of the following two types: (i) $r_A > 1$, $r_B < 1$ (or $r_A < 1$, $r_B > 1$) copolymerization and (ii) $r_A < 1$, $r_B < 1$ azeotropic copolymerization. This is evident from Table 2.13 which gives the reactivity ratios and their products for some representative free-radical copolymerizations.

The reactivity of a monomer is strongly dependent upon the ability of its substituent group(s) to stabilize the corresponding polymeric radical. This is because the greater is the stability of the polymeric radical, the more readily it is formed by reaction of the monomer. Thus reactive monomers have substituent groups which stabilize the polymeric radical by delocalization of the unpaired electron (i.e. by resonance). Hence a monomer of high reactivity yields a polymeric radical of low reactivity and vice versa.

TABLE 2.13 *Some typical values of reactivity ratios for free-radical copolymerization at 60°C*

Monomer A	Monomer B	r_A	r_B	$r_A r_B$
Styrene	Butadiene	0.78	1.39	1.08
Styrene	Methyl methacrylate	0.52	0.46	0.24
Styrene	Methyl acrylate	0.75	0.18	0.14
Styrene	Acrylonitrile	0.40	0.04	0.02
Styrene	Maleic anhydride	0.02	0	0
Styrene	Vinyl chloride	17	0.02	0.34
Vinyl acetate	Vinyl chloride	0.23	1.68	0.39
Vinyl acetate	Acrylonitrile	0.06	4.05	0.24
Vinyl acetate	Styrene	0.01	55	0.55
Methyl methacrylate	Methyl acrylate	1.69	0.34	0.57
Methyl methacrylate	n-Butyl acrylate	1.8	0.37	0.67
Methyl methacrylate	Acrylonitrile	1.20	0.15	0.18
Methyl methacrylate	Vinyl acetate	20	0.015	0.30
trans-Stilbene	Maleic anhydride	0.03	0.03	0.001

The order in which common substituent groups provide increasing resonance stabilization is

where R is an alkyl group. Vinyl monomers (CH_2=CHX) tend to be of lower reactivity than the corresponding 1,1-disubstituted monomers (CH_2=CXY) because the resonance effects of the substituents tend to be additive and because secondary radicals (i.e. ⌁CH_2—ĊHX) generally are of lower stability than tertiary radicals (i.e. ⌁CH_2—ĊXY) due to steric shielding of the active site being greater for the latter. In contrast, 1,2-disubstituted monomers (CHX=CHY) are of low reactivity due to steric hindrance of the propagation reaction by the 2-substituent in the monomer. Although such monomers homopolymerize only with great difficulty, they can be copolymerized with vinyl and 1,1-disubstituted monomers because of the reduced steric hindrance in the cross-propagation reactions. However, they always have low reactivity ratios.

In $r_A > 1$, $r_B < 1$ copolymerizations, the ratio r_A/r_B increases (or for $r_A < 1$, $r_B > 1$ copolymerization, the ratio r_B/r_A increases) as the difference between the reactivities of the monomers increases (cf. reactivity ratios for copolymerization of vinyl acetate with vinyl chloride, acrylonitrile and styrene). Thus the most simple copolymerizations tend to be between monomers with similar reactivities. However, this general statement requires qualification because of the influence of steric effects (such as those described above) and polar effects. The latter are of great importance and are most evident in azeotropic copolymerizations. In general, as the difference between the electron densities of the C=C bonds (i.e. polarities) of the two monomers increases, the alternating tendency increases (i.e. $r_A r_B$ decreases; cf. data in Table 2.13 for copolymerization of styrene with methyl methacrylate, methyl acrylate, acrylonitrile and maleic anhydride). Strong polar effects can negate the effects of steric hindrance, as is demonstrated by the alternating copolymerization of stilbene with maleic anhydride. Since specific steric and polar effects operate in most copolymerizations, there are relatively few copolymerizations which closely approximate to ideal behaviour (i.e. $r_A r_B = 1$), the majority have $r_A r_B < 1$.

The complexity of copolymerization makes theoretical prediction of reactivity ratios rather difficult. Nevertheless the semi-empirical Q–e scheme is often used for estimating reactivity ratios and also provides an approximate ranking of the reactivities and polarities of monomers. The basis of the scheme is that the rate constant k_{pm} for reaction of a polymeric radical (p) with a monomer (m) is given by

$$k_{pm} = P_p Q_m \exp\left(-e_p e_m\right) \tag{2.92}$$

where P_p and Q_m are measures of the reactivities of the polymeric radical and the monomer respectively, and e_p and e_m are measures of the 'electrostatic charges' associated with the polymeric radical and the monomer respectively. Expressions for k_{AA}, k_{AB}, k_{BB} and k_{BA} can be generated by application of Equation (2.92) to each propagation reaction. By taking the appropriate ratios of these expressions (see Equation (2.86)) and assuming that the charge associated with a polymeric radical is equal to the charge associated with the monomer from which it is derived, the following equations for the reactivity ratios are obtained

$$r_A = (Q_A/Q_B) \exp\left[-e_A(e_A - e_B)\right] \tag{2.93}$$

$$r_B = (Q_B/Q_A) \exp\left[-e_B(e_B - e_A)\right] \tag{2.94}$$

where $e_A = e_{pA} = e_{mA}$ and $e_B = e_{pB} = e_{mB}$. Thus the terms in polymeric radical reactivity have divided out and the equations essentially relate reactivity ratios to the reactivities (Q_A and Q_B) and electrostatic charges (e_A and e_B) of the C=C bonds of the two monomers. The prediction of alternating tendency

TABLE 2.14 *Values of Q and e for some important monomers*

Monomer	Q	e
Isoprene	3.33	−1.22
Butadiene	2.39	−1.05
Styrene	1.00	−0.80
Methyl methacrylate	0.74	0.40
Acrylonitrile	0.60	1.20
Ethyl acrylate	0.52	0.22
Maleic anhydride	0.23	2.25
Vinyl chloride	0.044	0.20
Vinyl acetate	0.026	−0.22

$$r_A r_B = \exp\left[-(e_A - e_B)^2\right] \qquad (2.95)$$

is in accord with experimental observations since $r_A r_B$ decreases as $(e_A - e_B)$ increases. Equation (2.95) also predicts that ideal copolymerization will occur when $e_A = e_B$ since this gives $r_A r_B = 1$.

In order to develop the Q–e scheme, styrene was chosen as the reference monomer and arbitrarily assigned values of Q and e. Initially the Q and e values of other monomers were evaluated from pairs of experimentally-determined r_A, r_B values for copolymerization with styrene. These values have been refined after consideration of experimentally-determined r_A, r_B values for copolymerizations not involving styrene and the 'best-fit' Q–e values for each monomer recorded for prediction of pairs of r_A, r_B values using Equations (2.93) and (2.94).

Table 2.14 gives Q–e values for some important monomers and includes those currently accepted for the reference monomer (styrene). The values of Q and e increase with increasing monomer reactivity and increasing electron deficiency of the C=C bond respectively. Negative values of e indicate that the C=C bond is electron rich.

The theoretical basis of the Q–e scheme has been criticized in several respects, the most important being: (i) the assumption of equal charges for a monomer and its corresponding polymeric radical is unrealistic, (ii) alternating tendency results from polarization phenomena in the transition state of propagation and not from interactions of permanent charges (e.g. free-radical copolymerizations are not affected by changes in the dielectric constant of the reaction medium), and (iii) steric effects are not directly taken into account, though they are incorporated into the experimentally-evaluated Q–e values. Thus the Q–e scheme cannot be used to give rigorous quantitative predictions. Nevertheless it does provide a useful guide to free-radical copolymerization behaviour.

TABLE 2.15 *Some typical values of reactivity ratios for cationic and anionic copolymerization of styrene (monomer A)*

Type of copolymerization	Temperature (°C)	Initiator	Solvent	Monomer B	r_A	r_B	$r_A r_B$
Cationic	−90	$AlCl_3$	Dichloromethane	Isobutylene	0.24	1.79	0.43
	0	BF_3	Nitroethane	Chloroprene	33.0	0.15	4.95
	0	$TiCl_4$	Carbon tetrachloride	para-Methoxystyrene	0.05	46	2.30
Anionic	−78	^sBuLi	Tetrahydrofuran	Butadiene	11.0	0.04	0.44
	25	^sBuLi	Tetrahydrofuran	Butadiene	4.0	0.3	1.20
	25	^sBuLi	Benzene	Butadiene	0.04	10.8	0.43
	25	^nBuLi	Tetrahydrofuran	Isoprene	9.0	0.10	0.90
	30	^nBuLi	Benzene	Isoprene	0.26	10.6	2.76
	20	^sBuLi	Benzene	para-Methylstyrene	0.74	1.10	0.81

2.16.6 *Ionic copolymerization*

In comparison to free-radical copolymerizations, there are relatively few ionic copolymerizations which yield copolymers containing significant proportions of repeat units from each of the monomers involved. This is because the ionic charge associated with the active centre emphasizes substituent group effects and gives rise to much greater differences in monomer reactivities. In most cases one monomer is much more reactive than the other and so has a strong tendency to homopolymerize, i.e. $r_A \gg 1$, $r_B \ll 1$ or vice versa (see Table 2.15). The formation of copolymers is best accomplished using comonomers which are of very similar structure (e.g. anionic copolymerization of styrene with *para*-methylstyrene).

A further distinguishing feature of ionic copolymerizations is the strong dependence of reactivity ratios upon the nature of the solvent and the counter-ion (and hence the initiator). These effects can be dramatic and can lead to a reversing of the order of relative monomer reactivities (cf. reactivity ratios for anionic copolymerization of styrene with butadiene and isoprene in different solvents).

With the exception of block copolymers (Section 2.16.9), butyl rubber is the only copolymer of major industrial importance prepared by ionic copolymerization.

2.16.7 *Ziegler–Natta coordination copolymerization*

For many years Ziegler–Natta coordination copolymerization of ethylene with propylene and non-conjugated dienes (such as hexa-1,4-diene and 5-ethylidene-2-norbornene) has been used to prepare an important class of rubbers known collectively as EPDM rubbers. More recently, copolymerization of ethylene with small proportions of higher α-olefins (such as but-1-ene, hex-1-ene and oct-1-ene) has become important and is used to prepare a range of copolymers known as linear low-density polyethylenes (LLDPE).

Whilst the theory presented in Section 2.16.1 can be applied to copolymerizations performed using Ziegler–Natta catalysts, there are several features of such reactions which make it less than appropriate. The theory assumes there to be only one type of active centre (for example, an unpaired electron for free-radical copolymerization), whereas heterogeneous Ziegler–Natta catalysts possess a range of active surface sites with different activities and stereochemical selectivities. Thus the reactivity ratios obtained are average values and are dependent upon the precise methods and conditions employed for preparation of the catalyst and for copolymerization. Furthermore, attempts to predict copolymer composition drift using the simple theory often are thwarted by time-dependent changes in the catalytic activities of the active surface sites.

Despite the limited applicability of simple copolymerization theory, it is useful for establishing relative monomer reactivities and the effects of changing catalyst composition. For example, it is found that the relative reactivity of α-olefins decreases as the size of the substituent group increases and that the relative reactivity of ethylene in copolymerization with propylene generally is increased by changing from vanadium- to titanium-based catalysts. Often, very large differences in reactivity ratios are observed for copolymerization of ethylene with higher α-olefins, e.g. r_A(ethylene) > 50 with r_B(but-1-ene) < 0.1.

2.16.8 *Other types of copolymerization*

The other types of chain polymerization described earlier in this chapter can be used to prepare copolymers but such copolymerizations have received relatively little attention. In view of this, they will not be considered here and the reader is referred to more advanced texts.

2.16.9 *Block copolymers*

The most convenient way of preparing block copolymers by chain polymerization is through the use of living polymers. This involves the formation of living homopolymer molecules by polymerization of the first monomer (A) under carefully controlled conditions. The terminal active sites of these molecules are then used to initiate polymerization of a different monomer (B), thus extending the molecules by formation of a second homopolymer block to produce an AB di-block copolymer. This general synthetic strategy has been applied to the preparation of block copolymers with well-defined structures using ionic, Ziegler–Natta, ring-opening and group transfer methods of chain polymerization. In each case the order in which the monomers are polymerized is of crucial importance because the living polymer formed from the first monomer must be capable of efficiently initiating polymerization of the second monomer. Hence, the less reactive monomer is polymerized first. For example, in order to prepare block copolymers from methyl methacrylate and styrene by anionic living polymerization, the styrene must be polymerized first.

Anionic living polymerization provides the most important chain polymerization route to block copolymers. By careful control of the reaction conditions it is possible to produce block copolymers with blocks of pre-defined molar mass, narrow molar mass distribution and controlled stereochemistry. In particular, anionic living polymerization is used to prepare block copolymers suitable for use as thermoplastic elastomers (Section 4.5.1), e.g. ABA tri-block copolymers in which homopolymer A is glassy (e.g. polystyrene) and homopolymer B is rubbery (e.g. polyisoprene). Such block copolymers can be prepared by polymerizing styrene using ⁵BuLi, then adding isoprene to form living di-block copolymer molecules which finally are coupled together by their reaction with a dihalide compound (e.g. phosgene or dichlorodimethylsilane). The preparation is performed in a non-polar solvent (e.g. benzene) in order to ensure that styrene is the less reactive of the two monomers and so can be used in the first stage. A schematic representation of the sequence of reactions is shown below

$$^{s}BuLi \xrightarrow{\text{styrene}} {^{s}Bu} \text{---} {^{-}Li^{+}} \xrightarrow{\text{isoprene}} {^{s}Bu} \text{\wedge\wedge\wedge\wedge\wedge} {^{-}Li^{+}}$$

$$^{s}Bu \text{\wedge\wedge\wedge} R \text{\wedge\wedge\wedge} {^{s}Bu} + 2LiX \longleftarrow X\text{---}R\text{---}X$$

where — represents a polystyrene block and ∿ represents a polyisoprene block. More complex block copolymer structures can be produced by coupling with polyfunctional halide compounds (e.g. RX_4 would yield a four-armed star-shaped block copolymer). An alternative method for preparation of ABA tri-block copolymers involves the use of an electron transfer initiator (e.g. sodium naphthalide) in a polar solvent, such as tetrahydrofuran, in which isoprene is the monomer of lower reactivity. This allows the isoprene to be polymerized first to produce dicarbanionic living polyisoprene molecules which then are used to polymerize styrene before terminating the reaction

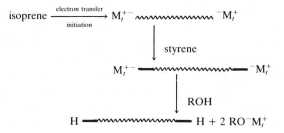

where M_t^{+} is a metal ion. However, for these particular block copolymers, this route is less satisfactory because a high proportion of the undesirable 1,2- and 3,4-additions occur upon anionic polymerization of isoprene in a polar solvent (Section 2.8.2).

Methods involving transformation from one type of chain polymerization to another have been developed for the preparation of block copolymers when the second monomer to be polymerized is not susceptible to the type of chain polymerization used to form the first block. For example, anionic living polymers of monomer A can be terminated by reaction with an excess of a dibromide to yield polymer molecules with terminal bromide groups which can be reacted with an appropriate silver salt (e.g. $AgPF_6$) and used to initiate cationic polymerization of monomer B

$$n A \xrightarrow{\text{}^sBuLi} \text{}^sBu \wedge\wedge\wedge A^- Li^+ \xrightarrow[\text{Br—R—Br}]{\text{Excess}} \text{}^sBu \wedge\wedge\wedge A\text{—}R\text{—}Br + LiBr$$

$$\downarrow AgPF_6$$

$$\text{}^sBu \wedge\wedge\wedge A\text{—}R \blacksquare\blacksquare B^+ PF_6^- \xleftarrow{n\,B} \text{}^sBu \wedge\wedge\wedge A\text{—}R^+ PF_6^- + AgBr$$

where $\wedge\wedge\wedge$ represents poly(A) and $\blacksquare\blacksquare$ represents poly(B).

Several different transformations have been used, e.g. anionic to cationic, cationic to anionic, and anionic to free-radical polymerization. An up-to-date review should be consulted for details of these and other methods (e.g. coupling of terminally-functionalized polymers) of block copolymer preparation.

2.16.10 *Graft copolymers*

Graft copolymers are the branched equivalents of block copolymers, and most commonly are prepared from prepolymers which possess groups along the chain that can be activated to initiate polymerization of a second monomer, thus forming branches on the prepolymer. The simplest of such procedures involve either exposure of the prepolymer to high-energy radiation in presence of the second monomer, or heating of the prepolymer in the presence of a suitable free-radical initiator and the second monomer. In this way free-radical sites are produced along the prepolymer chain either by direct interaction of the atoms in the chain with the radiation or via abstraction of atoms (e.g. hydrogen atoms) from the prepolymer by other free-radical species. The prepolymer free-radicals then initiate polymerization of the second monomer. Whilst these simple methods allow a very wide range of polymers to be used as prepolymers, they are not suited to the formation of well-defined graft copolymers and inevitably yield graft copolymers that are contaminated by some homopolymer formed from the second monomer. Methods involving the use of prepolymers with side groups which can be activated to initiate either cationic or anionic polymerization of the second monomer allow better control of graft copolymer structure and of homopolymer formation. Additionally, graft copolymers can be prepared by coupling living polymers to reactive side groups on a prepolymer.

An alternative approach to the preparation of graft copolymers involves the use of *macromonomers*, the most important of which are prepolymers with terminal polymerizable C=C bonds. For example, anionic living polymerization can be used to prepare a prepolymer with a terminal hydroxyl group which can then be reacted with methacryloyl chloride to produce a macromonomer

$$\text{wwOH} + \underset{\substack{\| \quad |}}{\text{Cl-C-C=CH}_2} \rightarrow \text{wwO-C-C=CH}_2 + \text{HCl}$$

$$\underset{\text{O CH}_3}{} \qquad \underset{\text{O CH}_3}{}$$

macromonomer

Graft copolymers are produced by chain copolymerization of the macromonomer with another ethylenic monomer. This enables more well-defined graft copolymers to be prepared, especially with regard to the molar mass and molar mass distribution of the branches since these properties are defined by those of the macromonomer.

For more details concerning the preparation of graft copolymers the reader should consult an up-to-date review.

Further reading

Allen, G. and Bevington, J.C. (eds) (1989), *Comprehensive Polymer Science*, Vols 3–6, Pergamon Press, Oxford.

Allport, D.C. and Janes, W.H. (eds) (1973), *Block Copolymers*, Applied Science, London.

Elias, H-G. (1984), *Macromolecules*, Vol. 2, 2nd edn, John Wiley, New York.

Flory, P.J. (1953), *Principles of Polymer Chemistry*, Cornell University Press, Ithaca.

Lenz, R.W. (1967), *Organic Chemistry of Synthetic High Polymers*, Interscience, New York.

Mark, H.F., Bikales, N.M., Overberger, C.G. and Menges, G. (Eds) (1985–89), *Encyclopedia of Polymer Science and Engineering*, Wiley-Interscience, New York.

Noshay, A. and McGrath, J.E. (1977), *Block Copolymers: An Overview and Critical Survey*, Academic Press, New York.

Odian, G. (1981), *Principles of Polymerization*, 2nd edn, Wiley-Interscience, New York.

Saunders, K.J. (1988), *Organic Polymer Chemistry*, 2nd edn, Chapman and Hall, London.

Stevens, M.P. (1975), *Polymer Chemistry—An Introduction*, Addison-Wesley, Reading, MA.

Problems

2.1 Write down the reaction scheme for the polymerization of an ω-amino carboxylic acid.

The following data were obtained during the condensation of 12-hydroxy stearic acid at 433.5 K in the molten state by sampling the reaction mixture at various times. [COOH] was determined for each sample by titrating with ethanolic sodium hydroxide.

t(h)	[COOH] (mol dm^{-3})
0	3.10
0.5	1.30
1.0	0.83
1.5	0.61
2.0	0.48
2.5	0.40
3.0	0.34

Determine the rate constant for the reaction under these conditions and the order of the reaction. Also suggest whether or not a catalyst was used.

What would be the extent of the reaction after 1 hour and 5 hours?

2.2 Neglecting the contribution of end groups to polymer molar mass, calculate the percentage conversion of functional groups required to obtain a polyester with a number-average molar mass of $24000\,\mathrm{g\,mol^{-1}}$ from the monomer $HO(CH_2)_{14}COOH$.

2.3 A polyamide was prepared by bulk polymerization of hexamethylene diamine (9.22 g) with adipic acid (11.68 g) at 280 °C. Analysis of the whole reaction product showed that it contained $2.6 \times 10^{-3}\,\mathrm{mol}$ of carboxylic acid groups. Evaluate the number-average molar mass, $\bar{M}_n$, of the polyamide, and also estimate its weight-average molar mass, $\bar{M}_w$, by assuming that it has the most probable distribution of molar mass.

2.4 A polycondensation reaction takes place between 1.2 moles of a dicarboxylic acid, 0.4 moles of glycerol (a triol) and 0.6 moles of ethylene glycol (a diol).

Calculate the critical extents of reaction for gelation using the Statistical Theory of Flory and the Carothers Theory.

Comment on the observation that the measured value of the critical extent of reaction is 0.866.

2.5 1 kg of a polyester of number-average molar mass $10000\,\mathrm{g\,mol^{-1}}$ is mixed with 1 kg of another polyester of number-average molar mass $30000\,\mathrm{g\,mol^{-1}}$. The mixture then is heated to a temperature at which it undergoes an ester interchange reaction.

Assuming that the two original polymer samples and the polymer produced by the ester interchange reaction have the most probable distribution of molar mass, calculate $\bar{M}_n$ and $\bar{M}_w$ for the mixture before and after the ester interchange reaction.

2.6 Write down a reaction scheme for polymerization of styrene ($CH_2{=}CHPh$) initiated by thermolysis of azobisisobutyronitrile (AIBN), including both combination and disproportionation as possible modes of termination.

A sample of polystyrene prepared by bulk polymerization at $60\,°C$ using radioactive (^{14}C) AIBN as initiator, was found to have $\bar{M}_n = 1000\,kg\,mol^{-1}$ and a radioactivity of 6×10^3 counts $s^{-1}\,g^{-1}$ (measured using a liquid-scintillation counter). Given that the AIBN has a radioactivity of 6×10^9 counts $s^{-1}\,mol^{-1}$, determine the mode of termination which operated in the preparation of the polystyrene sample.

2.7 Calculate the half-life of benzoyl peroxide in benzene at $333\,K$ given that the rate constant, k_d, for dissociation of this initiator under these conditions is $3.4 \times 10^{-6}\,s^{-1}$. What would be the change in initiator concentration after 1 hour at $333\,K$? Comment on the values obtained.

2.8 In a free-radical polymerization reaction what would be the effect of (a) increasing $[M]_0$ four times at constant $[I]_0$ and (b) increasing $[I]_0$ four times at constant $[M]_0$ upon:

(i) the total radical concentration at steady state,
(ii) the rate of polymerization, and
(iii) the number-average degree of polymerization.

2.9 Methyl methacrylate was polymerized at a mass concentration of $200\,g\,dm^{-3}$ in toluene using AIBN as initiator at a mass concentration of $1.64 \times 10^{-2}\,g\,dm^{-3}$ and a reaction temperature of $60\,°C$. Calculate the initial rate of polymerization and the molar mass of the poly(methyl methacrylate) formed in the initial stages of the reaction given that the relevant rate constants at $60\,°C$ are:

(i) initiator dissociation, $k_d = 8.5 \times 10^{-6}\,s^{-1}$,
(ii) propagation, $k_p = 367\,dm^3\,mol^{-1}\,s^{-1}$,
(iii) termination, $k_t = 9.3 \times 10^6\,dm^3\,mol^{-1}\,s^{-1}$,
(iv) transfer to monomer, $k_{trM} = 3.93 \times 10^{-3}\,dm^3\,mol^{-1}\,s^{-1}$,
(v) transfer to solvent, $k_{trS} = 7.34 \times 10^{-3}\,dm^3\,mol^{-1}\,s^{-1}$.

Assume that the initiator efficiency $f = 0.7$, that the termination parameter $q = 1$ (i.e. termination by combination is negligible), and that the density of the initial solution of methyl methacrylate in toluene is $860\,g\,dm^{-3}$.

2.10 Assuming that in free-radical polymerization the rate constants can be replaced by the appropriate Arrhenius expressions:

$$k_d = A_d \exp(-E_d/RT)$$
$$k_p = A_p \exp(-E_p/RT)$$
$$k_t = A_t \exp(-E_t/RT)$$

calculate the changes in the rate of polymerization and the degree of polymerization caused by increasing the temperature of polymerization of styrene in benzene initiated by AIBN from $60\,°C$ to $70\,°C$ given that:

$E_p = 34 \, \text{kJ mol}^{-1}$
$E_t = 10 \, \text{kJ mol}^{-1}$
$E_d = 126 \, \text{kJ mol}^{-1}$

Assume that the concentration of monomer and initiator and the values of f and q remain unchanged when the temperature is increased.

2.11 Styrene was polymerized at a mass concentration of $208 \, \text{g dm}^{-3}$ in ethylene dichloride using sulphuric acid as the initiator at 25 °C. Given that the rate constants for propagation, ion-pair rearrangement and transfer to monomer are $7.6 \, \text{dm}^3$ $\text{mol}^{-1} \text{s}^{-1}$, $4.9 \times 10^{-2} \, \text{s}^{-1}$ and $0.12 \, \text{dm}^3 \, \text{mol}^{-1} \text{s}^{-1}$ respectively, calculate the molar mass of the polystyrene formed early in the reaction.

2.12 For polymerization of styrene in tetrahydrofuran at 25 °C using sodium naphthalide as initiator, the rate constant for propagation is $550 \, \text{dm}^3 \, \text{mol}^{-1} \, \text{s}^{-1}$. If the initial mass concentration of styrene is $156 \, \text{g dm}^{-3}$ and that of sodium naphthalide is $3.02 \times 10^{-2} \, \text{g dm}^{-3}$, calculate the initial rate of polymerization and, for complete conversion of the styrene, the number-average molar mass of the polystyrene formed. Comment upon the expected value of the polydispersity index ($\overline{M}_w / \overline{M}_n$) and the stereoregularity of the polystyrene produced.

2.13 A $0.16 \, \text{mol dm}^{-3}$ solution ($3.50 \, \text{cm}^3$) of s-butyllithium in toluene was added to a solution of styrene ($8.40 \, \text{g}$) in toluene ($200 \, \text{cm}^3$). After complete conversion of the styrene, isoprene ($28.00 \, \text{g}$) was added. When the isoprene had completely polymerized, the reaction was completed by addition of a $0.10 \, \text{mol dm}^{-3}$ solution ($2.80 \, \text{cm}^3$) of dichloromethane in toluene.

Write down the reactions occurring in each stage of this reaction sequence and show the structure of the final polymer. Evaluate the relevant degrees of polymerization and the corresponding molar masses for the final polymer.

2.14 Outline the methods and conditions of polymerization you would use to prepare the following polymers, giving reasons for your choices:

   (i)   isotactic poly(but-1-ene);
   (ii) isotactic poly(methyl methacrylate);
   (iii) poly(oxytetramethylene) with hydroxyl end-groups;
   (iv) poly(methyl methacrylate) for use in high-quality optical lenses;
   (v) poly(vinyl acetate) for use in formulating water-based adhesives;
   (vi) polystyrene with carboxylic acid end-groups.

2.15 Table 2.13 (Section 2.16.5) gives the reactivity ratios for free-radical copolymerization of styrene with (a) butadiene, (b) methyl methacrylate, (c) methyl acrylate, (d) acrylonitrile, (e) maleic anhydride, (f) vinyl chloride and (g) vinyl acetate. For each of these copolymerizations calculate:

   (i) the composition of the copolymer formed at low conversions of an equimolar mixture of the two monomers;
   (ii) the comonomer composition required to form a copolymer consisting of 50 mol% styrene repeat units.

For the azeotropic copolymerizations, calculate the azeotropic composition.

2.16 The initial concentrations of styrene, $[S]_0$, and acrylonitrile, $[AN]_0$, employed in a series of low conversion free-radical copolymerizations are given in the table below together with the nitrogen contents (%N by weight) of the corresponding poly(styrene-*stat*-acrylonitrile) samples produced.

$[S]_0$/mol dm^{-3}	3.45	2.60	2.10	1.55
$[AN]_0$/mol dm^{-3}	1.55	2.40	2.90	3.45
%N in copolymer	5.69	7.12	7.77	8.45

By use of an appropriate plot, evaluate the reactivity ratios for free-radical copolymerization of styrene with acrylonitrile.

Explain why it was necessary to restrict the copolymerizations to low conversions.

3 Characterization

3.1 Introduction

Relationships between the synthesis and molecular properties of polymers (Chapter 2), and between their molecular and bulk properties (Chapters 4 and 5), provide the foundations of Polymer Science. In order to establish these relationships, and to test theories, it is essential to accurately and thoroughly characterize the polymers under investigation. Furthermore, use of these relationships to predict and understand the in-use performance of a particular polymer depends upon the availability of good characterization data for that polymer. Thus polymer characterization is of great importance, both academically and commercially. The current chapter is concerned with *molecular characterization* of polymer samples, by which is meant the determination of their average molar masses, molar mass distributions, molecular dimensions, overall compositions, basic chemical structures and detailed molecular microstructures. Since most methods of molecular characterization involve analysis of polymers in dilute solution ($< 20\,\mathrm{g\,dm^{-3}}$), the relevant theories for polymers in solution will be introduced before considering the individual methods.

Polymers in solution

3.2 Thermodynamics of polymer solutions

A *solution* can be defined as a homogeneous mixture of two or more substances, i.e. the mixing is on a molecular scale. Under the usual conditions of constant temperature T and pressure P, the thermodynamic requirement for formation of a two-component solution is that the Gibbs free energy G_{12} of the mixture must be less than the sum of the Gibbs free energies G_1 and G_2 of the pure components in isolation. This requirement is defined in terms of the Gibbs free energy of mixing

$$\Delta G_m = G_{12} - (G_1 + G_2)$$

which must be negative (i.e. $\Delta G_m < 0$) for a solution to form. Since Gibbs free energy is related to enthalpy H and entropy S by the standard thermodynamic equation

$$G = H - TS \tag{3.1}$$

a more useful expression for ΔG_m is

$$\Delta G_m = \Delta H_m - T\Delta S_m \qquad (3.2)$$

where ΔH_m is the enthalpy (or heat) of mixing and ΔS_m is the entropy of mixing.

3.2.1 *Thermodynamics of ideal solutions*

Ideal solutions are mixtures of molecules (i) that are identical in size and (ii) for which the energies of like (i.e. 1–1 or 2–2) and unlike (i.e. 1–2) molecular interactions are equal. The latter condition leads to *athermal mixing* (i.e. $\Delta H_m = 0$), which also means that there are no changes in the rotational, vibrational and translational entropies of the components upon mixing. Thus ΔS_m depends only upon the *combinatorial* (or configurational) entropy change, ΔS_m^{comb}, which is positive because the number of distinguishable spatial arrangements of the molecules increases when they are mixed. Hence ΔG_m is negative and formation of an ideal solution always is favourable. The methods of *statistical mechanics* can be used to derive an equation for ΔS_m^{comb} by assuming that the molecules are placed randomly into cells which are of molecular size and which are arranged in the form of a three-dimensional lattice (represented in two dimensions for cubic cells in Fig. 3.1(a)).

From statistical mechanics, the fundamental relation between the entropy S of an assembly of molecules and the total number Ω of distinguishable degenerate (i.e. of equal energy) arrangements of the molecules is given by Boltzmann's equation

$$S = k \ln \Omega \qquad (3.3)$$

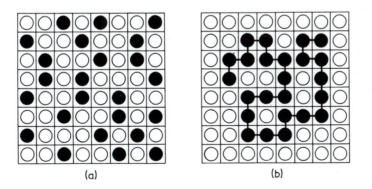

(a)   (b)

Fig 3.1 *Schematic representation of a liquid lattice. (a) mixture of molecules of equal size, (b) mixture of solvent molecules with a polymer molecule showing the connectivity of polymer segments.*

where **k** is the Boltzmann constant. Application of this equation to formation of an ideal solution gives

$$\Delta S_m^{comb} = \mathbf{k}[\ln \Omega_{12} - (\ln \Omega_1 + \ln \Omega_2)] \tag{3.4}$$

where Ω_1, Ω_2 and Ω_{12} are respectively the total numbers of distinguishable spatial arrangements of the molecules in the pure solvent, the pure solute and the ideal mixture. Since all the molecules of a pure substance are identical, there is only one distinguishable spatial arrangement of them (i.e. of arranging N identical molecules in a lattice containing N cells). Thus $\Omega_1 = 1$ and $\Omega_2 = 1$, allowing Equation (3.4) to be reduced to

$$\Delta S_m^{comb} = \mathbf{k} \ln \Omega_{12} \tag{3.5}$$

For ideal mixing of N_1 molecules of solvent with N_2 molecules of solute in a lattice with $(N_1 + N_2)$ cells, the total number of distinguishable spatial arrangements of the molecules is equal to the number of permutations of $(N_1 + N_2)$ objects which fall into two classes containing N_1 identical objects of type 1 and N_2 identical objects of type 2 respectively

$$\text{i.e. } \Omega_{12} = \frac{(N_1 + N_2)!}{N_1! N_2!} \tag{3.6}$$

Substitution of Equation (3.6) into Equation (3.5) gives

$$\Delta S_m^{comb} = \mathbf{k} \ln \left[\frac{(N_1 + N_2)!}{N_1! N_2!} \right] \tag{3.7}$$

and introducing Stirling's approximation, $\ln N! = N \ln N - N$ (for large N), leads to

$$\Delta S_m^{comb} = -\mathbf{k}\{N_1 \ln [N_1/(N_1 + N_2)] + N_2 \ln [N_2/(N_1 + N_2)]\} \tag{3.8}$$

It is more usual to write thermodynamic equations in terms of numbers of moles, n, and mole fractions, X, which here are defined by $n_1 = N_1/\mathbf{N_A}$, $n_2 = N_2/\mathbf{N_A}$, $X_1 = n_1/(n_1 + n_2)$ and $X_2 = n_2/(n_1 + n_2)$, where $\mathbf{N_A}$ is the Avogadro constant. Thus Equation (3.8) becomes

$$\Delta S_m^{comb} = -\mathbf{R}[n_1 \ln X_1 + n_2 \ln X_2] \tag{3.9}$$

since the universal gas constant $\mathbf{R} = \mathbf{k N_A}$. Hence for the formation of an ideal solution

$$\Delta G_m = \mathbf{R}T[n_1 \ln X_1 + n_2 \ln X_2] \tag{3.10}$$

This important equation provides the fundamental basis from which standard thermodynamic relationships for ideal solutions can be derived (e.g. Raoult's Law). However, since relatively few solutions behave ideally, the simple lattice theory requires modification to make it more generally applicable. For mixtures of small molecules, non-ideality invariably is due to non-athermal mixing (i.e. $\Delta H_m \neq 0$) and so requires the effects of non-equivalent intermolecular interactions to be taken into account. In contrast, polymer solutions show major deviations from ideal solution behaviour, even when $\Delta H_m = 0$. For example, the vapour pressure of solvent above a polymer solution invariably is very much lower than predicted from Raoult's Law. The failure of the simple lattice theory to give realistic predictions of the thermodynamic properties of polymer solutions arises from the assumption that the solvent and solute molecules are identical in size. Paul Flory and Maurice Huggins independently proposed a modified lattice theory which takes account of (i) the large differences in size between solvent and polymer molecules, and (ii) intermolecular interactions. This theory is described in an up-to-date form in the following section.

3.2.2 Flory–Huggins theory

The theory sets out to predict ΔG_m for the formation of polymer solutions by considering the polymer molecules to be chains of *segments*, each segment being equal in size to a solvent molecule. The number, x, of segments in the chain defines the size of a polymer molecule and is given by the ratio of the molecular volumes of polymer and solvent (hence x is not necessarily equal to the degree of polymerization). On this basis it is possible to place solvent molecules and/or polymer molecules in a three-dimensional lattice consisting of identical cells, each of which is the same size as a solvent molecule and has z cells as first neighbours. Each lattice cell is occupied by either a solvent molecule or a chain segment, and each polymer molecule is placed in the lattice so that its chain segments occupy a continuous sequence of x cells (as indicated in Fig. 3.1(b)).

The first stage in the development of the Flory–Huggins theory is to derive an expression for ΔS_m^{comb} when $\Delta H_m = 0$. As for mixing of simple molecules this involves application of Equation (3.4), but this time with $\Omega_2 > 1$ because each molecule in a pure amorphous polymer can adopt many different *conformations* (i.e. distinguishable spatial arrangements of the chain of segments). Hence, for polymer solutions

$$\Delta S_m^{comb} = \mathbf{k} \ln (\Omega_{12}/\Omega_2) \qquad (3.11)$$

The procedures and mathematical manipulations used to evaluate Ω_{12}, Ω_2 and ΔS_m^{comb} will only be considered in outline because they are relatively complex. The basis of the calculations is the formation of a theoretical polymer solution by mixing N_2 polymer molecules, each consisting of x chain segments, with N_1 solvent molecules in a lattice containing $N_1 + xN_2 \, (= N)$ cells. The N_2 polymer molecules are added one-by-one to the lattice before adding the solvent molecules. When adding a polymer molecule, the first segment of the chain can be placed in any empty cell. However, for the connectivity required to represent a polymer molecule, each successive segment of the remaining $x - 1$ segments must be placed in an empty cell adjacent to the previously-placed segment. The individual numbers of possible placements are calculated for each segment of the chain, and by multiplying these numbers together the total number, v, of possible conformations of the polymer molecule in the lattice is obtained. This series of calculations is carried out for each of the N_2 polymer molecules as they are added to the lattice. The restrictions imposed by partial occupancy of the lattice are taken into account using a *mean-field approximation* whereby the segments of the previously-added polymer molecules are assumed to be distributed uniformly in the lattice. Since the N_1 solvent molecules are identical, there is only one distinguishable spatial arrangement of them, namely one solvent molecule in each of the remaining N_1 empty cells. Thus the total number of distinguishable spatial arrangements of the mixture is given by

$$\Omega_{12} = \frac{1}{N_2!} \prod_{i=1}^{N_2} v_i \tag{3.12}$$

where $\prod$ is the sign for a continuous product (in this case $v_1 \times v_2 \times v_3 \times v_4 \times \ldots \times v_{N_2}$). The factor of $1/N_2!$ takes account of the fact that the N_2 polymer molecules are identical and therefore indistinguishable from each other. The resulting expression for Ω_{12} is given below in a useful form presented more recently by Flory

$$\Omega_{12} = [(z-1)^{(x-1)N_2}] \left[\left(\frac{x}{e^{x-1}} \right)^{N_2} \right] \left[\left(\frac{N}{N_1} \right)^{N_1} \left(\frac{N}{xN_2} \right)^{N_2} \right] \tag{3.13}$$

The first quantity in square brackets is the total number of conformations of the N_2 polymer molecules when each one is given the complete freedom of the lattice (i.e. the lattice is assumed to be empty when adding each polymer molecule). The second quantity is the fraction of these conformations that are allowed when the N_2 polymer molecules compete for cells in a lattice of xN_2 cells (i.e. in the pure amorphous polymer). The third quantity is the factor by which the total number of conformations of the N_2 polymer molecules is increased when they are given greater spatial freedom by dilution with N_1 solvent molecules. Thus Ω_2 is given by the product of the first two factors in Equation (3.13) and so Equation (3.11) simplifies to

$$\Delta S_m^{comb} = k \ln \left[\left(\frac{N}{N_1} \right)^{N_1} \left(\frac{N}{xN_2} \right)^{N_2} \right]$$

which simplifies further to give

$$\Delta S_m^{\text{comb}} = -\mathbf{k}[N_1 \ln \phi_1 + N_2 \ln \phi_2] \qquad (3.14)$$

where ϕ_1 and ϕ_2 are the volume fractions of solvent and polymer respectively, and are given by $\phi_1 = N_1/(N_1 + xN_2)$ and $\phi_2 = xN_2/(N_1 + xN_2)$. Writing Equation (3.14) in terms of the number of moles, n_1 and n_2, of solvent and polymer in the mixture gives

$$\Delta S_m^{\text{comb}} = -\mathbf{R}[n_1 \ln \phi_1 + n_2 \ln \phi_2] \qquad (3.15)$$

which may be compared to the corresponding expression for ideal mixing, i.e. Equation (3.9). Thus the only difference is that for polymer solutions, the mole fractions in Equation (3.9) are replaced by the corresponding volume fractions. Equation (3.15) is in fact a more general expression for athermal mixing and reduces to Equation (3.9) when $x = 1$.

Having derived an expression for ΔS_m^{comb}, the second stage in the development of Flory–Huggins theory is to calculate a term which accounts for the effects of intermolecular interactions. In the original theory, this was considered only in terms of an enthalpy change. However, the term subsequently was modified in recognition that there must be an entropy change associated with the non-randomness induced by interactions. Thus the effects of interactions will be treated here in terms of a contact Gibbs free energy change, $\Delta G_m^{\text{contact}}$. The calculation is simplified by restriction to first neighbour interactions on the basis that the forces between uncharged molecules are known to decrease rapidly with their distance of separation. Three types of contact need to be considered, each with its own Gibbs free energy of interaction:

Type of contact	Gibbs free energy of interaction
Solvent–solvent	g_{11}
Segment–segment	g_{22}
Solvent–segment	g_{12}

For every two solvent–segment contacts formed on mixing, one solvent–solvent and one segment–segment contact of the pure components are lost. Thus the Gibbs free energy change, Δg_{12}, for the formation of a single solvent–segment contact is given by

$$\Delta g_{12} = g_{12} - \tfrac{1}{2}(g_{11} + g_{22})$$

If p_{12} is the total number of solvent–segment contacts in the solution, then

$$\Delta G_m^{\text{contact}} = p_{12} \Delta g_{12} \qquad (3.16)$$

With the exception of the segments at the chain ends, each chain segment has two connected segmental neighbours. Hence there are $(z - 2)x + 2$ lattice sites adjacent to each polymer molecule. When x is large, $(z-2)x \gg 2$ and the total number of lattice sites adjacent to all the polymer molecules can be taken as $N_2(z-2)x$. On the basis of the mean-field approximation, a fraction ϕ_1 of these sites will be occupied by solvent molecules. Thus

$$p_{12} = N_2(z-2)x\phi_1$$

from which x can be eliminated to give

$$p_{12} = (z-2)N_1\phi_2 \tag{3.17}$$

because $xN_2\phi_1 = N_1\phi_2$. Combining Equations (3.16) and (3.17) leads to

$$\Delta G_m^{\text{contact}} = (z-2)N_1\phi_2\Delta g_{12} \tag{3.18}$$

The lattice parameter z and the Gibbs free energy change Δg_{12} are not easily accessible, and so are eliminated by introducing a single new parameter, χ, which is commonly known as the *Flory–Huggins polymer–solvent interaction parameter* and is defined by

$$\chi = (z-2)\Delta g_{12}/kT \tag{3.19}$$

Thus χ is a temperature-dependent dimensionless quantity which characterizes polymer–solvent interactions and can be expressed more simply in the form

$$\chi = a + b/T$$

where a and b are temperature-independent quantities. More generally χ is given as the sum of enthalpy, χ_H, and entropy, χ_S, components

$$\chi = \chi_H + \chi_S \tag{3.20}$$

from which it is simple to show that $\chi_H = b/T$ and $\chi_S = a$, and that

$$\chi_H = -T\left(\frac{\partial \chi}{\partial T}\right) \text{ and } \chi_S = \frac{\partial(T\chi)}{\partial T}$$

Combination of Equations (3.18) and (3.19) together with the relations $n_1 = N_1/N_A$ and $\mathbf{R} = \mathbf{k}N_A$ leads to

$$\Delta G_m^{\text{contact}} = \mathbf{R}Tn_1\phi_2\chi \tag{3.21}$$

Now all that is required to obtain the *Flory–Huggins equation for the Gibbs free energy of mixing* is to combine ΔS_m^{comb} (Equation (3.15)) and $\Delta G_m^{\text{contact}}$ (Equation (3.21)) as follows

$$\Delta G_m = \Delta G_m^{\text{contact}} - T\Delta S_m^{\text{comb}}$$

to give

$$\Delta G_m = \mathbf{R}T[n_1 \ln\phi_1 + n_2 \ln\phi_2 + n_1\phi_2\chi] \tag{3.22}$$

Using the Flory–Huggins equation it is possible to account for the equilibrium thermodynamic properties of polymer solutions, particularly the large negative deviations from Raoult's Law, phase-separation and fractionation behaviour, melting-point depressions in crystalline polymers, and swelling of networks. However, whilst the theory is able to predict general trends, precise agreement with experimental data is not achieved. The deficiencies of the theory result from the limitations both of the model and of the assumptions employed in its derivation. The use of a single type of lattice for pure solvent, pure polymer and their mixtures is unrealistic and requires there to be no volume change upon mixing. The mathematical procedure used to calculate the total number of possible conformations of a polymer molecule in the lattice does not exclude self-intersections of the chain, which clearly is physically unrealistic. Furthermore, the use of a mean-field approximation to facilitate this calculation for placement of a polymer molecule in a partly-filled lattice, is satisfactory only when the volume fraction, ϕ_2, of polymer is high. In view of this, Flory–Huggins theory is least satisfactory for dilute polymer solutions because the polymer molecules in such solutions are isolated from each other by regions of pure solvent, i.e. the segments are not uniformly distributed in the lattice. Additionally, the interaction parameter, χ, introduced to account for the effects of contact interactions, is not a simple parameter. Equation (3.20) resulted from a recognition that χ contains both enthalpy and entropy contributions, and χ also has been shown to depend upon ϕ_2. Despite these shortcomings, Flory–Huggins lattice theory was a major step forward towards understanding the thermodynamics of polymer solutions, and is the basis of many other theories. The theory will be developed further in subsequent sections since the relationships obtained are instructive, though the above limitations must be borne in mind when applying the resulting equations. Some of the more refined theories developed since the advent of Flory–Huggins lattice theory will be given brief consideration, but in general are beyond the scope of this book.

3.2.3 *Partial molar quantities and chemical potential*

In thermodynamics it often is necessary to distinguish between *extensive* and *intensive* properties. The value of an extensive property changes with the amount of material in the system, whereas the value of an intensive property is independent of the amount of material. Examples of extensive properties are mass, volume, and *total* free energy, enthalpy and entropy, whereas intensive properties include pressure, temperature, density and *molar* free energy, enthalpy and entropy. Generally, systems at equilibrium are defined in terms of their intensive properties.

For multicomponent systems such as polymer solutions, it is important to know how thermodynamic parameters (represented by Z) vary when a small amount of one component (represented by i) is added to the system whilst maintaining the pressure, temperature and amounts of all other components (represented by j) of the system constant. This requires the use of *partial molar quantities* such as that represented by $\overline{Z}_i$ which gives the rate at which Z changes with the number, n_i, of moles of component i and is defined by the partial differential

$$\overline{Z}_i = \left(\frac{\partial Z}{\partial n_i}\right)_{P,T,n_j} \tag{3.23}$$

The partial molar quantities most commonly encountered in the thermodynamics of polymer solutions are partial molar volumes $\overline{V}_i$ and partial molar Gibbs free energies $\overline{G}_i$. The latter quantities are called *chemical potentials*, μ_i, which therefore are defined by

$$\mu_i = \left(\frac{\partial G}{\partial n_i}\right)_{P,T,n_j} \tag{3.24}$$

In single-component systems such as pure solvents, the partial molar quantities are identical to the corresponding molar quantities.

For a system at equilibrium, the chemical potential, μ_i, of component i is related to its activity, a_i, in the system by the standard thermodynamic relation

$$\mu_i - \mu_i^0 = \mathbf{R}T \ln a_i \tag{3.25}$$

where μ_i^0 is the chemical potential of component i in its standard state, and $\mu_i - \mu_i^0$ is the partial molar Gibbs free energy change, $\overline{\Delta G}_i = (\partial \Delta G/\partial n_i)_{P,T,n_j}$, for formation of the system. Partial differentiation of the Flory–Huggins expression for ΔG_m (Equation (3.22)) leads to the following relations for the solvent

$$\mu_1 - \mu_1^0 = \mathbf{R}T\left[\ln \phi_1 + \left(1 - \frac{1}{x}\right)\phi_2 + \chi\phi_2^2\right] \tag{3.26}$$

and for the polymer

$$\mu_2 - \mu_2^0 = \mathbf{R}T[\ln \phi_2 - (x - 1)\phi_1 + x\chi\phi_1^2] \tag{3.27}$$

and per polymer chain segment, i.e. $(\mu_2 - \mu_2^0)/x$

$$\mu_s - \mu_s^0 = \mathbf{R}T\left[(\ln \phi_2)/x - \left(1 - \frac{1}{x}\right)\phi_1 + \chi\phi_1^2\right] \tag{3.28}$$

These equations can be combined with Equation (3.25) to obtain expressions for the corresponding activities of the components, which can then be used to calculate theoretical activities for comparison with those determined by experiment. Flory has shown that x should be replaced by its number-average value when performing calculations for polydisperse polymers.

3.2.4 Dilute polymer solutions

When analysing the thermodynamic properties of polymer solutions it is sufficient to consider only one of the components, which for reasons of simplicity normally is the solvent. Thus application of Equation (3.25) gives

$$\mu_1 - \mu_1^0 = \mathbf{R}T\ln a_1 \tag{3.29}$$

which can be separated into *ideal* and *non-ideal* (or *excess*) contributions by recognizing that $a_1 = \gamma_1 X_1$ where γ_1 and X_1 are respectively the activity coefficient and mole fraction of the solvent. Hence

$$(\mu_1 - \mu_1^0)^{\text{ideal}} = \mathbf{R}T\ln X_1 \tag{3.30}$$

$$(\mu_1 - \mu_1^0)^E = \mathbf{R}T\ln \gamma_1 \tag{3.31}$$

where the superscript E indicates an *excess* contribution.

The methods for molar mass characterization use dilute polymer solutions, typically $<20\,\text{g}\,\text{dm}^{-3}$, for which $n_1 \gg xn_2$ and $n_1 \gg n_2$ leading to

$$\phi_2 = xn_2/(n_1 + xn_2) \approx xn_2/n_1 \tag{3.32}$$

$$X_2 = n_2/(n_1 + n_2) \approx n_2/n_1 \tag{3.33}$$

Hence for dilute polymer solutions both ϕ_2 and X_2 are small and related to each other by $X_2 = \phi_2/x$. Under these conditions the logarithmic terms in Equations (3.26) and (3.30) can be approximated by series expansion to give

$$\ln \phi_1 = \ln(1 - \phi_2) = -\phi_2 - \phi_2^2/2 - \phi_2^3/3 - \dots \tag{3.34}$$

and

$$\ln X_1 = \ln(1 - X_2) = -X_2 - X_2^2/2 - X_2^3/3 - \dots$$

which upon substitution of $X_2 = \phi_2/x$ gives

$$\ln X_1 = -(\phi_2/x) - (\phi_2/x)^2/2 - (\phi_2/x)^3/3 - \dots \tag{3.35}$$

Since ϕ_2 is small and x is large, only the first two terms of Equation (3.34) and the first term of Equation (3.35) need to be retained. This leads to

$$\mu_1 - \mu_1^0 = -\mathbf{R}T\phi_2/x + \mathbf{R}T(\chi - \tfrac{1}{2})\phi_2^2 \tag{3.36}$$

from Equation (3.26) and

$$(\mu_1 - \mu_1^0)^{\text{ideal}} = -\mathbf{R}T\phi_2/x \tag{3.37}$$

from Equation (3.30). Comparison of these equations shows that for dilute polymer solutions, Flory–Huggins theory predicts the excess contribution to be

$$(\mu_1 - \mu_1^0)^E = \mathbf{R}T(\chi - \tfrac{1}{2})\phi_2^2 \tag{3.38}$$

in which the quantity $-\mathbf{R}T\phi_2^2/2$ arises from the connectivity of polymer chain segments and the quantity $\mathbf{R}T\chi\phi_2^2$ results from contact interactions. When $\chi = \tfrac{1}{2}$ these two effects just exactly compensate each other, $(\mu_1 - \mu_1^0)^E = 0$ and the dilute polymer solution behaves ideally, i.e. as if $\Delta H_m = 0$ and the polymer chain segments were not connected. Flory introduced the term *theta conditions* to describe this ideal state of a dilute polymer solution and developed the concept further by defining the excess partial molar enthalpy, $\overline{\Delta H_1^E}$, and entropy, $\overline{\Delta S_1^E}$, of mixing as follows

$$\overline{\Delta H_1^E} = \mathbf{R}T\varkappa\phi_2^2 \tag{3.39}$$

$$\overline{\Delta S_1^E} = \mathbf{R}\psi\phi_2^2 \tag{3.40}$$

where $\varkappa$ and ψ are respectively enthalpy and entropy parameters. Since

$$(\mu_1 - \mu_1^0)^E = \overline{\Delta G_1^E} = \overline{\Delta H_1^E} - T\overline{\Delta S_1^E} \tag{3.41}$$

it is easy to show that

$$\varkappa - \psi = \chi - \tfrac{1}{2} \tag{3.42}$$

and that theta conditions occur at a particular temperature, θ, known as the *theta* (or *Flory*) *temperature*, for which $(\mu_1 - \mu_1^0)^E = 0$. Thus

$$\overline{\Delta H_1^E} = \theta\overline{\Delta S_1^E} \tag{3.43}$$

This equation suggests that the enthalpic interactions produce a proportionate change in the excess entropy, and reveals that $\overline{\Delta H_1^E}$ and $\overline{\Delta S_1^E}$ (or alternatively $\varkappa$ and ψ) must have the same sign (i.e. either both positive or both negative). Equation (3.43) also provides an alternative definition of the theta temperature, as being the proportionality constant relating $\overline{\Delta H_1^E}$ to $\overline{\Delta S_1^E}$. On this basis the origin of the variation of θ from one polymer–solvent system to another is clearly evident.

Substituting Equations (3.39) and (3.40) into (3.43) gives

$$\varkappa = \left(\frac{\theta}{T}\right)\psi \tag{3.44}$$

which when combined with Equations (3.39), (3.40) and (3.41) leads to

$$(\mu_1 - \mu_1^0)^E = \mathbf{R}T\psi\left[\left(\frac{\theta}{T}\right) - 1\right]\phi_2^2 \tag{3.45}$$

TABLE 3.1 *Values of theta temperature for different polymer–solvent systems*[*]

Polymer	Solvent	$\theta/°C$
Polyethylene	Biphenyl	125
Polystyrene	Cyclohexane	34
Poly(vinyl acetate)	Methanol	6
Poly(methyl methacrylate)	Butyl Acetate	−20
Poly(methyl methacrylate)	Pentyl Acetate	41
Poly(vinyl alcohol)	Water	97
Poly(acrylic acid)	1,4-Dioxan	29

[*] The entropy parameter ψ is positive for all of the systems except poly(vinyl alcohol): water and poly(acrylic acid): 1,4-dioxan for which it is negative.

Equations (3.38) and (3.45) reveal that $(\mu_1 - \mu_1^0)^E$ promotes mixing (i.e. is negative) when $\chi < \frac{1}{2}$ which corresponds to $T > \theta$ for positive ψ and $T < \theta$ for negative ψ. Furthermore, they show that under theta conditions $\chi = \frac{1}{2}$, $T = \theta$ and the dilute polymer solution behaves as an ideal solution. Some values of θ are given in Table 3.1 for a series of different polymer–solvent systems.

The theory presented above for dilute polymer solutions is based upon the Flory–Huggins Equation (3.22) which strictly is not valid for such solutions because of the mean-field approximation. Nevertheless, whilst Equations (3.38) and (3.45) do not accurately predict $(\mu_1 - \mu_1^0)^E$, they are of the correct functional form, i.e. the relationships

$$(\mu_1 - \mu_1^0)^E \propto (\chi - \tfrac{1}{2})$$

and

$$(\mu_1 - \mu_1^0)^E \propto \psi \left[\left(\frac{\theta}{T} \right) - 1 \right]$$

also are obtained from the more satisfactory excluded-volume theories (Section 3.3.3). Thus the Flory–Huggins theory leads to correct conclusions about the general thermodynamic behaviour of dilute polymer solutions, especially with regard to the prediction of ideal behaviour when $\chi = \frac{1}{2}$ and $T = \theta$. The accuracy of the Flory–Huggins dilute polymer solution theory improves as theta conditions are approached and so the predictions of $(\mu_1 - \mu_1^0)^E$ afforded by Equations (3.38) and (3.45) become better as $\chi \to \frac{1}{2}$ and $T \to \theta$.

3.2.5 *The solubility parameter approach*

In many practical situations all that is required is a reasonable guide to the compatibility (i.e. miscibility) of specific polymer–solvent systems. Examples include the selection of a solvent for dissolving a particular polymer, the specification of an elastomer which will not swell (i.e. absorb liquid) in applications known to involve contact with certain liquids, and the prediction of possible environmental crazing of a polymer during service (Section 5.6.7). The solubility parameter approach is most useful for this purpose and was first developed by Hildebrand for calculating estimates of the enthalpy of mixing, $\Delta H_m^{\text{contact}}$, for mixtures of liquids. The equation employed is

$$\Delta H_m^{\text{contact}} = V_m\phi_1\phi_2(\delta_1 - \delta_2)^2 \tag{3.46}$$

where V_m is the molar volume of the mixture, and δ_1 and δ_2 are the solubility parameters of components 1 and 2 respectively. The *solubility parameter*, δ, of a liquid is the square root of the energy of vaporization per unit volume and is given by

$$\delta = [(\Delta H_v - \mathbf{R}T)/V]^{1/2} \tag{3.47}$$

where ΔH_v is its molar enthalpy of vaporization and V is its molar volume. The quantity δ^2 is called the *cohesive energy density* (CED) since it characterizes the strength of attraction between the molecules in unit volume. For a volatile liquid CED, and hence δ, can be determined experimentally by measuring ΔH_v and V.

Equation (3.46) yields only zero or positive values for ΔH_m and predicts that mixing becomes more favourable (i.e. ΔH_m becomes less positive) as the difference between the solubility parameters of the two components decreases, with $\Delta H_m = 0$ when $\delta_1 = \delta_2$. Specific effects such as hydrogen bonding and charge transfer interactions can lead to negative ΔH_m but are not taken into account by Equation (3.46) and so a separate qualitative judgement must be made to predict their effect upon miscibility.

Extension of the solubility parameter approach to the prediction of polymer–solvent and polymer–polymer miscibility requires knowledge of δ values for polymers. Because polymers are not volatile, their δ values must be obtained indirectly. Most commonly the δ value is taken as being that of the solvent which gives the maximum degree of swelling for network polymers and the maximum intrinsic viscosity (Section 3.14) for soluble polymers. Methods also are available for calculating estimates of δ from a knowledge of the repeat unit structure and polymer density.

Some values of δ for common solvents and polymers are given in Table 3.2. Thus the solubility parameter differences correctly indicate that with the exception of methanol and water, polystyrene can be dissolved in each of the solvents listed. In general, however, the solubility parameter approach must be treated as a guide to miscibility which is most accurate for mixtures of non-polar amorphous polymers with non-polar solvents. The limitations of the theory have already been mentioned with respect

TABLE 3.2 *Solubility parameters for common solvents and polymers. (Taken from the 'Polymer Handbook' edited by Brandrup and Immergut, reproduced with permission)*

Solvent	$\delta/10^3 J^{1/2} m^{-3/2}$
Acetone	20.3
Carbon tetrachloride	17.6
Chloroform	19.0
Cyclohexane	16.8
Methanol	29.7
Toluene	18.2
Water	47.9
Xylene	18.0
Polymer	$\delta/10^3 J^{1/2} m^{-3/2}$
Polyethylene	16.4
Polystyrene	18.5
Poly(methyl methacrylate)	19.0
Polypropylene	17.2
Poly(vinyl chloride)	20.0

to contributions from hydrogen-bonding interactions and other specific effects. Additionally, since the theory is based upon liquid–liquid mixtures it does not take into account the need to overcome the enthalpy of crystallization when dissolving crystalline polymers. The latter contribution is the reason why crystalline polymers are generally less soluble than amorphous polymers, and more specifically why polyethylene dissolves in toluene and xylene *only* at high temperatures (i.e. above about 80 °C).

By combining Equation (3.46) with Equations (3.20) and (3.21) it is possible to obtain the following expression for χ_H in terms of solubility parameters

$$\chi_H = V_1(\delta_1 - \delta_2)^2 / \mathbf{R}T \tag{3.48}$$

where V_1 is the molar volume of the solvent. Together with χ_S, which typically is about 0.2, Equation (3.48) allows estimates of χ to be calculated, though obviously the estimates are subject to the limitations of solubility parameter theory.

3.3 Chain dimensions

The subject of *chain dimensions* is concerned with relating the sizes and shapes of individual polymer molecules to their chemical structure, chain length and molecular environment. In the analysis of polymer solution thermodynamics given in the previous sections it was not necessary to consider in detail chain dimensions. However, in order to fully interpret the properties of dilute polymer solutions it is crucial to consider the behaviour of isolated polymer molecules, in particular their chain dimensions. In this and subsequent sections some of the more elementary aspects of chain dimensions will be introduced.

The shape of a polymer molecule is to a large extent determined by the effects of its chemical structure upon chain stiffness. Since most polymer molecules have relatively flexible backbones, they tend to be highly coiled and can be represented as random coils. However, as the backbone becomes stiffer the chains begin to adopt a more elongated worm-like shape and ultimately become rod-like. The theories which follow are concerned only with the chain dimensions of linear flexible polymer molecules. More advanced texts should be consulted for treatments of worm-like and rod-like chains.

3.3.1 *Freely-jointed chains*

The simplest measure of chain dimensions is the length of the chain along its backbone and is known as the *contour length*. For a chain of n backbone bonds each of length l, the contour length is nl. However, for linear flexible chains it is more usual, and more realistic, to consider the dimensions of the molecular coil in terms of the distance separating the chain ends, i.e. the *end-to-end distance r* (Fig. 3.2).

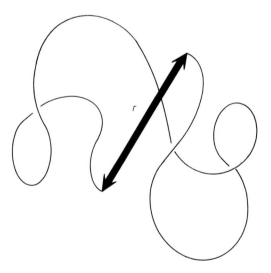

Fig. 3.2 *Schematic representation of a coiled polymer molecule showing the end-to-end distance.*

When considering an isolated polymer molecule it is not possible to assign a unique value of r because the chain conformation (and hence r) is continuously changing due to rotation of backbone bonds. However, whilst each conformation has a characteristic value of r, certain different conformations give rise to the same value of r. Thus some values of r are more probable than others and the probability distribution of r can be represented by the *root mean square (RMS) end-to-end distance*, $\langle r^2 \rangle^{1/2}$, where $\langle \ \rangle$ indicates that the quantity is averaged over time. The techniques of statistical mechanics can be used to calculate $\langle r^2 \rangle^{1/2}$ but in order to do so a model for the polymer molecule must be assumed. The simplest model is that of a *freely-jointed chain* of n links (i.e. backbone bonds) of length l for which there are no restrictions upon either bond angle or bond rotation. The analysis of this model is a simple extension of the well-known *random-walk* calculation which was first developed to describe the movement of molecules in an ideal gas. The only difference is that for the freely-jointed chain, each step is of equal length l. A three-dimensional rectangular co-ordinate system is used as shown in Fig. 3.3. One end of the chain is fixed at the origin O and the probability, $P(x,y,z)$, of finding the other end within a small volume element $dx.dy.dz$

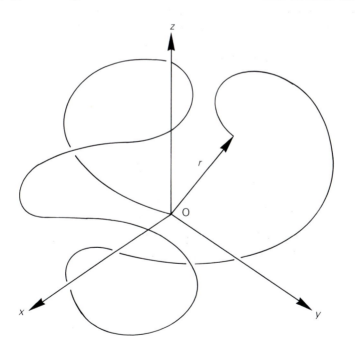

Fig. 3.3 *Diagram showing a coiled polymer molecule of end-to-end distance r in a rectangular co-ordinate system with one chain end fixed at the origin O.*

at a particular point with co-ordinates (x,y,z) is calculated. The calculation leads to an equation of the form

$$P(x,y,z) = W(x,y,z)dxdydz \qquad (3.49)$$

where $W(x,y,z)$ is a *probability density function*, i.e. a probability per unit volume. When $r \ll nl$ and n is large, $W(x,y,z)$ is given by

$$W(x,y,z) = (\beta/\pi^{1/2})^3\exp[-\beta^2(x^2 + y^2 + z^2)] \qquad (3.50)$$

where $\beta = [3/(2nl^2)]^{1/2}$. Since r^2 is equal to $(x^2 + y^2 + z^2)$, Equation (3.50) can be simplified to

$$W(x,y,z) = (\beta/\pi^{1/2})^3\exp(-\beta^2 r^2) \qquad (3.51)$$

This equation reveals that $W(x,y,z)$ is a *Gaussian* distribution function which has a maximum value at $r = 0$, as is shown by the plot of $W(x,y,z)$ given in Fig. 3.4(a). The co-ordinates (x,y,z) specified by $W(x,y,z)$ define a particular direction from the origin and so correspond only to one of many such co-ordinates each of which gives rise to an end-to-end distance equal to r but in a different direction. Thus a more important probability is that of finding one chain end at a distance r in any direction from the other chain end located at the origin. This is equal to the probability $W(r)dr$ of finding the chain end in a spherical shell of thickness dr positioned a radial distance r from the origin. The volume of the shell is $4\pi r^2 dr$ and so

$$W(r)dr = W(x,y,z).\ 4\pi r^2 dr \qquad (3.52)$$

which upon substituting Equation (3.51) for $W(x,y,z)$ leads to the following expression for the *radial distribution function $W(r)$*

$$W(r) = 4\pi(\beta/\pi^{1/2})^3 r^2 \exp(-\beta^2 r^2) \qquad (3.53)$$

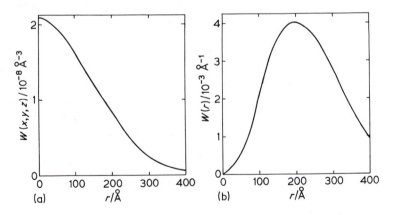

Fig. 3.4 *Plot of (a) Equation (3.51) and (b) Equation (3.53), for a chain of 10^4 links, each link of length 2.5 Å.*

This function is plotted in Fig. 3.4(b) and can be shown by simple differentiation (see problems) to have a maximum value at $r = 1/\beta$. It also is simple to show that $W(r)$ normalizes to unity, i.e.

$$\int_0^\infty W(r)\mathrm{d}r = 1 \tag{3.54}$$

though this highlights a deficiency of the theory since $W(r)$ is finite when $r > nl$, a condition which physically is not possible (i.e. the end-to-end distance cannot exceed the contour length). However, since $W(r)$ is small for $r > nl$ the errors introduced are negligible.

The mean square end-to-end distance, $\langle r^2 \rangle$, is the second moment of the radial distribution function and so is defined by the integral

$$\langle r^2 \rangle = \int_0^\infty r^2 W(r)\mathrm{d}r \tag{3.55}$$

Combining Equations (3.53) and (3.55) and integrating gives the result (see problems)

$$\langle r^2 \rangle = 3/2\beta^2 \tag{3.56}$$

Since $\beta^2 = 3/2nl^2$ the RMS end-to-end distance is given by

$$\langle r^2 \rangle_f^{1/2} = n^{1/2}l \tag{3.57}$$

where the subscript f indicates that the result is for a freely-jointed chain. This rather simple equation is very important and reveals that $\langle r^2 \rangle_f^{1/2}$ is a factor of $n^{1/2}$ smaller than the contour length. Since n is large, this highlights the highly coiled nature of flexible Gaussian polymer chains. For example, a freely-jointed chain of 10 000 segments each of length 3Å will have a fully-extended contour length of 30 000Å whereas its RMS end-to-end distance will be only 300Å.

In addition to the RMS end-to-end distance, the dimensions of linear chains often are characterized in terms of the RMS distance of a chain segment from the centre of mass of the molecule, which is known simply as the RMS *radius of gyration*, $\langle s^2 \rangle^{1/2}$. This quantity has the advantage that it also can be used to characterize the dimensions of branched macro-molecules (which have more than two chain ends) and cyclic macro-molecules (which have no chain ends). Furthermore, properties of dilute polymer solutions which are dependent upon chain dimensions are controlled by $\langle s^2 \rangle^{1/2}$ rather than $\langle r^2 \rangle^{1/2}$. However, for linear Gaussian chains $\langle s^2 \rangle^{1/2}$ is related to $\langle r^2 \rangle^{1/2}$ by

$$\langle s^2 \rangle^{1/2} = \langle r^2 \rangle^{1/2}/6^{1/2} \tag{3.58}$$

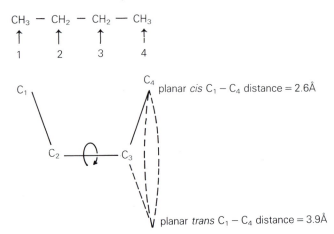

Fig. 3.5 *Effects of bond rotation upon the conformation of a n-butane molecule.*

and so in the theoretical treatment of linear flexible chains it is usual to consider only $\langle r^2 \rangle^{1/2}$.

3.3.2 *Effects of bond angle and short-range steric restrictions*

Whilst the freely-jointed chain is a simple model from which to begin prediction of chain dimensions, it is physically unrealistic. The links in real polymer chains are subject to bond angle restrictions and do not rotate freely because of short-range steric interactions. Both of these effects cause $\langle r^2 \rangle^{1/2}$ to be larger than $\langle r^2 \rangle_f^{1/2}$, and are most easily demonstrated by considering the conformations of simple alkanes.

The possible conformations of a n-butane molecule are illustrated schematically in Fig. 3.5. Each carbon atom in the molecule is sp^3 hybridized and so is tetrahedral with bond angles of 109.5°. The different conformations result from rotation about the C_2—C_3 bond and have stabilities which depend upon the steric interactions between the methyl groups and hydrogen atoms bonded to C_2 and C_3. The methyl groups are relatively bulky and so the *planar cis* conformation, corresponding to their distance of closest approach, is the least stable. Accordingly, the *planar trans* conformation is the most stable. The potential energy of a n-butane molecule is given in Fig. 3.6 as a function of the angle ϕ through which the C_3—C_4 bond is rotated from the *planar trans* conformation about the plane of the C_1—C_2—C_3 bonds ($\phi = 0$ for *planar trans* and $\phi = 180°$ for *planar cis*). The minima correspond to the staggered conformations (*planar trans* and *gauche* ±) in which there is maximum separation of the substituents on C_2 and C_3, i.e. the methyl groups and hydrogen atoms are at 60° to each

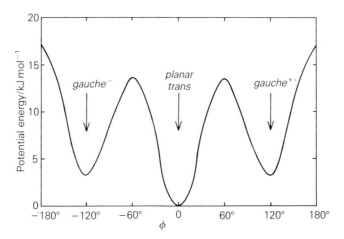

Fig. 3.6 *Potential energy of a n-butane molecule as a function of the angle ϕ of bond rotation.*

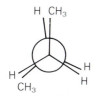 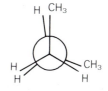

eclipsed conformations

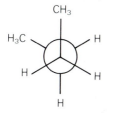 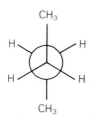 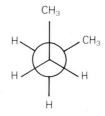

staggered conformations

Fig. 3.7 *Newman projections of the eclipsed and staggered conformations of a n-butane molecule.*

other. These conformations are shown as Newman projections in Fig. 3.7 together with the eclipsed conformations which give rise to the maxima in Fig. 3.6. Thus short-range steric interactions are important in determining the relative probabilities of existence of different conformations of a molecule.

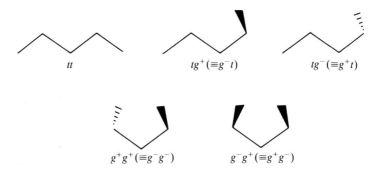

Fig. 3.8 *The five distinct conformations of n-pentane.*

The restrictions imposed by short-range steric interactions upon the conformations of a polymer molecule occur at a local level in short sequences of chain segments. The interactions are more complex than in n-butane because the conformation about a given chain segment is dependent upon the conformations about the segments to which it is directly connected, i.e. there is an interdependence of local chain conformations. This is most easily appreciated by considering the conformations of n-pentane in terms of sequences of *planar trans* (*t*) and *gauche* ± (*g*±) conformations. If the three central carbon atoms in n-pentane are fixed in the plane of the paper, then the five distinct conformations of the molecule can be represented in skeletal form as shown in Fig. 3.8. The g^-g^+ and g^+g^- conformations bring the methyl groups into close proximity (with their carbon atoms about 2.5Å apart) and so are of high energy and low probability in comparison to the other conformations. Such interdependent steric restrictions affect the local chain conformation all along a polymer chain and have a significant effect upon chain dimensions.

The simplest modification to the freely-jointed chain model is the introduction of bond angle restrictions whilst still allowing free rotation about the bonds. This is known as the *valence angle model* and for a polymer chain with backbone bond angles all equal to θ, leads to the following equation for the mean square end-to-end distance

$$\langle r^2 \rangle_{fa} = nl^2(1 - \cos\theta)/(1 + \cos\theta) \tag{3.59}$$

where the subscript *fa* indicates that the result is for chains in which the bonds rotate freely about a fixed bond angle. Since $180° > \theta > 90°$, $\cos\theta$ is negative and $\langle r^2 \rangle_{fa}$ is greater than $\langle r^2 \rangle_f$. Equation (3.59) is directly applicable to polymers derived from ethylenic monomers because they have C—C backbone bonds with $\theta \approx 109.5°$ for which $\cos\theta \approx -1/3$ and the equation becomes

$$\langle r^2 \rangle_{fa} = 2nl^2 \qquad (3.60)$$

Thus for such polymers (e.g. polyethylene) bond angle restrictions cause the RMS end-to-end distance to increase by a factor of $\sqrt{2}$ (i.e. *c.* 1.4) from that of the freely-jointed chain.

The restrictions upon bond rotation arising from short-range steric interactions are more difficult to quantify theoretically. The usual procedure is to assume that the conformations of each sequence of three backbone bonds are restricted to discrete *rotational isomeric states* corresponding to potential energy minima such as those shown for n-butane in Fig. 3.6. The effects of interdependent steric interactions of the backbone atoms also can be taken into account in this way (e.g. by excluding g^-g^+ and g^+g^- sequences). For the simplest case of polyethylene, rotational isomeric state theory leads to the following equation

$$\langle r^2 \rangle_o = nl^2 \left(\frac{1 - \cos\theta}{1 + \cos\theta} \right) \left(\frac{1 + \overline{\cos\phi}}{1 - \overline{\cos\phi}} \right) \qquad (3.61)$$

where the subscript *o* indicates that the result is for a polymer chain with hindered rotation about a fixed bond angle. The quantity $\overline{\cos\phi}$ is the average value of $\cos\phi$ where ϕ is the angle of bond rotation as defined for n-butane (Figs. 3.6 and 3.7). When bond rotation is unrestricted, $\overline{\cos\phi} = 0$ because all values of ϕ are equally probable causing the positive and negative $\cos\phi$ contributions to cancel each other out and resulting in Equations (3.61) and (3.59) becoming identical. However, due to short-range steric restrictions, values of $|\phi| < 90°$ are favoured so that $\overline{\cos\phi}$ is positive and $\langle r^2 \rangle_o$ is greater than $\langle r^2 \rangle_{fa}$. The presence of side groups on the polymer chain (e.g. phenyl groups for polystyrene) tends to further hinder bond rotation and introduces additional interdependent steric interactions involving the side groups. These effects are very difficult to treat theoretically and so Equation (3.61) usually is written in the more general form

$$\langle r^2 \rangle_o = \sigma^2 nl^2 (1 - \cos\theta)/(1 + \cos\theta) \qquad (3.62)$$

where σ is a *steric parameter* the value of which normally exceeds that of the square root of the $(1 + \overline{\cos\phi})/(1 - \overline{\cos\phi})$ term in Equation (3.61), and is the factor by which $\langle r^2 \rangle_o^{1/2}$ exceeds $\langle r^2 \rangle_{fa}^{1/2}$. Since values of σ are rather difficult to calculate, they usually are evaluated from values of $\langle r^2 \rangle_o^{1/2}$ measured experimentally.

An idea of the stiffness of a polymer chain can be gained from the ratio $\langle r^2 \rangle_o^{1/2}/\langle r^2 \rangle_f^{1/2}$ which is the square-root of the *characteristic ratio*, $C_\infty = \langle r^2 \rangle_o/nl^2$. Values of σ and C_∞ for some common polymers are given in Table 3.3. It is found that σ typically has values between 1.5 and 2.5, and

TABLE 3.3 *Typical values of σ and C∞ for some common polymers*

Polymer	Temperature/°C	σ	C_∞
Polyethylene	140	1.8	6.8
Isotactic polypropylene	140	1.6	5.2
Poly(vinyl chloride)	25	1.8	6.7
Polystyrene	25	2.3	10.8
Polystyrene	70	2.1	9.2
Poly(methyl methacrylate)	25	2.1	8.6
Poly(methyl methacrylate)	72	1.8	6.6

that the combination of fixed bond angles and short-range steric interactions causes the RMS end-to-end distances of real polymer chains to be greater than those of freely-jointed chains by factors of 2–3.

It is possible to represent a real polymer chain by an *equivalent freely-jointed chain* of N links each of length b such that the chains have the same contour length (i.e. $nl = Nb$) and the same RMS end-to-end distance (i.e. $\langle r^2 \rangle_o = Nb^2$). Thus, in terms of the characteristic ratio

$$N = n/C_\infty \qquad (3.63)$$

and

$$b = C_\infty l \qquad (3.64)$$

3.3.3 *Effects of long-range steric interactions: chains with excluded volume*

The mathematical model for evaluating $\langle r^2 \rangle_o^{1/2}$ is that of a series of connected vectors (representing the backbone bonds) which are restricted at a local level to certain allowed orientations relative to each other. The bond vectors are volumeless lines in space and the model places no restrictions upon the relative positions of two bond vectors widely separated in the chain. Thus the model does not prevent remotely-connected bond vectors occupying the same volume in space, and allows self-intersections of the chain. In a real isolated polymer molecule, each part of the molecule excludes other more remotely-connected parts from its volume. These long-range steric interactions cause the true RMS end-to-end distance, $\langle r^2 \rangle^{1/2}$, to be greater than $\langle r^2 \rangle_o^{1/2}$ and usually are considered in terms of an *excluded volume*. Chain dimensions which correspond to $\langle r^2 \rangle_o^{1/2}$ are unperturbed by the effects of volume exclusion and so are called the *unperturbed dimensions*. The extent to which these dimensions are perturbed in real chains is defined by an *expansion parameter* α_r such that

$$\langle r^2 \rangle^{1/2} = \alpha_r \langle r^2 \rangle_o^{1/2} \qquad (3.65)$$

However, expansion of the molecular coil due to volume exclusion is not uniform, and is greatest where the segment density is highest. Thus a different expansion parameter α_s is required to account for expansion from the unperturbed RMS radius of gyration

$$\langle s^2 \rangle^{1/2} = \alpha_s \langle s^2 \rangle_o^{1/2} \tag{3.66}$$

Theoretical calculations give the expansion parameters as power series in the *excluded volume parameter z* which is given by

$$z = (3/2\pi \langle r^2 \rangle_o)^{3/2} N^2 \beta_e \tag{3.67}$$

where N is defined by Equation (3.63) and β_e, which is called the *excluded volume integral*, is the volume excluded by one chain segment to another. For small expansions (i.e. small z) the following relationships are obtained

$$\alpha_r^2 = 1 + (4/3)z \tag{3.68}$$

$$\alpha_s^2 = 1 + (134/105)z \tag{3.69}$$

and so $\alpha_r > \alpha_s$.

Approximate theories relating expansion parameters, α, to z give equations of the general form

$$\alpha^5 - \alpha^3 = Kz \tag{3.70}$$

where K is a constant. For large expansions $\alpha^5 \gg \alpha^3$ and so $\alpha^5 \approx Kz$. Under these conditions α is dependent upon $n^{1/10}$ because z is proportional to $N^{1/2}$ which in turn is proportional to $n^{1/2}$. Since the unperturbed dimensions are dependent upon $n^{1/2}$, the perturbed dimensions of a highly expanded chain are proportional to $n^{3/5}$.

The expansion parameters for a real chain also embody the effects of interactions of the chain with its molecular environment (e.g. with solvent molecules or other polymer molecules). The simplest case is that of a *pure amorphous polymer*. Each polymer molecule is surrounded by other polymer molecules of the same type. Expansion of a given chain to relieve long-range *intramolecular* steric interactions only serves to create an equal number of *intermolecular* steric interactions with neighbouring chains. These opposing volume exclusion effects exactly counteract each other and so in a pure amorphous polymer the polymer molecules adopt their unperturbed dimensions (i.e. $\alpha_r = \alpha_s = 1$), as proven more recently by neutron scattering experiments (Section 3.13). Since the net energy of interaction is zero, the individual molecular coils interpenetrate each other and become highly entangled.

In *dilute solution*, polymer molecules do not interact with each other and chain expansion depends upon the balance between the intramolecular segment–segment and intermolecular segment–solvent and solvent–solvent interactions. When a polymer is dissolved in a *good solvent*, the isolated polymer molecules expand from their unperturbed dimensions so as to increase the number of segment–solvent contacts (i.e. $\alpha > 1$). In a *poor solvent*, however, the segment–solvent interactions are weak and their Gibbs free energy of interaction may be positive (i.e. unfavourable); the polymer chains contract in order to reduce the number of segment–solvent contacts and so are relatively compact (i.e. values of α are close to unity). Thus in poor solvents, isolated polymer chains are subjected to two opposing influences: (i) expansion due to unfavourable segment–segment interactions (i.e. volume exclusion), and (ii) contraction due to unfavourable segment–solvent interactions. Under specific conditions in a poor solvent, the energy of segment–solvent interaction is just sufficiently positive to exactly counteract the energy of segment–segment interaction and each polymer molecule assumes its unperturbed dimensions. These conditions correspond to the *theta conditions* introduced in Section 3.2.4 to describe the situation when a dilute polymer solution behaves ideally. Under these conditions, the solvent is said to be a *theta solvent* for the polymer and unperturbed chain dimensions can be measured. If the solvency conditions deteriorate further, the polymer molecules will 'precipitate' to form a separate phase with a very high polymer concentration ($\alpha \approx 1$) rather than undergo extensive contraction ($\alpha < 1$) in the dilute solution.

The link between chain dimensions and the thermodynamic behaviour of dilute polymer solutions is of great importance and has its origin in the excluded volume integral β_e. This arises because β_e is the volume excluded by one segment to any other segment which therefore can be part of the same or a different molecule. Theories for β_e lead to equations of the form

$$\beta_e = C_e \left[\left(\frac{\theta}{T} \right) - 1 \right] \tag{3.71}$$

where C_e is a constant and θ is the theta temperature. Thus $\beta_e \to 0$ as $T \to \theta$.

Theories of chain expansion use β_e to account for *intramolecular* segment–segment volume exclusion and relate expansion parameters to β_e through the excluded volume parameter z. These theories show that $\alpha_r \to 1$ and $\alpha_s \to 1$ as $\beta_e \to 0$.

In dilute polymer solutions each polymer molecule excludes all others from its volume. Thus mean-field theories, such as Flory–Huggins theory, are inappropriate and more exact theories of dilute solutions calculate ΔG_m from the volume excluded by one polymer molecule to another, which in turn is calculated using β_e to account for *intermolecular* segment–segment volume exclusion. These theories show that $(\mu_1 - \mu_1^0)^E \to 0$ as $\beta_e \to 0$.

It may therefore be concluded that when $\beta_e = 0$ polymer molecules assume their unperturbed dimensions and dilute polymer solutions behave ideally.

3.4 Frictional properties of polymer molecules in dilute solution

Processes such as diffusion, sedimentation and viscous flow involve the motion of individual molecules. When a polymer molecule moves through a dilute solution it undergoes frictional interactions with solvent molecules. The nature and effects of these frictional interactions depend upon the size and shape of the polymer molecule as modified by its thermodynamic interactions with solvent molecules. Thus the chain dimensions can be evaluated from measurements of the frictional properties of a polymer molecule.

3.4.1 Frictional coefficients of polymer molecules

Two extremes of the frictional behaviour of polymer molecules can be identified, namely free-draining and non-draining. A polymer molecule is said to be *free-draining* when solvent molecules are able to flow past each segment of the chain, and *non-draining* when solvent molecules within the coiled polymer chain move with it. These two extremes of behaviour lead to different dependences of the frictional coefficient, f_o, of a polymer molecule upon chain length.

Free-draining polymer molecules are considered by dividing them into identical segments each of which has the same frictional coefficient ζ. Since solvent molecules permeate all regions of the polymer coil with equal ease (or difficulty), each segment makes the same contribution to f_o which therefore is given by

$$f_o = x\zeta \tag{3.72}$$

where x is number of segments in the chain.

A *non-draining polymer molecule* can be represented by an equivalent impermeable hydrodynamic particle, i.e. one which has the same frictional coefficient as the polymer molecule. Thus a non-draining random coil can be represented by an equivalent impermeable hydrodynamic sphere of radius R_h. From Stokes' Law

$$f_o = 6\pi\eta_0 R_h \tag{3.73}$$

where η_0 is the viscosity of the pure solvent. By making the reasonable assumption that R_h is proportional to $\langle s^2 \rangle^{1/2}$, then Equation (3.73) can be re-written in the form

$$f_o = K_o \, \alpha_\eta \langle s^2 \rangle_o^{1/2} \tag{3.74}$$

where K_o is a constant for a given system and α_η is the expansion parameter for the hydrodynamic chain dimensions. Theory shows that for small chain expansions $\alpha_\eta = \alpha_s^{0.81}$, i.e. $\alpha_\eta < \alpha_s$. Since $\langle s^2 \rangle_o^{1/2}$ is proportional to $x^{1/2}$ and for highly expanded coils α_η is approximately proportional to $x^{1/10}$ (Equation (3.70)), Equation (3.74) predicts that

$$f_o = K'_o x^{a_o} \tag{3.75}$$

where K'_o is another constant and $0.5 \leqslant a_o \leqslant 0.6$. This dependence of f_o upon x can now be contrasted with that for the free-draining polymer molecule, for which f_o is directly proportional to x.

The frictional behaviour of real polymer molecules comprises both free-draining and non-draining contributions. The free-draining contribution is dominant for very short chains and for highly elongated rod-like molecules, but for flexible (i.e. coiled) chains it decreases rapidly as the chain length increases. Since most polymers consist of long flexible chain molecules, their frictional properties closely approximate to those associated with polymer molecules in the non-draining limit. Therefore, in the following section, this limit will be considered further, specifically in relation to the viscosity of a dilute polymer solution.

3.4.2 *Hydrodynamic volume and intrinsic viscosity in the non-draining limit*

The viscosity of a suspension of rigid non-interacting spheres is given by the Einstein equation

$$\eta = \eta_0 \left[1 + \left(\frac{5}{2} \right) \phi_2 \right] \tag{3.76}$$

where η and η_0 are the viscosities of the suspension and the suspension medium respectively, and ϕ_2 is the volume fraction of the spheres. If the spheres are considered to be impermeable polymer coils of hydrodynamic volume V_h then

$$\phi_2 = (c/M) N_A V_h$$

where c is the polymer concentration (mass per unit volume), M is the molar mass of the polymer (mass per mol) and N_A is the Avogadro constant. Substituting this equation into Equation (3.76) and simplifying gives

$$\eta_{sp} = \left(\frac{5}{2} \right) (c/M) N_A V_h \tag{3.77}$$

where $\eta_{sp} = (\eta - \eta_0)/\eta_0$ and is known as the *specific viscosity*. Non-interaction of the polymer coils requires infinite dilution and this is achieved mathematically by defining a quantity called the *intrinsic viscosity*, $[\eta]$, according to the equation

$$[\eta] = \lim_{c \to 0} (\eta_{sp}/c) \tag{3.78}$$

Thus a more satisfactory form of Equation (3.77) is

$$[\eta] = \left(\frac{5}{2}\right) N_A V_h / M \tag{3.79}$$

which can be rearranged to give an equation for the hydrodynamic volume of an impermeable polymer molecule

$$V_h = \left(\frac{2}{5}\right) [\eta] M / N_A \tag{3.80}$$

Thus the quantity $[\eta]M$ is proportional to V_h and is often, *though incorrectly*, referred to as the hydrodynamic volume.

Assuming that V_h is proportional to $(a_\eta \langle s^2 \rangle_0^{1/2})^3$ then from Equation (3.79)

$$[\eta] = \Phi_o^s a_\eta^3 \left(\frac{\langle s^2 \rangle_o^{3/2}}{M} \right) \tag{3.81}$$

where Φ_o^s is a constant. For non-draining polymer molecules, Φ_o^s is independent of chain structure and chain length, and is dependent only upon the distribution of segments in the molecular coil. The value of Φ_o^s currently recommended for non-draining Gaussian polymer chains is $3.67 \times 10^{24} \, \text{mol}^{-1}$ and is based upon theoretical calculations and experimental measurements.

Since $\langle s^2 \rangle_o$ is directly proportional to the number, x, of chain segments, it is also directly proportional to M. Thus $\langle s^2 \rangle_o / M$ is a constant and Equation (3.81) is more commonly written in the form of the *Flory–Fox Equation*

$$[\eta] = K_\theta a_\eta^3 M^{1/2} \tag{3.82}$$

where

$$K_\theta = \Phi_o^s \left(\frac{\langle s^2 \rangle_o}{M} \right)^{3/2} \tag{3.83}$$

Furthermore, since a_η is approximately proportional to $x^{1/10}$, and hence to $M^{1/10}$, for highly-expanded flexible chains, the Flory–Fox Equation (3.82) suggests a general relationship of the form

$$[\eta] = KM^a \tag{3.84}$$

where K and a are constants for a given system, and a increases in the range $0.5 \leqslant a \leqslant 0.8$ with the degree of expansion of the molecular coils from their unperturbed dimensions ($a = 0.5$ under theta conditions). This important relationship between $[\eta]$ and M is most commonly called the *Mark–Houwink Equation* and was first proposed on the basis of experimental data.

Number-Average Molar Mass

3.5 Methods for measurement of number-average molar mass

The number-average molar mass, $\overline{M}_n$, of a polymer is of great importance and there are several methods for its measurement. The common feature underlying each of these methods, is that they measure the number of polymer molecules in a given mass of polymer. The methods which involve measurement of the colligative properties of dilute polymer solutions are applicable to most polymers and will be considered in Sections 3.6–3.8. Additionally, for certain polymers it is possible to measure $\overline{M}_n$ by end-group analysis and examples of such measurements will be described in Section 3.9.

Colligative properties are those properties of a solution of a non-volatile solute which in the limit of infinite dilution, depend only upon the number of solute species present in unit volume of the solution and not upon the nature or size of those species. Thus the colligative properties of polymer solutions enable $\overline{M}_n$ to be measured for linear and branched homopolymers and copolymers with equal ease. The four important colligative effects are the osmotic pressure of the solution, the lowering of solvent vapour pressure, the elevation of solvent boiling point and the depression of

TABLE 3.4 *The relative magnitudes of the colligative effects for polymers dissolved in benzene at a concentration of 10 g dm^{-3}. The values are given in practical units and have been calculated assuming ideal solution behaviour*

$\overline{M}_n$ (g mol^{-1})	Osmotic pressure* (mm benzene)	Lowering of vapour pressure (mm Hg)	Elevation of boiling point (K)	Depression of freezing point (K)
10 000	288	799×10^{-5}	272×10^{-5}	577×10^{-5}
100 000	29	80×10^{-5}	27×10^{-5}	58×10^{-5}
500 000	6	16×10^{-5}	5×10^{-5}	12×10^{-5}
1 000 000	3	8×10^{-5}	3×10^{-5}	6×10^{-5}

* At 25°C.

solvent freezing point. The relative magnitudes of these effects are indicated in Table 3.4. Thus osmotic pressure is the only colligative property which is accurately measurable for polymers of high molar mass.

3.6 Membrane osmometry

When a solution is separated from the pure solvent by a *semi-permeable membrane*, i.e. a membrane that permits the passage of solvent molecules but not of solute molecules, the solvent molecules always tend to pass through the membrane into the solution. This general phenomenon is known as *osmosis*, and the flow of solvent molecules leads to the development of an *osmotic pressure* which at equilibrium just prevents further flow. The *equilibrium osmotic pressure*, π, can be measured using a capillary osmometer such as that shown schematically in Fig. 3.9.

3.6.1 Osmosis and chemical potential

Osmosis occurs when the pressure, P_0, above the solution and solvent compartments is equal, because the chemical potential, μ_1^0, of the pure solvent is higher than the chemical potential, μ_1, of the solvent in the

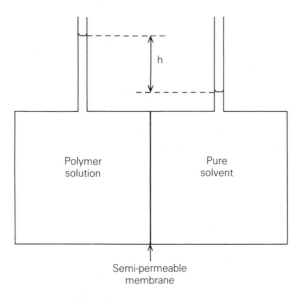

Fig. 3.9 *Schematic representation of a membrane osmometer in which a polymer solution and pure solvent separated by a semi-permeable membrane are in equilibrium. The osmotic pressure, π, is equal to $h\rho_0 g$ where ρ_0 is the solvent density and $\mathbf{g}$ is the acceleration due to gravity.*

solution. Thus the solvent molecules flow through the membrane in order to reduce their chemical potential. The increase in pressure on the solution side of the membrane caused by this flow, has the effect of increasing the chemical potential of the solvent in the solution. At equilibrium the chemical potential of the solvent in the solution at pressure P is equal to μ_1^0 and so

$$\mu_1^0 = \mu_1 + \int_{P_0}^{P} \left(\frac{\partial \mu_1}{\partial P}\right)_{T,n_1,n_2} dP \tag{3.85}$$

From Equation (3.24) it follows that

$$\left(\frac{\partial \mu_1}{\partial P}\right)_{T,n_1,n_2} = \frac{\partial}{\partial n_1}\left(\frac{\partial G}{\partial P}\right)_{T,n_1,n_2}$$

Since $(\partial G/\partial P)_{T,n_1,n_2} = V$ (i.e. the solution volume) then

$$\left(\frac{\partial \mu_1}{\partial P}\right)_{T,n_1,n_2} = \left(\frac{\partial V}{\partial n_1}\right)_{T,P,n_2} = \overline{V}_1 \tag{3.86}$$

where $\overline{V}_1$ is by definition (see Equation (3.23)) the partial molar volume of the solvent. Substituting Equation (3.86) into (3.85) and solving the integral gives

$$\mu_1^0 = \mu_1 + (P - P_0)\overline{V}_1 \tag{3.87}$$

The pressure difference $(P - P_0)$ is the equilibrium osmotic pressure π and so Equation (3.87) can be re-written in the standard form

$$\mu_1 - \mu_1^0 = -\pi \overline{V}_1 \tag{3.88}$$

which is analogous to Equation (3.29). Thus Equation (3.88) is applicable to any solute/solvent system, regardless of whether or not the system is ideal.

An equation applicable to dilute polymer solutions can be obtained by substituting the Flory–Huggins expression for $\mu_1 - \mu_1^0$ (i.e. Equation (3.36)) into Equation (3.88)

$$\pi = RT\phi_2/x\overline{V}_1 + RT(\tfrac{1}{2} - \chi)\phi_2^2/\overline{V}_1 \tag{3.89}$$

Since the solution is dilute, $\overline{V}_1$ is approximately equal to the molar volume V_1 of the solvent. Also $\phi_2 = xn_2/(n_1 + xn_2) \approx xn_2/n_1$ and the total solution volume $V \approx n_1 V_1$. Thus for a dilute polymer solution

$$\phi_2/x\overline{V}_1 = n_2/V \tag{3.90}$$

which upon substitution into Equation (3.89) gives

$$\pi = RT(n_2/V) + RT(\tfrac{1}{2} - \chi)x^2 V_1(n_2/V)^2 \tag{3.91}$$

This equation shows clearly that π depends upon the number of moles of polymer molecules per unit volume of solution (n_2/V).

Assuming that the polymer is polydisperse, then $n_2 = \sum n_i$ and the total mass of polymer in the solution is given by $m = \sum n_i M_i$ where n_i is the total number of moles of polymer molecules of molar mass M_i present in the solution. Thus from Equation (1.3) it follows that

$$\bar{M}_n = \sum n_i M_i \Big/ \sum n_i = m/n_2 \tag{3.92}$$

and hence that

$$n_2/V = (m/V)(n_2/m) = c/\bar{M}_n \tag{3.93}$$

where c is the polymer concentration (mass per unit volume). Substitution of Equation (3.93) into (3.91) and rearrangement of the resulting equation leads to the following expression for the *reduced osmotic pressure* (π/c)

$$\pi/c = \mathbf{R}T/\bar{M}_n + (\mathbf{R}T/V_1)(\tfrac{1}{2} - \chi)(xV_1/\bar{M}_n)^2 c \tag{3.94}$$

By definition (Section 3.2.2) the number of chain segments is given by

$$x = V_2/V_1 \tag{3.95}$$

where V_2 is the molar volume of the polymer. Thus

$$xV_1/\bar{M}_n = V_2/\bar{M}_n = 1/\rho_2 \tag{3.96}$$

where ρ_2 is the density of the polymer. Substitution of Equation (3.96) into (3.94) gives

$$\pi/c = \mathbf{R}T/\bar{M}_n + (\mathbf{R}T/V_1\rho_2^2)(\tfrac{1}{2} - \chi)c \tag{3.97}$$

on the basis of Flory–Huggins dilute solution theory. Equation (3.97) shows that for an ideal solution, i.e. under theta conditons ($\chi = \tfrac{1}{2}$)

$$(\pi/c)_\theta = \mathbf{R}T/\bar{M}_n \tag{3.98}$$

and the measurement of π enables an *absolute value* of $\bar{M}_n$ to be calculated (i.e. no simplifying model or instrument calibration is required). For most measurements on dilute polymer solutions, non-ideality is eliminated by extrapolating π/c data to $c = 0$ since

$$(\pi/c)_{c\to 0} = \mathbf{R}T/\bar{M}_n \tag{3.99}$$

A typical plot of experimental data is shown in Fig. 3.10. Thus $\bar{M}_n$ can be evaluated from the intercept, and from the slope it is possible to calculate an estimate of χ using Equation (3.97). However, since Flory–Huggins theory strictly is not valid for dilute polymer solutions, Equation (3.97) usually is written in the form of a *virial equation* such as

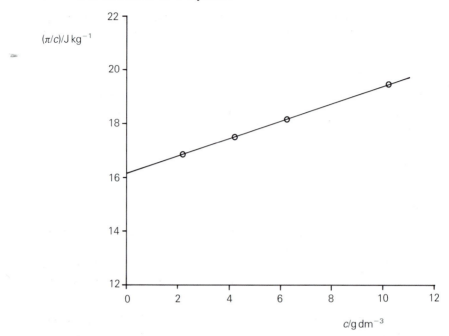

Fig. 3.10 *Plot of π/c against c obtained from measurements at 310 K on solutions in toluene of a sample of poly(vinyl acetate) with $\bar{M}_n = 160\,kg\,mol^{-1}$.*

$$\pi/c = \mathbf{R}T\left[\frac{1}{\bar{M}_n} + A_2c + A_3c^2 + \ldots\right] \tag{3.100}$$

or alternatively

$$\pi/c = (\pi/c)_{c\to 0}[1 + \Gamma_2c + \Gamma_3c^2 + \ldots] \tag{3.101}$$

where A_2 and Γ_2 and A_3 and Γ_3 are known as the second and third virial coefficients respectively, and quite generally $\Gamma_i = A_i\bar{M}_n$. Virial equations are often used to describe the thermodynamic behaviour of real solutions and real gases, and have the property that at infinite dilution $(c \to 0)$ the behaviour of the solutions or gases becomes ideal. The Flory–Huggins theory in the form of Equation (3.97) gives

$$A_2 = (\tfrac{1}{2} - \chi)/V_1\rho_2^2 \tag{3.102}$$

whereas the more satisfactory excluded-volume theories lead to slight modification of this equation

$$A_2 = F(z)(\tfrac{1}{2} - \chi)/V_1\rho_2^2 \tag{3.103}$$

where $F(z)$ is a function of the excluded-volume parameter z and is equal to unity under theta conditions (i.e. when $z = 0$). For *small z*, $F(z)$ takes the form

$$F(z) = 1 - b_1 z + b_2 z^2 \qquad (3.104)$$

in which b_1 and b_2 are positive constants the values of which have been calculated from theory and depend upon chain architecture (e.g. for linear chains $b_1 = 2.8654$ and $b_2 = 13.928$).

3.6.2 *Measurement of osmotic pressure*

The membrane osmometers used for measurement of osmotic pressures can be divided into two categories:

(i) *static osmometers* in which the equilibrium osmotic pressure is attained naturally by diffusion of solvent through the membrane;

(ii) *dynamic osmometers* in which flow of solvent through the membrane is detected and just prevented by application of pressure to the solution cell.

In order to minimize equilibration times (typically 30–60 min), modern osmometers are designed so that the solution and solvent cells are small (typically $0.3–2\,cm^3$) and the membrane surface area is high. Additionally, since osmotic pressures are sensitive to temperature, good temperature control is essential and should be $\pm 0.01°C$. The simplest static osmometers are of the basic form indicated schematically in Fig. 3.9 (e.g. Pinner–Stabin osmometer) and have relatively large cell volumes (typically $3–20\,cm^3$). The volume of solvent transported across the membrane at equilibrium is reduced by using capillaries for the pressure head. However, these capillary osmometers are cumbersome to use and give relatively long equilibration times (typically several hours). Modern static osmometers have much smaller cell volumes and pressure sensors which monitor the pressure in the solution cell, usually indirectly (e.g. Wescan and Knauer osmometers). It normally is not necessary to correct the solution concentration for dilution due to movement of solvent into the solution cell because the effect is usually negligible.

In dynamic osmometers the solvent cell is connected to a solvent reservoir. The vertical position of this reservoir relative to that of the solution cell controls the pressure head and is adjusted by a servo-motor responding to a signal from a sensor that monitors movement of solvent across the membrane. In the Hallikainen (Shell or Stabin) osmometer the sensor is a diaphragm which forms part of the sealed solution side of the cell and is one half of a capacitor. Displacement of the diaphragm due to ingress of solvent into the solution cell causes a change in capacitance which automatically is translated into an increase in the pressure head by servo-motor controlled movement of the solvent reservoir. In the Mechrolab osmometer, solvent flow is monitored by an optical system as movement of an air bubble in a capillary tube connected to the solvent cell, an automatic signal again being sent to the servo-motor. For both instruments, the solvent reservoir is moved to a position such that the flow of solvent across the membrane is just counteracted by the increase in pressure head (i.e. there is no net flow of solvent across the membrane).

In order to evaluate $\bar{M}_n$ for a particular polymer sample it is necessary to measure the osmotic pressure, π, of a series of dilute solutions with different concentrations (usually $c < 20\,\mathrm{g\,dm^{-3}}$). A plot of π/c against c gives an intercept at $c = 0$ from which $\bar{M}_n$ can be calculated using Equation (3.99). The slope of the initial linear region of the plot can be used with Equation (3.100) to evaluate the second virial coefficient A_2 from which it is possible to calculate the Flory–Huggins polymer–solvent interaction parameter χ, preferably by use of Equation (3.103) rather than Equation (3.102). In good solvents A_3 is significant and the plots (see Fig. 3.11) show distinct upward curvature at higher c due to the c^2 term in Equation (3.100). The linear region of the plots extends to higher c and A_2 becomes smaller as the solvency conditions become poorer (see Fig. 3.11). Under

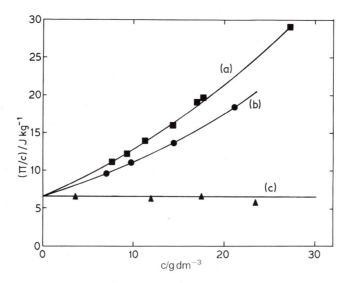

Fig. 3.11 *Plot of π/c versus c for poly(methyl methacrylate) dissolved in three different solvents, (a) toluene, (b) acetone and (c) acetonitrile (data taken from Fox, Kinsinger, Mason and Schuele (1962) Polymer **3**, 71).*

theta conditions $A_2 = 0$, and so by determining A_2 from measurements at different temperatures and extrapolating the data to $A_2 = 0$ it is possible to estimate the theta temperature of a polymer–solvent system.

The upper limit for accurate measurement of $\overline{M}_n$ is determined by the sensitivity of the pressure-detecting system and for most osmometers is in the range 5×10^5 to $1 \times 10^6\,\mathrm{g\,mol^{-1}}$. The lower limit for $\overline{M}_n$ is determined by the extent to which the membrane is permeated by low molar mass polymer molecules. Most membranes are cellulose-based and permeable to molecules with molar masses below about $5000\,\mathrm{g\,mol^{-1}}$, though more retentive membranes are now available. Thus low molar mass polymer molecules move through the membrane into the solvent cell and at equilibrium *do not contribute* to the osmotic pressure. Even when these molecules constitute only a small weight fraction of the polymer sample, their number fraction can be significant (because they are of low molar mass) and so their equilibration on either side of the membrane can lead to serious overestimation of $\overline{M}_n$. Therefore, the lower limit for accurate evaluation of $\overline{M}_n$ depends upon the molar mass distribution of the polymer sample and the permeability of the membrane, and typically is in the range 5×10^4 to $1 \times 10^5\,\mathrm{g\,mol^{-1}}$.

3.7 Vapour pressure osmometry

The technique of *vapour pressure osmometry* (VPO) has its basis in the lowering of solvent vapour pressure by a polymeric solute, and involves measuring the temperature difference between a polymer solution and pure solvent in vapour-phase equilibrium with each other. The principal features of a vapour pressure osmometer are shown schematically in Fig. 3.12. Thus two matched thermistor beads are positioned in a sealed thermostatted ($\pm 0.001°C$) chamber which is saturated with solvent vapour from the solvent contained in the base of the chamber. Using pre-positioned syringes, a drop of pure solvent is placed onto one thermistor bead and a drop of solution onto the other. Due to the lower chemical potential of the solvent in the polymer solution, solvent molecules from the vapour phase condense onto the solution drop giving up their enthalpy of vaporization and causing the temperature of the solution to rise above the temperature, T_o, of the chamber and solvent drop. The thermistors form part of a Wheatstone bridge circuit and the temperature difference, ΔT, between the solution and solvent drops is measured as a resistance difference, ΔR. At equilibrium the vapour pressure P_{1,T_s} of the solution drop at temperature T_s is exactly equal to the vapour pressure P^0_{1,T_o} of the pure solvent at temperature T_o. The equilibrium value of $\Delta T_e = T_s - T_o$ is a measure of the vapour pressure lowering caused by the polymeric solute in the solution at T_o which has vapour pressure P_{1,T_o}. Assuming that the

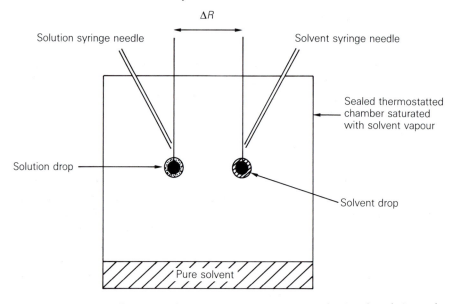

Fig. 3.12 *Schematic illustration of a vapour pressure osmometer showing the solution and solvent drops on matched thermistor beads. The temperature difference between the solution and solvent is measured as the resistance difference, ΔR, between the thermistor beads.*

vapour behaves as an ideal gas, the variation of vapour pressure with temperature is given by the Clausius–Clapeyron Equation

$$\frac{1}{P}\left(\frac{dP}{dT}\right) = \frac{\Delta H_v}{RT^2} \tag{3.105}$$

where ΔH_v is the molar enthalpy of vaporization of the solvent. Thus

$$\int_{P_{1,T_o}}^{P_{1,T_s}}\left(\frac{1}{P}\right)dP = \int_{T_0}^{T_s}\left(\frac{\Delta H_v}{RT^2}\right)dT$$

Since ΔT is very small, ΔH_v can be assumed constant and solution of the integrals gives

$$\ln(P_{1,T_s}/P_{1,T_o}) = \Delta H_v \Delta T_e/RT_sT_o$$

Furthermore, at equilibrium $P_{1,T_s} = P^0_{1,T_o}$ and $T_s \approx T_o$ so that the equation simplifies to

$$\ln(P^0_{1,T_o}/P_{1,T_o}) = \Delta H_v \Delta T_e/RT_o^2 \tag{3.106}$$

The activity of the solvent in the solution at T_o is given by $a_{1,T_o} = P_{1,T_o}/P^0_{1,T_o}$ and so Equation (3.106) may be re-written as

$$\Delta T_e = -\left(\frac{\mathbf{R}T_o^2}{\Delta H_v}\right)\ln a_{1,T_o} \tag{3.107}$$

Substitution of a dilute solution expression for $\ln a_{1,T_o}$ (e.g. Flory–Huggins), followed by a series of approximations similar to those used in deriving the equations for the reduced osmotic pressure, leads to a virial equation of the form

$$\Delta T_e/c = K_e\left[\frac{1}{\overline{M}_n} + A_2 c + A_3 c^2 + \ldots\right] \tag{3.108}$$

where $K_e = V_1\mathbf{R}T_o^2/\Delta H_v$ and depends only upon the solvent and temperature, V_1 is the molar volume of the solvent, A_2 and A_3 are the second and third virial coefficients defined previously for Equation (3.100), and c is the polymer concentration (mass per unit volume). The dilution effect upon c of the condensed solvent vapour is negligible for most practical values of c.

In practice, ΔT_e is not attained due to heat losses from the solution drop to the thermistor bead and its stem, and to the saturated vapour. Instead, a steady-state value ΔT_s is achieved and is given by

$$\Delta T_s/c = K_s\left[\frac{1}{\overline{M}_n} + A_2^v c + A_3^v c^2 + \ldots\right] \tag{3.109}$$

where $K_s = K_e/\varkappa$ in which $\varkappa$ depends upon the heat transfer processes, A_2^v and A_3^v are the VPO second and third virial coefficients and are given by $A_2^v = A_2 + \varkappa_2$ and $A_3^v = A_3 + \varkappa_3$ where $\varkappa_2$ and $\varkappa_3$ depend upon the ratio K_s/K_e. Thus unlike membrane osmometry (MO), VPO is not an absolute method for measurement of $\overline{M}_n$.

Vapour pressure osmometers measure the resistance difference between the two thermistor beads and this is assumed to be proportional to the temperature difference, i.e. $\Delta R_s = k_R\Delta T_s$. Thus, in order to be of practical use, Equation (3.109) must be modified to the form

$$\Delta R_s/c = K_{Rs}\left[\frac{1}{\overline{M}_n} + A_2^v c + A_3^v c^2 + \ldots\right] \tag{3.110}$$

where $K_{Rs} = k_R K_s$. There is experimental evidence which shows that through $\varkappa$, K_{Rs} is slightly dependent upon the size of the solution drop and upon the molar mass of the solute. However, for most purposes it is reasonable to assume that K_{Rs} is a constant for a given solvent/temperature/osmometer system. Nevertheless it is good practice to use a consistent procedure and to evaluate K_{Rs} by performing measurements on several samples (rather than one) of accurately known but different $\overline{M}_n$. In this way a mean value for K_{Rs} can be obtained and any variation of K_{Rs} with $\overline{M}_n$ detected.

Measurements of ΔR_s are made for a series of dilute solutions of the polymer in a solvent which has a relatively high vapour pressure (preferably 100–400 mmHg) at the temperature of analysis. A plot of $\Delta R_s/c$ against c is extrapolated to $c = 0$ to give an intercept from which $\bar{M}_n$ can be calculated

$$(\Delta R_s/c)_{c \to 0} = K_{Rs}/\bar{M}_n \tag{3.111}$$

The initial linear slopes of such plots give a relative indication of solvency conditions but, unlike MO, cannot yield absolute information.

Despite the non-absolute nature of VPO, it is used extensively to determine $\bar{M}_n$ for polymers of low molar mass. Most commercial instruments are able to detect values of ΔT_s of the order of 10^{-4} K, giving an upper limit for accurate measurement of $\bar{M}_n$ of about 1.5×10^4 g mol^{-1}. Thus VPO and MO are complementary techniques.

3.8 Ebulliometry and cryoscopy

Measurements of the elevation of solvent boiling point (*ebulliometry*) and of the depression of solvent freezing point (*cryoscopy*) caused by the presence of a polymeric solute enable $\bar{M}_n$ to be evaluated in a similar way to VPO. Nowadays, however, such measurements are rarely used to evaluate $\bar{M}_n$ for polymers and so will be given only brief consideration here.

Theoretical treatments lead to equations that are identical to Equation (3.108) derived for VPO except that:

(i) for ebulliometry $\Delta T_e = T_b - T_o$ where T_b and T_o are the boiling points of the solution and pure solvent respectively, and

(ii) for cryoscopy $\Delta T_e = T_o - T_f$ where T_f and T_o are the freezing points of the solution and pure solvent respectively, and ΔH_v is replaced by the molar enthalpy of melting of the pure solvent, ΔH_m.

Non-ideality is eliminated in the usual way by extrapolating $\Delta T_e/c$ data to $c = 0$ to give an intercept from which $\bar{M}_n$ can be calculated.

Accurate measurements of ΔT_e in ebulliometry and cryoscopy of polymer solutions are made difficult by the small magnitudes of ΔT_e since superheating and supercooling can introduce serious errors. Other practical problems often encountered include separate measurements of T_f and T_o in cryoscopy, and foaming of the boiling polymer solution in ebulliometry. With current commercial instruments the upper limit for accurate measurement of $\bar{M}_n$ is about 5×10^3 g mol^{-1}.

3.9 End-group analysis

The non-thermodynamic methods for evaluation of $\bar{M}_n$ have their basis in the determination of the number of moles of end-groups of a particular type in a given mass of polymer and therefore are methods of *end-group analysis*. There are four essential requirements which must be satified for end-group analysis to be applicable:

(i) The end-group(s) on the polymer molecules must be amenable to quantitative analysis (usually by titrimetry or spectroscopy).

(ii) The number of analysable end-groups per polymer molecule must be accurately known.

(iii) Other functional groups which interfere with the end-group analysis either must be absent or their effects must be capable of being corrected for.

(iv) The concentration of end-groups must be sufficient for accurate quantitative analysis.

Thus end-group analysis is restricted to low molar mass polymers with well-defined structures and distinguishable end-groups. The upper limit for accurate measurement of $\overline{M}_n$ is dependent upon the sensitivity of the technique used to determine the end-group concentration, but typically is 1×10^4 to $1.5 \times 10^4 \, \text{g mol}^{-1}$.

End-group analysis is most appropriate to polymers prepared by step polymerization since they tend to be of relatively low molar mass and have characteristic end-groups suitable for analysis, e.g. polyesters (—OH and —CO_2H end-groups), polyamides (—NH_2 and —CO_2H end-groups) and polyurethanes (—OH and —NCO end-groups). Low molar mass polyesters, polyamides and polyethers (—OH end-groups) prepared by ring-opening polymerization similarly are suitable for end-group analysis. The methods most commonly employed are titrimetry (often involving back titrations) and nuclear magnetic resonance spectroscopy (Section 3.20). Certain low molar mass polymers prepared from ethylenic monomers by chain polymerization, in particular ionic polymerization, also are amenable to end-group analysis. However, in most cases polymers prepared by chain polymerization have structures which are insufficiently well-defined and have molar masses which are too high for accurate end-group analysis.

End-group analysis yields the *equivalent mass* (sometimes called the equivalent weight) of the polymer, M_e, which is the mass of polymer per mol of end-groups. If for a polydisperse polymer sample, n_i is the number of moles of polymer molecules of molar mass M_i, and f is the number of analysable end-groups per polymer molecule then

$$M_e = \sum n_i M_i \bigg/ \sum f n_i \qquad\qquad (3.112)$$

Since f is a constant and $\overline{M}_n = \Sigma n_i M_i / \Sigma n_i$, Equation (3.112) can be re-written as

$$M_e = \overline{M}_n / f \qquad\qquad (3.113)$$

TABLE 3.5 *Theoretical effects of low molar mass impurities upon the value of* $\overline{M}_n$ *as calculated using Equation (1.6)*

Impurity	Weight fraction of impurity	$\overline{M}_n$ of pure polymer/g mol^{-1}	$\overline{M}_n$ of impure polymer/g mol^{-1}
Water ($M = 18$ g mol^{-1})	0.010	10 000	1 528
	0.001	10 000	6 433
	0.010	200 000	1 784
	0.001	200 000	16 515
Toluene ($M = 92$ g mol^{-1})	0.010	10 000	4 815
	0.001	10 000	9 028
	0.010	200 000	8 799
	0.001	200 000	63 033

showing that $\overline{M}_n$ can be evaluated from M_e and f.

The functionalities, f, of low molar mass functionalized prepolymers (e.g. polyether polyols used in the preparation of polyurethanes) are often determined from Equation (3.113) by combining the results from VPO (i.e. $\overline{M}_n$) and functional-group analysis (i.e. M_e). In this respect the functional groups do not have to be end groups.

3.10 Effects of low molar mass impurities upon $\overline{M}_n$

Purity is of great importance when evaluating $\overline{M}_n$ because low molar mass impurities can give rise to major errors in the measured value of $\overline{M}_n$, as is shown by the data in Table 3.5. These errors are reduced in MO due to equilibration of the impurity on either side of the membrane and in VPO by evaporation of volatile impurities from the solution drop. Thus the effects of low molar mass impurities upon measurement of $\overline{M}_n$ can be negligible but should never be ignored.

Scattering Methods

3.11 Static light scattering

The phenomenon of light scattering is encountered widely in everyday life. For example, light scattering by airborne dust particles causes a beam of light coming through a window to be seen as a shaft of light, the poor visibility in a fog results from light scattering by airborne water droplets, and laser beams are visible due to scattering of the radiation by atmospheric particles. Also, light scattering by gas molecules in the atmosphere gives rise to the blue colour of the sky and the spectacular colours that can sometimes be seen at sunrise and sunset. These are all examples of *static light scattering* since the time-averaged intensity of scattered light is observed.

In general, interaction of electromagnetic radiation with a molecule results either in absorption or scattering of the radiation. Absorption of radiation forms the basis of the spectroscopic techniques discussed in Sections 3.19–3.21. Scattering results from interaction of the molecules with the oscillating electric field of the radiation, which forces the electrons to move in one direction and the nuclei to move in the opposite direction. Thus a dipole is induced in the molecules, which for isotropic scatterers is parallel to, and oscillates with, the electric field. Since an oscillating dipole is a source of electromagnetic radiation, the molecules emit light, the *scattered light*, in all directions. Almost all of the scattered radiation has the same wavelength (and hence frequency) as the incident radiation and results from *elastic (or Rayleigh) scattering* (i.e. with zero energy change). Additionally, a small amount of the scattered radiation has a higher, or lower, wavelength than the incident radiation and arises from *inelastic (or Raman) scattering* (i.e. with non-zero energy change). Inelastically-scattered light carries information relating to bond vibrations and is the basis of Raman spectroscopy, a technique which increasingly is being used in studies of polymer structure and polymer deformation micromechanics. However, in this section it is the elastically-scattered light from dilute polymer solutions which is of interest since it enables the weight-average molar mass to be determined and can also yield values for the Flory–Huggins interaction parameter and the radius of gyration of the polymer molecules.

3.11.1 *Light scattering by small molecules*

The theory of light scattering was first developed by Lord Rayleigh in 1871 during his studies on the properties of gases. In his theory the molecular (or particle) dimensions are assumed to be very much smaller than the wavelength, λ, of the incident monochromatic light. Also, in common with all of the theories which will be discussed here, it is implicitly assumed that the scattering is perfectly elastic. The latter assumption is satisfactory because the intensity of the inelastically-scattered light is insignificant compared to that of the elastically-scattered light. However, the former assumption is not always satisfactory and the Rayleigh theory requires modification when the molecular dimensions approach or are comparable

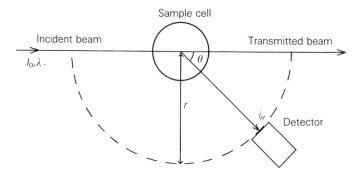

Fig. 3.13 *Main features of the apparatus required to measure the light scattered from a polymer solution (after Billingham).*

to λ (which is the situation for many polymer molecules when using visible incident radiation).

A schematic diagram which shows the basic principles of a light scattering experiment is given in Fig. 3.13. The basic Rayleigh equation gives the angular variation of the intensity, I_θ, of the scattered radiation arising from interaction of a single molecule (or particle) with unpolarized incident radiation of intensity I_0

$$\frac{I_\theta}{I_0} = \frac{8\pi^4\alpha^2(1 + \cos^2\theta)}{\lambda^4 r^2} \tag{3.114}$$

where r and θ are as defined in Fig. 3.13, and α is the polarizability of the molecule (i.e. the proportionality constant relating the magnitude of the induced dipole to the magnitude of the electric field strength of the incident radiation). Now consider a volume V of a dilute gas in which there are N gas molecules. Assuming that the *total intensity, i_θ, of scattered radiation per unit volume* of the dilute gas is the sum of that due to the individual gas molecules present

$$\frac{i_\theta}{I_0} = \left(\frac{N}{V}\right)\frac{8\pi^4\alpha^2(1 + \cos^2\theta)}{\lambda^4 r^2} \tag{3.115}$$

Since α is not easily determined, use is made of its relationship to refractive index, n, according to the equation

$$4\pi(N/V)\alpha = n^2 - 1 \tag{3.116}$$

The refractive index of a vacuum is unity and so for a dilute gas

$$n = 1 + \left(\frac{dn}{dc}\right)c \tag{3.117}$$

where dn/dc is the linear rate of increase of n with gas concentration c (mass per unit volume) and is known as the *refractive index increment*. Since c is small, from Equation (3.117)

$$n^2 \approx 1 + 2\left(\frac{\mathrm{d}n}{\mathrm{d}c}\right)c$$

which upon substitution into Equation (3.116) with rearrangement gives

$$\alpha = (\mathrm{d}n/\mathrm{d}c)c/2\pi(N/V)$$

Substitution of this expression into Equation (3.115) yields

$$\frac{i_\theta}{I_0} = \frac{2\pi^2(\mathrm{d}n/\mathrm{d}c)^2 c^2(1 + \cos^2\theta)}{\lambda^4 r^2(N/V)} \tag{3.118}$$

Now $N/V = c/(M/N_A)$ where M is the molar mass of the gas molecules and N_A is the Avogadro constant. Thus Equation (3.118) reduces to

$$\frac{i_\theta}{I_0} = \frac{2\pi^2(\mathrm{d}n/\mathrm{d}c)^2 Mc(1 + \cos^2\theta)}{\lambda^4 r^2 N_A} \tag{3.119}$$

This is the *Rayleigh Equation for ideal elastic scattering* of unpolarized incident radiation. The scattering envelope (i.e. variation of i_θ with θ) described by this equation is symmetrical (i.e. $i_\theta = i_{180-\theta}$), as is shown in two dimensions in Fig. 3.14.

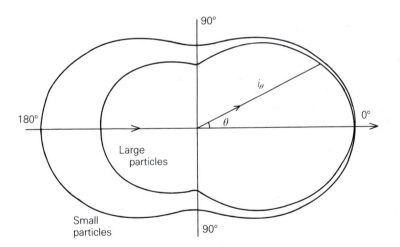

Fig. 3.14 *Two-dimensional scattering envelopes showing the effect of molecular size upon the intensity distribution of light scattered at different angles. The distance from the origin of the scattering to the perimeter of the scattering envelope represents the relative magnitude of i_θ. The intensity distribution is symmetrical for small particles, but for larger particles is unsymmetrical with the intensities reduced at all angles except zero.*

It is convenient at this stage to introduce a quantity called the *Rayleigh ratio*, *R*, which is the reduced relative scattering intensity defined by

$$R = i_\theta r^2 / I_0 (1 + \cos^2\theta) \tag{3.120}$$

and is independent of both *r* and *θ* on the basis of the Rayleigh Equation (3.119) (N.B. the Rayleigh ratio sometimes is defined as the quantity $i_\theta r^2 / I_0$ which is dependent upon *θ*; beware!). Hence the *Rayleigh Equation* may be simplified to

$$R = \frac{2\pi^2 (dn/dc)^2 Mc}{\lambda^4 N_A} \tag{3.121}$$

which shows that *M* can be determined from measurements of *R* at a given gas concentration when *λ* and *dn/dc* are known.

3.11.2 *Light scattering by liquids and solutions of small molecules*

The concentration of molecules in a liquid is very much greater than in an ideal gas. However, the increase in scattering is not proportionate to the increase in concentration because the molecules have some degree of order with respect to one another, and this order gives rise to partial destructive interference of the light scattered from different points within the liquid. It is usual to consider the liquid to be divided into a perfect lattice of volume elements which can contain many molecules but which are very much smaller than *λ*, and so can be considered as point scatterers. If the liquid were perfectly ordered, then each volume element would contain the same number of molecules and complete destructive interference of the scattered light would take place at all angles except *θ* = 0. Clearly this is not the case, and light scattering by liquids results from local density fluctuations within the liquid, i.e. at a given instant in time the volume elements do not necessarily contain the same number of molecules (though their time-averaged densities must be equal). Thus at a given instant in time, there is an excess of scattering from certain volume elements which is not lost by destructive interference.

For dilute solutions, there is an additional contribution resulting from local solute concentration fluctuations. In the study of dilute polymer solutions, only the scattering due to the polymer molecules is required. Thus the scattering arising from the local solvent density fluctuations is eliminated by taking the difference, Δ*R*, between the Rayleigh ratios of the solution and the pure solvent

$$\Delta R = R_{\text{solution}} - R_{\text{solvent}} \tag{3.122}$$

Δ*R* is commonly known as the *excess Rayleigh ratio* and for static light scattering depends upon the time-averaged mean-square excess polarizability of a volume element arising from the local solute concentration fluctuations, i.e. $\langle(\Delta a)^2\rangle$. If there are N volume elements each of volume δV then $N/V = 1/\delta V$ and from Equations (3.115), (3.120) and (3.122)

$$\Delta R = \left(\frac{1}{\delta V}\right)\frac{8\pi^4\langle(\Delta a)^2\rangle}{\lambda^4} \tag{3.123}$$

At a particular instant in time $\Delta a = \overline{\Delta a} + \delta\Delta a$ where $\overline{\Delta a}$ is the average excess polarizability (corresponding to the solute concentration c) and $\delta\Delta a$ is the fluctuation about $\overline{\Delta a}$ (due to the solute concentration fluctuation δc). Thus

$$\langle(\Delta a)^2\rangle = \langle[(\overline{\Delta a})^2 + 2\overline{\Delta a}\delta\Delta a + (\delta\Delta a)^2]\rangle$$

The first term in this relationship is the same for all volume elements at every instant. It therefore makes no overall contribution to scattering because it corresponds to perfect order and complete destructive interference. Additionally, the time-average of the second term is zero because equivalent positive and negative values of $\delta\Delta a$ are equally probable. On this basis Equation (3.123) can be modified to

$$\Delta R = \left(\frac{1}{\delta V}\right)\frac{8\pi^4\langle(\delta\Delta a)^2\rangle}{\lambda^4} \tag{3.124}$$

Now from Equation (3.116)

$$4\pi(1/\delta V)\Delta a = n^2 - n_0^2 \tag{3.125}$$

where n and n_0 are the refractive indices of the solution and solvent respectively. Using the same reasoning as for Equation (3.117)

$$n = n_0 + \left(\frac{dn}{dc}\right)c$$

and since the solution is dilute

$$n^2 \approx n_0^2 + 2n_0\left(\frac{dn}{dc}\right)c$$

Substitution of this relation into Equation (3.125) and differentiating gives the relationship between $\delta\Delta a$ and δc

$$\delta\Delta a = \left(\frac{n_0(dn/dc)\delta V}{2\pi}\right)\delta c \tag{3.126}$$

Thus the time-average $\langle(\delta\Delta a)^2\rangle$ is given by

$$\langle(\delta\Delta a)^2\rangle = \left(\frac{n_0^2(dn/dc)^2\delta V^2}{4\pi^2}\right)\langle(\delta c)^2\rangle$$

which can be substituted into Equation (3.124) to give

$$\Delta R = \frac{2\pi^2 n_0^2 (dn/dc)^2 \delta V \langle (\delta c)^2 \rangle}{\lambda^4} \quad (3.127)$$

The fluctuation theories of Einstein and Smoluchowski lead to the following general expression

$$\langle (\delta c)^2 \rangle = \frac{RTc}{\delta V N_A (\partial \pi / \partial c)} \quad (3.128)$$

where $\partial \pi / \partial c$ is the first derivative of the osmotic pressure with respect to the solute concentration. From Equation (3.100)

$$\partial \pi / \partial c = RT \left[\frac{1}{M} + 2A_2 c + 3A_3 c^2 + \ldots \right] \quad (3.129)$$

for a monodisperse sample. Thus combining Equations (3.127), (3.128) and (3.129) yields

$$\Delta R = \frac{2\pi^2 n_0^2 (dn/dc)^2 c}{N_A \lambda^4 [(1/M) + 2A_2 c + 3A_3 c^2 + \ldots]} \quad (3.130)$$

which shows that as interactions become more significant (i.e. as A_2 and A_3 increase), ΔR decreases. This is because interactions increase the degree of order in the solution.

It is usual to define an *optical constant*, K, as follows

$$K = \frac{2\pi^2 n_0^2 (dn/dc)^2}{N_A \lambda^4} \quad (3.131)$$

and to rearrange Equation (3.130) into the form

$$\frac{Kc}{\Delta R} = \frac{1}{M} + 2A_2 c + 3A_3 c^2 + \ldots \quad (3.132)$$

which gives the concentration dependence of the quantity $Kc/\Delta R$ and shows that at $c = 0$ it is equal to $1/M$. Thus the molar mass of the solute can be determined by extrapolation of experimental $Kc/\Delta R$ data to $c = 0$.

3.11.3 *Light scattering by large molecules in solution*

In deriving Equation (3.132) it was implicitly assumed that the solvent and solute molecules act as point scatterers, i.e. that their dimensions are very much smaller than λ. Whilst in general this assumption is satisfactory for the solvent molecules, it is often inappropriate for polymer solute molecules. The theory for ΔR begins to fail when the solute molecules have dimensions of the order of $\lambda'/20$, where λ' is the wavelength of the

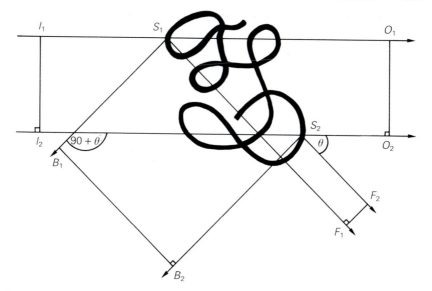

Fig. 3.15 *Schematic illustration of the origin of the interference effects which occur between light scattered by two different segments, S_1 and S_2, in the same molecule. The path-length difference $(I_2S_2B_2 - I_1S_1B_1)$ in the backward direction is greater than the path-length difference $(I_1S_1F_1 - I_2S_2F_2)$ in the forward direction. The path-length difference is zero at $\theta = 0$ $(I_1S_1O_1 = I_2S_2O_2)$.*

light in the medium $(\approx \lambda/n_0)$. Thus the limiting dimensions when using He–Ne laser light $(\lambda = 632.8\,\text{nm})$ are typically about 20 nm. Most polymer molecules of interest have dimensions which either are close to or exceed $\lambda'/20$. The consequence of this is that interference occurs between light scattered from different parts of the same molecule, and so ΔR is reduced. This effect is illustrated schematically in Fig. 3.15 which also shows that the path-length (i.e. phase) difference increases with θ and is zero only for $\theta = 0$. Thus the scattering envelope is unsymmetrical with i_θ greater for forward scattering than for backward scattering, as shown in Fig. 3.14. In order to account for such interference effects, a *particle scattering factor* $P(\theta)$ is introduced and is given by the ratio

$$P(\theta) = \frac{\Delta R_\theta}{\Delta R_{\theta=0}} \qquad (3.133)$$

where ΔR_θ is the measured value of ΔR at the scattering angle θ (i.e. including the effects of interference) and $\Delta R_{\theta=0}$ is the value of ΔR in the absence of interference effects. For small molecules $P(\theta) = 1$ for all values of θ, whereas for large molecules $P(\theta)$ can be very much smaller than unity but increases as θ decreases and at $\theta = 0$ becomes equal to unity. By combining Equations (3.132) and (3.133) a more general expression is obtained

$$\frac{Kc}{\Delta R_\theta} = \frac{1}{P(\theta)}\left[\frac{1}{M} + 2A_2c + 3A_3c^2 + \dots\right] \tag{3.134}$$

Since $\theta = 0$ is inaccessible experimentally (i_0 cannot be differentiated from I_0), analytical expressions for $P(\theta)$ are required so that an appropriate extrapolation to $\theta = 0$ can be performed in order to eliminate the effects of interference. Debye showed that the functional form of $P(\theta)$ depends upon the shape of the scatterers and derived the following expression for monodisperse Gaussian coils

$$P(\theta) = (2/u^2)(u - 1 + e^{-u}) \tag{3.135}$$

where $u = q^2\langle s^2\rangle$ in which $\langle s^2\rangle$ is the mean-square radius of gyration of the coil and q is the *scattering vector* which for dilute solutions is given by

$$q = \frac{4\pi n_0 \sin(\theta/2)}{\lambda} \tag{3.136}$$

Guinier recognized that $P(\theta)$ becomes independent of shape as $\theta \to 0$ and obtained the following general shape-independent expression for $P(\theta)$ when $q\langle s^2\rangle^{1/2} \leqslant 1$

$$P(\theta) = 1 - (q^2\langle s^2\rangle/3) \tag{3.137}$$

Since $1/(1 - x) \approx 1 + x$ when $x \ll 1$, then from Equation (3.137)

$$1/P(\theta) = 1 + (q^2\langle s^2\rangle/3) \tag{3.138}$$

Combining Equations (3.134), (3.138) and (3.136) leads to a general expression for static light scattering

$$\frac{Kc}{\Delta R_\theta} = \left[\frac{1}{M} + 2A_2c + 3A_3c^2 + \dots\right]\left[1 + \left(\frac{16\pi^2 n_0^2 \sin^2(\theta/2)}{3\lambda^2}\right)\langle s^2\rangle\right] \tag{3.139}$$

which is applicable at low solute concentrations and scattering angles corresponding to $q\langle s^2\rangle^{1/2} \leqslant 1$ for unpolarized incident radiation.

3.11.4 *Effect of polydispersity*

When using light scattering to characterize polymers, it is necessary to take into account the effect of polydispersity. From Equation (3.134), at infinite dilution

$$\Delta R_\theta = KcMP(\theta) \tag{3.140}$$

and so the contribution to ΔR_θ due to the n_i molecules of molar mass M_i present in the scattering volume V is given by

$$\Delta R_{\theta,i} = K(n_i M_i / V) M_i P_i(\theta)$$

where $P_i(\theta)$ is the particle scattering factor for species i. Since $\Delta R_\theta = \sum \Delta R_{\theta,i}$ then

$$\Delta R_\theta = (K/V) \sum n_i M_i^2 P_i(\theta)$$

This can be re-written as

$$\Delta R_\theta = K \left(\frac{\sum n_i M_i}{V} \right) \left(\frac{\sum n_i M_i^2}{\sum n_i M_i} \right) \left(\frac{\sum n_i M_i^2 P_i(\theta)}{\sum n_i M_i^2} \right)$$

which reduces to

$$\Delta R_\theta = Kc\overline{M}_w \overline{P(\theta)}_z \tag{3.141}$$

Thus for a polydisperse solute, ΔR_θ depends upon the weight-average molar mass and the z-average particle scattering factor. The latter can be examined further by applying Equation (3.137) and leads to

$$\overline{P(\theta)}_z = 1 - (q^2/3) \left[\frac{\sum n_i M_i^2 \langle s^2 \rangle_i}{\sum n_i M_i^2} \right]$$

which reduces to

$$\overline{P(\theta)}_z = 1 - (q^2/3) \langle \overline{s^2} \rangle_z \tag{3.142}$$

Thus for a *polydisperse solute* Equation (3.139) becomes

$$\frac{Kc}{\Delta R_\theta} = \left[\frac{1}{\overline{M}_w} + 2A_2 c + 3A_3 c^2 + \dots \right] \left[1 + \left(\frac{16\pi^2 n_0^2 \sin^2(\theta/2)}{3\lambda^2} \right) \langle \overline{s^2} \rangle_z \right]$$

$$\tag{3.143}$$

which enables $\overline{M}_w$, A_2 and, for polymer molecules of sufficient size, $\langle \overline{s^2} \rangle_z$ to be determined from measurements of ΔR_θ at different angles θ for each of several polymer solutions with different concentrations c.

Although Equation (3.143) describes a three-dimensional surface for the variation of $Kc/\Delta R_\theta$ with c and θ, it is usual to plot the experimental data in two dimensions using an elegant technique devised by Zimm. In the *Zimm plot*, $Kc/\Delta R_\theta$ is plotted against the composite quantity $\sin^2(\theta/2) + k'c$ where k' is an arbitrary constant the value of which is chosen so as to give a clear separation of the data points (usually k' is taken as 0.1 when the units of c are $g\,dm^{-3}$). A grid-like graph is obtained, such as that shown schematically in Fig. 3.16, and consists of two sets of lines, one set joining

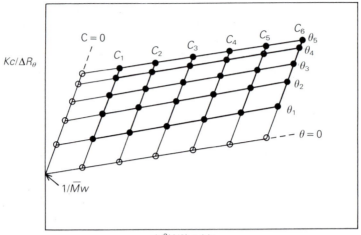

$$\text{sin}^2(\theta/2) + k'c$$

Fig. 3.16 *Schematic illustration of a Zimm plot for analysing light-scattering data. The solid points (●) represent experimental measurements, the open circles (○) are extrapolated points.*

points of constant c and the other joining points with the same value of θ. In order to evaluate A_2, each of the lines at constant c is extrapolated to $\theta = 0$ (i.e. to $k'c$) and the limiting points thus obtained further extrapolated to $c = 0$. The initial linear slope of this limiting extrapolation then gives A_2 from

$$A_2 = \left(\frac{k'}{2}\right)\left[\mathrm{d}[Kc/\Delta R_\theta]_{\theta=0}/\mathrm{d}(k'c)\right]_{c\to 0} \tag{3.144}$$

Similarly, each of the lines at constant θ are extrapolated to $c = 0$ (i.e. to $\text{sin}^2(\theta/2)$) and the limiting points obtained extrapolated to $\theta = 0$. In this case the initial linear slope gives $\langle s^2 \rangle_z$ in terms of $\overline{M}_w$, λ and n_0

$$\langle s^2 \rangle_z = \left(\frac{3\lambda^2 \overline{M}_w}{16\pi^2 n_0^2}\right)\left[\mathrm{d}[Kc/\Delta R_\theta]_{c=0}/\mathrm{d}[\text{sin}^2(\theta/2)]\right]_{\theta\to 0} \tag{3.145}$$

The common intercept of the two limiting extrapolations allows $\overline{M}_w$ to be evaluated

$$\left(\frac{Kc}{\Delta R_\theta}\right)_{\substack{c\to 0 \\ \theta\to 0}} = \frac{1}{\overline{M}_w} \tag{3.146}$$

and hence $\langle s^2 \rangle_z$ from Equation (3.145). Thus static light scattering measurements yield *absolute values* for $\overline{M}_w$, A_2 (from which χ can be evaluated, cf. Section 3.6.1) and $\langle s^2 \rangle_z$, the latter without reference to molecular shape. Furthermore, the form of the variation of $P(\theta)_{c=0}$ with $q^2\langle s^2 \rangle_z$ can be used to deduce the shape of the molecules in solution.

A simpler, though less satisfactory, procedure for evaluation of $\overline{M}_w$ and $\langle s^2 \rangle_z$ involves measurement of ΔR_θ at $90°$ and two other angles symmetrically-positioned about $\theta = 90°$ (i.e. θ and $180 - \theta$, usually $45°$ and $135°$). The *dissymmetry ratio*, Z, is calculated from

$$Z = \frac{\Delta R_\theta}{\Delta R_{180-\theta}} \qquad (3.147)$$

for each polymer solution and extrapolated to $c = 0$. This limiting value of Z is used to obtain $P(90)$ and $\langle s^2 \rangle_z/\lambda'$ from published tables which give the variation of $P(90)$ with Z and of Z with $\langle s^2 \rangle_z/\lambda'$ for scatterers with specific geometries (e.g. Gaussian coils, rigid rods, ellipsoids). Thus the shape of the polymer molecules must be either assumed or known. It is then a simple matter to calculate $\langle s^2 \rangle_z$ from $\langle s^2 \rangle_z/\lambda'$ and $\overline{M}_w$ from

$$\left(\frac{Kc}{\Delta R_{90}} \right)_{c \to 0} = \frac{1}{\overline{M}_w P(90)} \qquad (3.148)$$

3.11.5 Static light scattering measurements

The basic features of a light-scattering instrument are evident from Fig. 3.13, namely a light source and a thermostatted ($\pm 0.01°C$) sample cell positioned on a precision goniometer which enables a photomultiplier detector to be moved accurately to a wide range of scattering angles. Since $i_\theta \propto I_0\lambda^{-4}$, greatest sensitivity is achieved by using high-intensity visible radiation ideally of short wavelength. Modern instruments use monochromatic laser radiation (e.g. He–Ne: $\lambda = 632.8\,nm$, and Ar ion: $\lambda = 488.0\,nm$ or $514.5\,nm$) and are able to measure i_θ for θ from about $10°$ to $160°$. It is usual practice to record i_θ as the photomultiplier voltage output, and to evaluate the Rayleigh ratios of the solvent and the solutions from the ratios of their photomultiplier voltage outputs to that of a standard (e.g. benzene) for which the Rayleigh ratio and refractive index are accurately known for the incident radiation at the temperature of measurement. In this way the need to determine I_0 and to measure absolute values of i_θ are avoided. The Rayleigh ratios thus obtained then are corrected for the effects of refraction and for the variation of the scattering volume with θ.

In addition to the nature of the light source, the sensitivity of the measurements depends strongly upon the refractive index increment, dn/dc, which should be as large as possible. Thus solvents which have refractive indices substantially different from that of the polymer should be chosen (typically to give dn/dc of about $0.0001\,dm^3 g^{-1}$). The value of dn/dc at the wavelength of the incident radiation must be accurately known in order to perform the calculations of $\overline{M}_w$, A_2 and $\langle s^2 \rangle_z$, and if not available in the published literature, can be determined using a differential refractometer.

A further consideration which is of great practical importance is the need to ensure that all of the samples analysed are completely dust-free, otherwise the dust particles would contribute to the scattering intensity and lead to erroneous results. Thus the standard, solvent and solutions analysed must be clarified just prior to use, usually by micropore (e.g. $0.2–0.5\,\mu$m porosity) pressure filtration or by high-speed centrifugation.

Light scattering measurements have the potential to be accurate over a broad range of $\overline{M}_w$, typically 2×10^4 to $5 \times 10^6\,$g mol^{-1}. The lower limit of $\overline{M}_w$ is determined by the ability to detect small values of i_θ and by the variation of dn/dc with molar mass at very low molar masses. The upper limit corresponds to the molecular dimensions approaching $\lambda'/2$ when complete destructive interference of light scattered from different parts of the same molecule occurs.

3.11.6 *Light scattering by multicomponent systems*

The treatment of light scattering presented in the previous sections is applicable to a homopolymer dissolved in a single solvent. The modifications required to analyse light scattering by a homopolymer in mixed solvents or by a copolymer in a single solvent are beyond the scope of this book. However, it should be noted that it is necessary to take into account (i) the fact that a single value of dn/dc is no longer applicable, (ii) the effects of preferential solvent absorption into the molecular coil, and (iii) for copolymers, the variation of repeat unit sequence distribution and of overall composition that occurs from one molecule to another and for different copolymer types (e.g. random, block, graft).

3.12 **Dynamic light scattering**

In static light scattering experiments the time-averaged (or 'total') intensity of the scattered light is measured, and for solutions is related to the time-averaged mean-square excess polarizability which in turn is related to the time-averaged mean-square concentration fluctuation. Whilst such measurements provide a wealth of information, still more can be obtained by considering the real-time random (i.e. Brownian) motion of the solute molecules. This motion gives rise to a *Doppler effect* and so the scattered light possesses a range of frequencies shifted very slightly from the frequency of the incident light (the scattering is said to be *quasi-elastic*). These frequency shifts yield information relating to the movement (i.e. the *dynamics*) of the solute molecules, and can be measured using specialized interferometers and spectrum analysers provided that the incident light has a very narrow frequency band width (i.e. much smaller than the magnitude of the frequency shifts). Thus the availability of laser light sources has greatly facilitated such measurements.

An alternative, and more popular, means of monitoring the motion of the solute molecules is to record the real-time fluctuations in the intensity of the scattered light. The magnitude and frequency of the intensity fluctuations are at a maximum when light scattered by a single volume element is observed (i.e. corresponding to a specific point in the solution). They are reduced if the overall (i.e. 'mean') intensity of light scattered by several volume elements is measured by the detector. In view of this, highly sensitive photomultiplier or photodiode detectors with very small apertures are used so that the scattered light entering the detector can be considered to have derived from a single point in the solution. The total numbers of photons of scattered light entering the detector during each of a sequence of time intervals (typically variable from about 50 ns up to about 1 min) are recorded and analysed by a digital correlator interfaced to a computer. The time interval between successive photon countings is known as the *sample time* Δt, and the separation in time between two particular photon countings is known as their *correlation time* τ. It is essential to choose Δt so that it is much smaller than the timescale of the intensity fluctuations, which for polymers is typically in the range $1\,\mu s$ to $1\,ms$. Thus if τ is only a few multiples (e.g. one, two or three) of Δt, the corresponding photon counts will be closely related, and are said to be *correlated*. However, if τ is many multiples of Δt (i.e. larger than the timescale of the intensity fluctuations), the corresponding photon counts will not be correlated. The *autocorrelation function*, $G^{(1)}(\tau)$, of the intensity, i_θ, is defined by

$$G^{(1)}(\tau) = \lim_{T \to \infty} \left[\frac{1}{T} \int_0^T i_\theta(t) i_\theta(t + \tau) dt \right] \tag{3.149}$$

and its normalized value, $g^{(1)}(\tau)$, by

$$g^{(1)}(\tau) = G^{(1)}(\tau)/G^{(1)}(0) \tag{3.150}$$

where $G^{(1)}(0)$ is the time-averaged value of the square of the intensity, i.e. $\langle i_\theta(t)^2 \rangle$. In the timescale of the experimental measurements, the correlator evaluates $g^{(1)}(\tau)$ for a series of values of τ from the photon countings. Hence this technique of dynamic light scattering is known as *photon correlation spectroscopy*.

The decay of $g^{(1)}(\tau)$ with increasing τ carries information relating to the rate of movement of the solute molecules. For a monodisperse solute undergoing Brownian motion the decay curve is that of a single exponential and

$$g^{(1)}(\tau) = \exp(-\Gamma\tau) \tag{3.151}$$

where Γ is the *characteristic decay rate* which is related to the *translational diffusion coefficient*, D, of the solute by

$$\Gamma = q^2 D \tag{3.152}$$

where q is the scattering vector defined by Equation (3.136). Thus by fitting the experimental $g^{(1)}(\tau)$ data to an exponential curve, it is possible to evaluate Γ and hence D.

For polydisperse solutes such as polymers, $g^{(1)}(\tau)$ approximates to a weighted sum of exponentials

$$g^{(1)}(\tau) = \sum C_i \exp(-\Gamma_i \tau) \tag{3.153}$$

where C_i and Γ_i are respectively the fractional weighting factor and characteristic decay rate for species i. In practice, sophisticated algorithms are required to fit $g^{(1)}(\tau)$ data accurately to a sum of more than two exponential components. The arithmetic mean, $\overline{\Gamma}$, of the characteristic decay rates is given by $\overline{\Gamma} = \Sigma C_i \Gamma_i$, and is related to the *z-average translational diffusion coefficient*, D_z, through

$$\overline{\Gamma} = q^2 D_z \tag{3.154}$$

The translational diffusion coefficient of a molecule is related to its frictional coefficient, f_o, by the Einstein diffusion equation

$$D = \mathbf{k}T/f_0 \tag{3.155}$$

where $\mathbf{k}$ is the Boltzmann constant and T is the temperature. Thus by combining this equation with Stokes' Equation (3.73), it is possible to evaluate from D_z the *z-average hydrodynamic radius* of the polymer molecules. Additional information relating to the molar mass distribution can be extracted from the distribution of Γ_i, though at present only for relatively narrow distributions. Also, by performing measurements at a series of scattering angles, it is possible to assess the shape of the polymer molecules in solution.

Photon correlation spectroscopy (PCS) is particularly suitable for studies of the hydrodynamic behaviour of polymers, and for studies of the effects of solvency conditions (e.g. solvent, temperature) and skeletal structure (e.g. linear, branched, cyclic) upon chain dimensions. The development of PCS has been greatly assisted by developments in multi-bit processors (which have enabled $g^{(1)}(\tau)$ to be determined with ever greater accuracy) and in the analytical methods by which information is extracted from

$g^{(1)}(\tau)$. Nowadays, most light scattering instruments are designed so that they are capable of performing both static (i.e. 'total intensity') and dynamic light scattering measurements, the latter by PCS.

3.13 Small-angle X-ray and neutron scattering

The basic equations resulting from the theory presented for the scattering of visible radiation also can be applied to the scattering of X-rays and neutrons provided that the optical constant K is redefined to take account of the different origins of the scattering. In each case, K is proportional to the square of the appropriate property difference between the scatterers of interest (e.g. solute molecules) and their environment (e.g. solvent molecules). Scattering of visible radiation results from differences in polarizability which give rise to the $[n_0(dn/dc)]^2$ term in Equation (3.131) for K_{LS}. Scattering of X-rays results from *electron density differences* $\Delta\rho_e$, and $K_{XS} \propto (\Delta\rho_e)^2$. Finally, scattering of neutrons results from *neutron scattering length density differences* $\Delta\rho_n$, and $K_{NS} \propto (\Delta\rho_n)^2$.

The wavelengths of X-rays (typically about 0.1 nm) and neutrons (typically 0.1 nm to 2 nm) are very much smaller than those of visible radiation. Thus X-ray and neutron scattering measurements are able to provide size information on a much smaller dimensional scale than is possible using light scattering. However, in order to access this information from interference effects (i.e. from $P(\theta)$), it is necessary to make measurements at very small scattering angles ($\theta < 2°$) so that the path-length differences are less than $\lambda'/2$. These measurements are made possible by placing the detector a long distance (typically 1–40 m) from the sample so that at the detector there is a large separation between the incident beam ($\theta = 0$) and the scattered radiation. Since the measurements are made at small angles (SA), the acronyms SAXS and SANS are used to describe them.

It is possible to evaluate $\overline{M}_w$, A_2 and $\langle s^2 \rangle_z$ from SAXS and SANS data in essentially the same way as for light scattering. By using the Guinier approximation defined by Equation (3.137) and the corresponding SAXS and SANS equivalents to Equation (3.143), $\langle s^2 \rangle_z$ can be determined without having to assume a molecular shape. However, the Guinier expression is satisfactory only when $q\langle s^2 \rangle_z^{1/2} \lesssim 1$ and so its use to evaluate $\langle s^2 \rangle_z^{1/2}$ is valid only when the extrapolation of experimental data is over a range of θ for which $\langle s^2 \rangle_z^{1/2} \lesssim q^{-1}$. In other words, for a given value of λ and range of accessible θ, there is an upper limit for determination of $\langle s^2 \rangle_z^{1/2}$ using the Guinier approximation. The upper limits for SAXS, SANS and light scattering are respectively about 5 nm, 20 nm and 200 nm, as is indicated by the values of q^{-1} given in Table 3.6 for some typical λ, θ combinations. When $q\langle s^2 \rangle_z^{1/2} > 1$, complete expressions for $P(\theta)$ are

TABLE 3.6 *Values of the reciprocal scattering vector q^{-1} calculated from Equation (3.136) for some typical λ, θ combinations***

Scattering method		λ/nm	θ	q^{-1}/nm
SAXS	{	0.154	1.0°	1.4
		0.154	0.5°	2.8
		0.154	0.2°	7.0
SANS	{	1.0	1.0°	9.1
		1.0	0.5°	18.2
		1.0	0.2°	45.6
Light scattering	{	632.8	60.0°	67.1
		632.8	30.0°	129.7
		632.8	10.0°	385.2

* The refractive index, n_0, of the medium is arbitrarily assumed to be 1.0 for SAXS and SANS, and 1.5 for light scattering.

required in order to interpret the experimental data and necessitate the assumption of a particular model for the polymer molecules (e.g. Equation (3.135) for a Gaussian coil).

Synchrotron radiation sources enable X-ray radiation to be selected from wavelengths in the range $0.06 \leqslant \lambda \leqslant 0.3$ nm, whereas in laboratory instruments the Cu Kα line ($\lambda = 0.154$ nm) is most commonly used. Since $\langle s^2 \rangle_z^{1/2}$ is invariably greater than 5 nm for polymers, the Guinier region generally is inaccessible by SAXS and model assumptions are required in order to evaluate $\langle s^2 \rangle_z^{1/2}$.

The use of thermal neutrons (λ about 0.1 nm) direct from a nuclear reactor for SANS imposes similar restrictions to those encountered with SAXS. However, by reducing the velocities of the thermal neutrons it is possible to produce cold neutrons (λ about 1 nm) which enable $\langle s^2 \rangle_z^{1/2}$ up to about 20 nm to be determined by SANS in the Guinier region. Such measurements therefore complement similar measurements made by light scattering for which the lower limit of $\langle s^2 \rangle_z^{1/2}$ is about 15 nm using $\lambda = 488.0$ nm. In order to gain the necessary contrast (i.e. $\Delta\rho_n$) for SANS, use is made of the large difference between the neutron scattering lengths of hydrogen (^{1}H) atoms and deuterium (^{2}H) atoms. Thus, either a deutero-solvent or a deutero-solute is used. Since the ^{1}H and ^{2}H atoms act as individual point scatterers, it is possible to determine $\langle s^2 \rangle_z^{1/2}$ for specific blocks in molecules of block copolymers. This is achieved by preparing block copolymers in which only one block is of deutero-polymer and measuring the SANS in an appropriate ^{1}H solvent.

The high penetration of X-rays and neutrons enables SAXS and SANS measurements to be made on solid polymers. SAXS is commonly used to measure the size of dispersed phases in polymers and is particularly suitable for studies of polymer morphology. SANS measurements on solid polymers are achieved using mixtures of ^{1}H polymer with a small amount (i.e. low c) of the corresponding deutero-polymer (e.g. polystyrene with deutero-polystyrene). Such measurements gave the first irrefutable proof that in a pure amorphous polymer, the polymer molecules adopt their unperturbed dimensions (Section 3.3.3). Similarly the dimensions of individual blocks in phase-separated block copolymers, and of network chains in network polymers, can be determined using selective deuteration. Thus SANS is by far the most powerful technique for studies of chain dimensions and has many other applications beyond those mentioned here.

Frictional Properties

3.14 Dilute solution viscometry

A characteristic feature of a dilute polymer solution is that its viscosity is considerably higher than that of either the pure solvent or similarly dilute solutions of small molecules. This arises because of the large differences in size between polymer and solvent molecules, and the magnitude of the viscosity increase is related to the dimensions of the polymer molecules in solution. Therefore, measurements of the viscosities of dilute polymer solutions can be used to provide information concerning the effects upon chain dimensions of polymer structure (chemical and skeletal), molecular shape, degree of polymerization (hence molar mass) and polymer–solvent interactions. Most commonly, however, such measurements are used to determine the molar mass of a polymer.

3.14.1 *Intrinsic viscosity*

The quantities required and terminology used in dilute solution viscometry are summarized in Table 3.7. The terminology proposed by IUPAC was an attempt to eliminate the inconsistences associated with the common names, which define as viscosities, quantities that do not have the dimensions ($ML^{-1}T^{-1}$) of viscosity. However, the common names were well established when the IUPAC recommendations were published and the new terminology has largely been ignored.

The quantity of greatest importance for the purposes of polymer characterization is the *intrinsic viscosity* $[\eta]$ since it relates to the intrinsic ability of a polymer to increase the viscosity of a particular solvent at a given temperature. The *specific viscosity* of a solution of concentration c is related to $[\eta]$ by a power series in $[\eta]c$

TABLE 3.7 *Quantities and terminology used in dilute solution viscometry*

Common name	Name proposed by IUPAC	Symbol and definition*
Relative viscosity	Viscosity ratio	$\eta_r = \eta/\eta_0$
Specific viscosity	—	$\eta_{sp} = \eta_r - 1$
Reduced viscosity	Viscosity number	$\eta_{red} = \eta_{sp}/c$
Inherent viscosity	Logarithmic viscosity number	$\eta_{inh} = \ln{(\eta_r)}/c$
Intrinsic viscosity	Limiting viscosity number	$[\eta] = \lim_{c \to 0}{(\eta_{red})}$

* η_0 is the viscosity of the solvent and η is the viscosity of a polymer solution of concentration c (mass per unit volume).

$$\eta_{sp} = k_0[\eta]c + k_1[\eta]^2c^2 + k_2[\eta]^3c^3 + \ldots \tag{3.156}$$

where k_0, k_1, k_2, etc., are dimensionless constants and $k_0 = 1$. For dilute solutions, Equation (3.156) can be truncated and rearranged to the following form

$$\eta_{sp}/c = [\eta] + k_H[\eta]^2c \tag{3.157}$$

which is known as the *Huggins equation* and is valid for $[\eta]c \ll 1$. The Huggins constant $k_H(=k_1)$ is essentially independent of molar mass and has values which fall in the range 0.3 (for good polymer–solvent pairs) to 0.5 (for poor polymer–solvent pairs).

For dilute solutions with $\eta_{sp} \ll 1$

$$\ln(\eta_r) = \ln(1 + \eta_{sp}) \approx \eta_{sp} - \tfrac{1}{2}\eta_{sp}^2 \tag{3.158}$$

Substituting Equation (3.157) into (3.158), and retaining only the terms in c up to c^2 gives

$$\ln(\eta_r) = [\eta]c + (k_H - \tfrac{1}{2})[\eta]^2c^2$$

which can be rearranged into the form of the *Kraemer equation*

$$\ln(\eta_r)/c = [\eta] + k_K[\eta]^2c \tag{3.159}$$

where $k_K = k_H - \tfrac{1}{2}$ and so is negative (or zero).

Although there are other simplifications of Equation (3.156), the Huggins and Kraemer equations provide the most common procedure for evaluation of $[\eta]$ from experimental data. This involves a dual extrapolation according to these equations and gives $[\eta]$ as the mean intercept (Fig. 3.17).

3.14.2 *Interpretation of intrinsic viscosity data*

The intrinsic viscosity $[\eta]$ of a polymer is related to its viscosity-average molar mass M_v by the *Mark–Houwink equation*

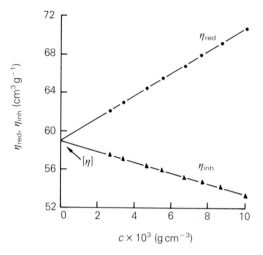

Fig. 3.17 *Evaluation of [η] for solutions in toluene of a sample of poly(tert-butyl acrylate) at 25°C using a dual Huggins–Kraemer plot.*

$$[\eta] = K\overline{M}_v^a \tag{3.160}$$

where K and a are characteristic constants for a given polymer/solvent/temperature system. Theoretical justification for the form of this equation is presented in Section 3.4.2. For Gaussian coils, it was shown that $a = 0.5$ under theta conditions, and that a increases to a limiting value of 0.8 with coil expansion (typically $a > 0.7$ for polymers in good solvents). The value of K tends to decrease as a increases and for flexible chains it is typically in the range 10^{-3} to 10^{-1} cm^3 g^{-1} (g mol^{-1})$^{-a}$.

The form of $\overline{M}_v$ can be deduced on the basis that specific viscosities are additive in the limit of infinite dilution, i.e.

$$\eta_{sp} = \sum(\eta_{sp})_i$$

where $(\eta_{sp})_i$ is the contribution to η_{sp} made by the n_i moles of polymer molecules of molar mass M_i present in unit volume of a dilute polymer solution. Since $c_i = n_i M_i$, $(\eta_{sp})_i = c_i[\eta]_i$ and $[\eta]_i = KM_i^a$ then

$$[\eta] = \left(\frac{\eta_{sp}}{c}\right)_{c \to 0} = K\left(\frac{\sum n_i M_i^{1+a}}{\sum n_i M_i}\right)$$

Comparison of this equation with Equation (3.160) shows that

$$\overline{M}_v = \left(\frac{\sum n_i M_i^{1+a}}{\sum n_i M_i}\right)^{1/a} \tag{3.161}$$

Thus for Gaussian chains $\overline{M}_v$ lies between $\overline{M}_n$ and $\overline{M}_w$, but is closer to $\overline{M}_w$. When $a = 1$ (as observed for certain inherently stiff or highly extended chains) $\overline{M}_v = \overline{M}_w$.

In order to evaluate $\overline{M}_v$ from $[\eta]$ using the Mark–Houwink equation, it is necessary to know the values of K and a for the system under study. These values most commonly are determined from measurements of $[\eta]$ for a series of polymer samples with known $\overline{M}_n$ or $\overline{M}_w$. Ideally the samples should have narrow molar mass distributions so that $\overline{M}_n \approx \overline{M}_v \approx \overline{M}_w$. If this is not the case, then provided that their molar mass distributions are of the same functional form (e.g. most probable distribution), the calibration is valid and yields equations that are similar to Equation (3.160) but in which $\overline{M}_v$ is replaced by $\overline{M}_n$ or $\overline{M}_w$. Generally a plot of $\log[\eta]$ against $\log M$ is fitted to a straight line from which K and a are determined. Theoretically this plot should not be linear over a wide range of M, so that K and a values should not be used for polymers with M outside the range defined by the calibration samples. However, in practice such plots are essentially linear over wide ranges of M, though curvature at low M is often observed due to the non-Gaussian character of short flexible chains. Some typical values of K and a are given in Table 3.8.

For small expansions of Gaussian polymer chains the expansion parameter, α_η, for the hydrodynamic chain dimensions is given by a closed expression of the form

$$\alpha_\eta^3 = 1 + bz \qquad (3.162)$$

where z is the excluded volume parameter defined by Equation (3.67) and b is a constant the value of which is uncertain (e.g. values of 1.55 and 1.05 have been obtained from different theories). Combining this relation with the Flory–Fox Equation (3.82) gives

$$[\eta] = K_\theta M^{1/2} + b K_\theta z M^{1/2}$$

which upon rearrangement, recognizing that from Equation (3.67) z is proportional to $M^{1/2}$, leads to

TABLE 3.8 *Some typical values of the Mark–Houwink constants K and a*

Polymer	Solvent	$T/°C$	K^*	a
Polyethylene	Decalin	135	6.2×10^{-2}	0.70
Polystyrene	Toluene	25	7.5×10^{-3}	0.75
Polystyrene	Cyclohexane	34[†]	8.2×10^{-2}	0.50
Poly(vinyl acetate)	Acetone	30	8.6×10^{-3}	0.74
Cellulose triacetate	Acetone	20	2.38×10^{-3}	1.0

* The units of K are $cm^3 g^{-1} (g\,mol^{-1})^{-a}$.
† Theta conditions, $T = \theta$.

$$[\eta]/M^{1/2} = K_\theta + BM^{1/2} \qquad\qquad\qquad (3.163)$$

where B depends upon the chain structure and polymer–solvent interactions, and is a constant for a given polymer/solvent/temperature system. Thus the corresponding $[\eta]$ and M data obtained for evaluation of Mark–Houwink constants from calibration samples with narrow molar mass distributions, can also be plotted as $[\eta]/M^{1/2}$ against $M^{1/2}$ to give K_θ as the intercept at $[\eta]/M^{1/2} = 0$. The value of K_θ can then be used to evaluate for the polymer to which it relates: (i) $\langle s^2 \rangle_0^{1/2}$ (i.e. unperturbed dimensions) for any M by assuming a theoretical value for Φ_0^s in Equation (3.83), and (ii) a_η for a given pair of corresponding $[\eta]$ and M values by using the Flory–Fox Equation (3.82).

The interpretation of $[\eta]$ for branched polymers and copolymers is considerably more complicated, and will not be dealt with here other than to highlight some of the additional complexities. The effect of branching is to increase the segment density within the molecular coil. Thus a branched polymer molecule has a smaller hydrodynamic volume and a lower intrinsic viscosity than a similar linear polymer of the same molar mass. For copolymers of the same molar mass, $[\eta]$ will differ according to the composition, composition distribution, sequence distribution of the different repeat units, interactions between unlike repeat units, and degree of preferential interaction of solvent molecules with one of the different types of repeat unit.

3.14.3 *Measurement of solution viscosity*

The viscosities of dilute polymer solutions most commonly are measured using capillary viscometers of which there are two general classes, namely U-tube viscometers and suspended-level viscometers (Fig. 3.18). A common feature of these viscometers is that a measuring bulb, with upper and lower etched marks, is attached directly above the capillary tube. The solution is either drawn or forced into the measuring bulb from a reservoir bulb attached to the bottom of the capillary tube, and the time required for it to flow back between the two etched marks is recorded.

In U-tube viscometers, the pressure head giving rise to flow depends upon the volume of solution contained in the viscometer, and so it is essential that this volume is exactly the same for each measurement. This normally is achieved after temperature equilibration by carefully adjusting the liquid level to an etched mark just above the reservoir bulb.

Most suspended-level viscometers are based upon the design due to Ubbelohde, the important feature of which is the additional tube attached just below the capillary tube. This ensures that during measurement the solution is suspended in the measuring bulb and capillary tube, with

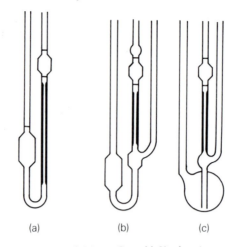

(a)   (b)   (c)

Fig. 3.18 *Schematic illustrations of (a) an Ostwald U-tube viscometer, (b) an Ubbelohde suspended-level viscometer, and (c) a modified Ubbelohde viscometer with a larger reservoir bulb for dilutions.*

atmospheric pressure acting both above and below the flowing column of liquid. Thus the pressure head depends only upon the volume of solution in and above the capillary, and so is independent of the total volume of solution contained in the viscometer. This feature is particularly useful because it enables solutions to be diluted in the viscometer by adding more solvent. When U-tube viscometers are used, they must be emptied, cleaned, dried and refilled with the new solution each time the concentration is changed.

Before use, it is essential to ensure that the viscometer is thoroughly clean and that the solvent and solutions are freed from dust by filtration, otherwise incorrect and erratic flow times can be anticipated. The viscometer is placed in a thermostatted water (or oil) bath with temperature control of ±0.01°C or better because viscosity generally changes rapidly with temperature. After allowing sufficient time for temperature equilibration of the solution, several measurements of flow time are made and should be reproducible to ±0.1% when measured visually using a stopwatch.

Under conditions of steady laminar Newtonian flow, the volume V of liquid which flows in time t through a capillary of length l and radius r is related to both the pressure difference P across the capillary and the viscosity η of the liquid by *Poiseuille's equation*

$$\frac{V}{t} = \frac{\pi r^4 P}{8\eta l} \tag{3.164}$$

The radial velocity profile corresponding to this equation is parabolic, with maximum velocity along the axis of the capillary tube and zero velocity at the wall. During the measurement of flow time, P continuously decreases and normally is given by

$$P = \langle h \rangle \rho \mathbf{g}$$

where $\langle h \rangle$ is the average pressure head, ρ is the density of the liquid and $\mathbf{g}$ is the acceleration due to gravity. Thus Poiseuille's Equation (3.164) can be rearranged to give

$$\eta = \frac{\pi r^4 \langle h \rangle \rho \mathbf{g} t}{8Vl} \qquad (3.165)$$

which has the form

$$\eta = A\rho t \qquad (3.166)$$

where A is a constant for a given viscometer. Poiseuille's equation does not take into account the energy dissipated in imparting kinetic energy to the liquid, but is satisfactory for most viscometers provided that the flow times exceed about $180\,\text{s}$.

Absolute measurements of viscosity are not required in dilute solution viscometry since it is only necessary to determine the viscosity of a polymer solution relative to that of the pure solvent. Application of Equation (3.166) leads to the following relation for the relative viscosity

$$\eta_r = \frac{\eta}{\eta_0} = \frac{\rho t}{\rho_0 t_0}$$

where ρ and ρ_0 are the densities, and t and t_0 are the flow times of a polymer solution of concentration c and of the pure solvent respectively. Since dilute solutions are used, it is common practice to assume that $\rho = \rho_0$ so that η_r is simply given by the ratio of the flow times t/t_0. The other quantities required are then calculated from η_r and c. In more accurate work kinetic energy and density corrections are applied, and when necessary the values of $[\eta]$ obtained are extrapolated to zero shear rate.

Although dilute solution viscometry is not an absolute method for characterization of polymers, in comparison to other methods it is simple, fast and uses relatively inexpensive equipment. It also has the advantage that it is applicable over the complete range of attainable molar masses. For these reasons dilute solution viscometry is widely used for routine measurements of molar mass.

3.15 Ultracentrifugation

Measurements of the sedimentation behaviour of polymer molecules in solution can provide a considerable amount of information, e.g. hydrodynamic volume, average molar masses and even some indication of molar mass distribution. Such measurements have been extensively used to characterize biologically-active polymers which often exist in solution as compact spheroids or rigid rods. However, sedimentation methods are rarely used to study synthetic polymers and so will be given only brief non-theoretical consideration here.

The normal gravitational force acting upon a polymer molecule in solution is insufficient to overcome its Brownian motion and does not cause sedimentation. Thus in order to make measurements of sedimentation it is necessary to subject the polymer solutions to much higher gravitational fields. This is achieved using *ultracentrifuges* which can attain rotation speeds of up to about 70 000 revolutions per minute (rpm) and generate centrifugal fields up to about 400 000 **g** where **g** is the acceleration due to gravity. The polymer solution is placed in a cell which fits into the rotor of the ultracentrifuge and has windows on either side. The cell is in the form of a truncated cone and is positioned so that its peak would be located at the axis of the rotor. This design ensures that sedimentation in radial directions is not restricted. The sedimentation process gives rise to a solvent phase and a concentrated polymer solution phase which are separated by a boundary layer in which the polymer concentration varies. There is, therefore, a natural tendency for backward diffusion of the molecules in order to equalize the chemical potentials of the components in the different regions of the cell, and this causes broadening of the boundary layer. The breadth of the boundary layer also increases with the degree of polydispersity because molecules of higher molar mass sediment at faster rates. The windows in the cell enable the radial variation in polymer concentration to be measured during ultracentrifugation, typically by monitoring refractive index differences (cf. light scattering, Sections 3.11.1 and 3.11.2). There are two general methods by which sedimentation experiments can be performed and they will now be briefly described.

In a *sedimentation equilibrium* experiment the cell is rotated at a relatively low speed (typically 5000–10 000 rpm) until an equilibrium is attained whereby the centrifugal force just balances the tendency of the molecules to diffuse back against the concentration gradient developed. Measurements are made of the equilibrium concentration profiles for a series of solutions with different initial polymer concentrations so that the results can be extrapolated to $c = 0$. A rigorous thermodynamic treatment is possible and enables absolute values of $\bar{M}_w$ and $\bar{M}_z$ to be determined. The principal restriction to the use of sedimentation equilibrium measurements is the long time required to reach equilibrium, since this is at least a few hours and more usually is a few days.

The *sedimentation velocity* method involves rotating the solution cell at very high speeds (typically 60 000–70 000 rpm) and gives results in much shorter timescales than sedimentation equilibrium measurements. The movement of the boundary layer is monitored as a function of time and its steady-state velocity used to calculate the mean sedimentation coefficient, S, for the polymer in solution. Measurements are made for a series of solution concentrations and enable the limiting sedimentation coefficient, S_0, to be obtained by extrapolation to $c = 0$. In order to calculate an average molar mass, it is necessary either to know the limiting diffusion coefficient of the polymer in the solvent or to calibrate the system by measuring S_0 for a series of similar polymers but which have narrow molar mass distributions and known molar masses. The latter procedure is more common and an equation similar in form to the Mark–Houwink Equation (3.160) is used to correlate S_0 data with molar mass for each specific polymer/solvent/temperature system. The resulting average molar mass usually is close to $\overline{M}_w$.

Molar Mass Distribution

3.16 Fractionation

In many instances, average molar masses and their ratios (i.e. polydispersity indices) are insufficient to describe the properties of a polymer and more complete information on the molar mass distribution is required. One way of obtaining this information is to separate (i.e. *fractionate*) the polymer into a number of fractions each of which has a narrow distribution of molar mass. The weight and molar mass of each polymer fraction are determined and enable the molar mass distribution to be constructed in the form of a histogram. However, such procedures are rarely used nowadays because much more rapid and powerful methods of size-exclusion chromatography (Section 3.17) are available for determining molar mass distributions. Nevertheless, fractionation itself is still practised, often for purposes of purification, and will be considered here in some detail because it introduces the important topic of phase-separation behaviour of polymers.

3.16.1 *Phase-separation behaviour of polymer solutions*

The simplest procedure for polymer fractionation is to dissolve the polymer at low concentration in a poor solvent and then to bring about stepwise phase separation (i.e. 'precipitation') of polymer fractions by either changing the temperature or adding a non-solvent. The highest molar mass species phase separate first and so the fractions are obtained in order of decreasing molar mass. Phase separation can be treated theoretically on the basis of Flory–Huggins theory. The effect of temperature upon phase separation of solutions of non-crystallizing polymers will be considered here since it is easier to analyse. It is usual to deal with molar quantities and so both sides of the Flory–Huggins Equation (3.22) must be divided by $(n_1 + n_2 x)$ where n_1 and n_2 are the numbers of moles of solvent and polymer present, and x is the *number of segments* in each of the polymer molecules (Section 3.2.2) which are assumed to be *monodisperse*. This gives the following equation for ΔG_m^*, the Gibbs free energy of mixing per mole of lattice sites (which therefore can be considered to be the Gibbs free energy of mixing per mol of segments)

$$\Delta G_m^* = RT[\phi_1 \ln \phi_1 + (\phi_2/x)\ln \phi_2 + \chi \phi_1 \phi_2] \tag{3.167}$$

This equation describes a series of curves for the variation of ΔG_m^* with ϕ_2, the volume fraction of polymer, one for each temperature. The curves

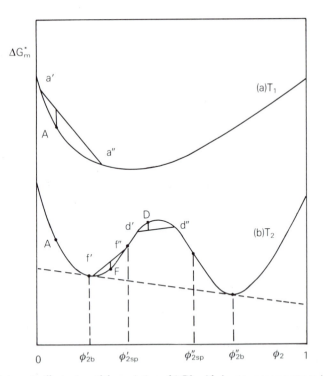

Fig. 3.19 *Schematic illustration of the variation of ΔG_m^* with ϕ_2 at two temperatures (a) T_1 and (b) T_2.*

have one of two general forms, as depicted in Fig. 3.19 which shows schematically curves that would be consistent with low values of x. At temperature T_1 (Fig. 3.19(a)) the polymer and solvent are miscible in all proportions, as is evident from consideration of any point on the curve. For example, if a homogeneous solution with ϕ_2 ($= \phi_{2A}$) corresponding to point A were to separate into two co-existing phases, conservation of matter demands that one should have $\phi_2 < \phi_{2A}$ and the other $\phi_2 > \phi_{2A}$, e.g. corresponding to points a' and a''. It is a relatively simple matter to show that the Gibbs free energy change associated with the phase separation process is given by the difference between (i) the value of ΔG_m^* corresponding to the point of intersection of the *tie-line* (joining points a' and a'') with the vertical $\phi_2 = \phi_{2A}$ line, and (ii) the value of ΔG_m^* on the curve at A. Clearly this difference is positive for all points on the curve and so the existence of a single homogeneous phase is favoured for all ϕ_2.

The situation at temperature T_2 (Fig. 3.19(b)) is rather more complex since two ΔG_m^* minima are present. Consider again phase separation from a homogeneous solution corresponding to ϕ_{2A}. It is easy to see that tie-lines joining any two points on the curve either side of ϕ_{2A} will intersect the vertical $\phi_2 = \phi_{2A}$ line above the curve. This is true for all compositions in the ranges $0 < \phi_2 < \phi_{2b}'$ and $\phi_{2b}'' < \phi_2 < 1$ and so homogeneous solutions with ϕ_2 in these ranges are stable at T_2. Now consider phase separation of a homogeneous solution with ϕ_2 ($= \phi_{2D}$) corresponding to point D. The tie-lines joining two points (such as d' and d'') immediately on either side of ϕ_{2D} intersect the vertical $\phi_2 = \phi_{2D}$ line *below* the curve. Thus the homogeneous solution is unstable and phase separation takes place until the system becomes stable when the *two co-existing phases* have the *binodal compositions* ϕ_{2b}' and ϕ_{2b}''. All homogeneous solutions with compositions in the range $\phi_{2sp}' < \phi_2 < \phi_{2sp}''$ are similarly unstable and separate into two phases corresponding to ϕ_{2b}' and ϕ_{2b}''. The general condition for equilibrium between two co-existing phases is that for each component, the chemical potential must be the same in both phases, i.e. $\mu_1' = \mu_1''$ and $\mu_2' = \mu_2''$. This condition is usually written in terms of chemical potential differences

$$\mu_1' - \mu_1^0 = \mu_1'' - \mu_1^0 \tag{3.168a}$$

$$\mu_2' - \mu_2^0 = \mu_2'' - \mu_2^0 \tag{3.168b}$$

since these are more directly related to ΔG_m^*. It can be shown that for any point on the curve, $(\mu_1 - \mu_1^0)$ and $(\mu_2 - \mu_2^0)/x$ are given by the values of ΔG_m^* which correspond to the intersections of the tangent to the curve at that point, with the vertical $\phi_2 = 0$ and $\phi_2 = 1$ lines respectively. Thus Equations (3.168a/b) are satisfied when two points on the curve have a *common tangent* as for ϕ_{2b}' and ϕ_{2b}'' in Fig. 3.19(b). Since the variation of

ΔG_m^* with ϕ_2 is unsymmetrical for polymer solutions, the binodal points do not correspond to the minima in the curve.

Finally, consider a homogeneous solution with ϕ_2 $(=\phi_{2F})$ corresponding to point F on the curve for T_2. Clearly phase separation into two phases corresponding to the binodal compositions is thermodynamically favoured. However, in order for this to occur an energy barrier must be overcome because the initial stages of phase separation about ϕ_{2F} (e.g. to points f' and f'') give rise to an increase in the Gibbs free energy. This is true for all homogeneous solutions with compositions in the ranges $\phi'_{2b} < \phi_2 < \phi'_{2sp}$ and $\phi''_{2sp} < \phi_2 < \phi''_{2b}$, where ϕ'_{2sp} and ϕ''_{2sp} are the *spinodal compositions* corresponding to the points of inflection in the curve. Such solutions are said to be *metastable* and will phase separate to the binodal compositions, but only if the energy barrier can be overcome. Since the spinodal compositions occur at points of inflection, they are located by application of the condition

$$\left(\frac{\partial^2 \Delta G_m^*}{\partial \phi_2^2}\right)' = \left(\frac{\partial^2 \Delta G_m^*}{\partial \phi_2^2}\right)'' = 0 \tag{3.169}$$

The existence of two minima in the variation of ΔG_m^* with ϕ_2 (and hence phase separation) results from the contribution to ΔG_m^* due to contact interactions (i.e. non-zero ΔH_m). This contribution decreases as the temperature changes from T_2 towards T_1, and the binodal and spinodal points get closer together until at a *critical temperature* T_c they just coincide at a single point corresponding to ϕ_{2c}. The curves defined by the binodal points and spinodal points as a function of temperature are known as the *binodal* and *spinodal* respectively. For most polymer solutions ΔH_m is positive, so that T_c $(>T_2)$ corresponds to the *common maximum* of the binodal and spinodal, and is known as the *upper critical solution temperature (UCST)* above which the polymer and solvent are miscible in all proportions (see Fig. 3.20(a)). For the less common situation when ΔH_m is negative, T_c $(<T_2)$ corresponds to the *common minimum* of the binodal and spinodal, and is known as the *lower critical solution temperature (LCST)* below which the polymer and solvent are completely miscible (see Fig. 3.20(b)). LCST behaviour usually is observed when there are specific favourable polymer–solvent interactions (e.g. hydrogen bonding, charge transfer). The last two examples given in Table 3.1, and poly(ethylene oxide) in water, are systems which show LCST behaviour due to hydrogen bonding interactions. LCST behaviour can also be caused by volume contraction upon mixing because this leads to a reduction in the entropy of mixing. Flory–Huggins theory assumes there to be no volume change and so more advanced theories are required to account quantitatively for the effects of volume changes. Such theories predict that all polymer–solvent systems should show both UCST and LCST behaviour,

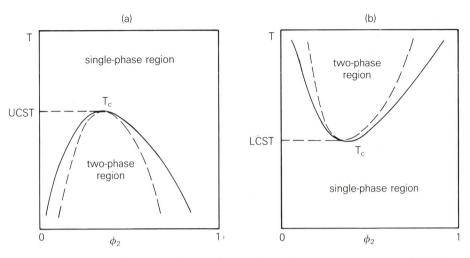

Fig. 3.20 *Schematic illustration of phase diagrams for polymer solutions showing (a) UCST behaviour and (b) LCST behaviour. The solid lines are the binodals and the dashed lines are the spinodals.*

though obviously in different temperature regimes. However, in all but a few cases (e.g. polystyrene in cyclohexane) only one of these regimes is experimentally accessible.

The regions outside the binodal correspond to stable homogeneous solutions, whereas the regions within the spinodal correspond to unstable solutions which will spontaneously phase-separate. The regions between the binodal and spinodal correspond to metastable solutions which only phase-separate if an energy barrier can be overcome.

For both UCST and LCST behaviour, T_c coincides with the turning point in the spinodal and so can be located by application of the condition

$$\left(\frac{\partial^2 \Delta G_m^*}{\partial \phi_2^2}\right) = \left(\frac{\partial^3 \Delta G_m^*}{\partial \phi_2^3}\right) = 0 \tag{3.170}$$

These derivatives can easily be evaluated from Equation (3.167) by recognizing that $\phi_1 = 1 - \phi_2$ and by assuming that χ is independent of ϕ_2. Thus $(\partial^2 \Delta G_m^*/\partial \phi_2^2) = 0$ leads to the following equation for the spinodal points

$$1/(1 - \phi_2) + 1/x\phi_2 - 2\chi = 0 \tag{3.171}$$

and so a theoretical spinodal curve can be constructed if the variation of χ with temperature is known. Application of the condition $(\partial^3 \Delta G_m^*/\partial \phi_2^3) = 0$ gives the critical composition as

$$\phi_{2c} = 1/(1 + x^{1/2}) \tag{3.172}$$

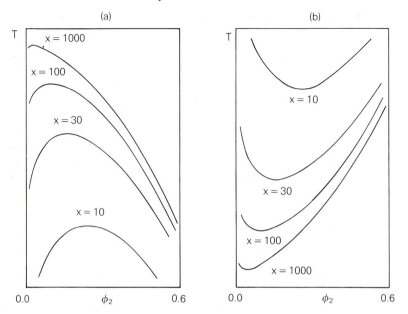

Fig. 3.21 *The effect of the number of chain segments, x, upon the binodals for (a) UCST behaviour and (b) LCST behaviour.*

The critical value χ_c of the Flory–Huggins interaction parameter is obtained by substituting Equation (3.172) into (3.171)

$$\chi_c = \tfrac{1}{2}[1 + 2/x^{1/2} + 1/x] \tag{3.173}$$

It should be noticed that as $x \rightarrow \infty$, $\phi_{2c} \rightarrow 0$ and $\chi_c \rightarrow \tfrac{1}{2}$. Thus there exist unique pairs of binodal and spinodal curves for each value of x. As x increases, these curves become increasingly skewed towards the $\phi_2 = 0$ axis and T_c moves either to higher temperatures for UCST behaviour or lower temperatures for LCST behaviour (Fig. 3.21). For phase separation of a solution of a polydisperse polymer, the binodal, spinodal and values of ϕ_{2c} and χ_c are obtained by replacing x by its number-average value, and are intermediate to those of the individual polymer species with specific values of x. The origin of fractionation by phase separation now is clearly evident, since preferential phase separation of the highest molar mass species can be expected.

The chemical potential difference $\mu_{2x} - \mu_{2x}^0$ for polymer species with x chain segments can be obtained from Equation (3.27) and is given by

$$\mu_{2x} - \mu_{2x}^0 = RT[\ln \phi_{2x} - (x - 1)\phi_1 + x\chi\phi_1^2] \tag{3.174}$$

and the equilibrium condition for their presence in two co-exiting phases may be written from Equation (3.168b) as

$$\mu'_{2x} - \mu^0_{2x} = \mu''_{2x} - \mu^0_{2x} \qquad (3.175)$$

The equilibrium conditions defined by Equations (3.168a) and (3.175) can be combined to give an equivalent single equilibrium condition

$$\mu'_{2x} - x\mu'_1 = \mu''_{2x} - x\mu''_1 \qquad (3.176)$$

Substituting into this equation expressions for μ'_1 and μ''_1 from Equation (3.26) (in which x is replaced by $\bar{x}_n$ and $\phi_2 = \Sigma\phi_{2x}$), and for μ'_{2x} and μ''_{2x} from Equation (3.174), after simplification gives

$$\phi''_{2x}/\phi'_{2x} = e^{\sigma x} \qquad (3.177)$$

where the parameter σ is given by

$$\sigma = \ln(\phi''_1/\phi'_1) + 2\chi(\phi''_2 - \phi'_2) \qquad (3.178)$$

For two particular co-existing phases the volume fractions in Equation (3.178) and χ have specific values, and σ is a constant. Since phase separation occurs when $\chi > \frac{1}{2}$ (because $\bar{x}_n < \infty$), σ is positive and so from Equation (3.177) $\phi''_{2x} > \phi'_{2x}$ for all values of x. Assuming that the polymer density is independent of x, then on the basis of Equation (3.177) the ratio of the mass fractions f''_{2x} and f'_{2x} of the x-mers is given by

$$f''_{2x}/f'_{2x} = Re^{\sigma x} \qquad (3.179)$$

where $R = V''/V'$, and V'' and V' are the volumes of the concentrated and dilute co-existing phases respectively. Efficient fractionation requires that $\sigma \ll 1$ and $R \ll 1$. The value of σ is not easily varied, but typically is about 0.01 and so is satisfactory. In contrast, the value of R can be altered and in order to ensure that $V' > V''$ it is necessary to begin with very dilute homogeneous solutions (typically with c about $2\,\mathrm{g\,dm^{-3}}$). The data given in Table 3.9 clearly show that whilst all species are present in each phase, the concentrated phase contains almost all of the high molar mass species. The volume fraction ratio ϕ''_{2x}/ϕ'_{2x} is close to unity for the low molar mass species so they are present predominantly in the dilute phase because of its

TABLE 3.9 *Some values of ϕ''_{2x}/ϕ'_{2x} and f''_{2x}/f'_{2x} calculated from Equations (3.177) and (3.179) with $\sigma = 0.01$ and $R = 0.005$*

x	10	100	500	1000	5000
ϕ''_{2x}/ϕ'_{2x}	1.1	2.7	148.4	2.2×10^4	5.2×10^{21}
f''_{2x}/f'_{2x}	0.006	0.014	0.74	$\lrcorner 10$	2.6×10^{19}

much larger volume. Thus the polymer present in the concentrated phase has a relatively narrow molar mass distribution, typically with $\bar{M}_w/\bar{M}_n$ in the range 1.1–1.3.

Flory–Huggins theory gives reasonable predictions for phase separation of dilute polymer solutions because the excluded volume is close to zero under the conditions of phase separation. The limitations of the theory presented here arise principally from the unsatisfactory assumptions that χ is independent of ϕ_2 and that the volume change upon mixing is zero. Whilst the prediction of T_c generally is good, experimentally-determined binodals tend to be less sharp with ϕ_{2c} larger than predicted by the theory. Better agreement can be gained by taking into account the dependence of χ upon ϕ_2 and the effects of volume changes, but such theories are beyond the scope of this book.

Measurements of T_c can be used to determine theta temperatures. Comparison of the Flory–Huggins dilute solution Equations (3.38) and (3.45) leads to

$$\chi - \tfrac{1}{2} = \psi\left[\left(\frac{\theta}{T}\right) - 1\right] \tag{3.180}$$

At $T = T_c$, χ can be substituted by χ_c from Equation (3.173) to give after rearrangement

$$1/T_c = 1/\theta + (1/\psi\theta)[1/x^{1/2} + 1/2x] \tag{3.181}$$

in which x is replaced by $\bar{x}_n$ for a polydisperse polymer. Thus a plot of $1/T_c$ against $(1/\bar{x}_n^{1/2} + 1/2\bar{x}_n)$ gives a straight line with intercept $1/\theta$. It is found that theta temperatures obtained in this way are in good agreement with those obtained from osmotic pressure measurements (Section 3.6.2). Inspection of Equation (3.181) reveals that $T_c \to \theta$ as $x \to \infty$, thus providing another alternative definition of θ as the critical temperature for miscibility in the limit of infinite molar mass.

3.16.2 *Procedures for fractionation*

The basic requirements for fractionation were established in the preceding section. Thus phase separation of a very dilute polymer solution is brought about by causing the solvency conditions to deteriorate (i.e. causing χ to increase). This can be achieved either by adding a non-solvent to the solution or by changing the temperature. The former procedure is preferred and involves addition of non-solvent to the polymer solution until phase separation is clearly evident. The addition of non-solvent is stopped at this point and the solution heated (assuming UCST behaviour) to redissolve the concentrated phase (i.e. χ is decreased). This solution then is cooled *slowly* to achieve equilibrium phase separation. After allowing the concentrated phase to settle at the bottom of the fractionation

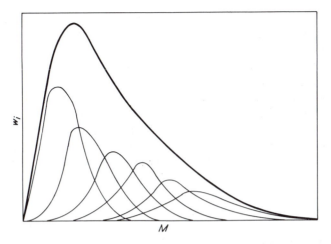

Fig. 3.22 *Schematic representation of typical molar mass distribution curves for the fractions obtained from a polydisperse polymer sample, the molar mass distribution of which is shown as the bold line.*

flask, it is removed and the remaining dilute solution phase subjected to further similar fractionations. The molar mass distributions of the fractions obtained are shown schematically in Fig. 3.22 in relation to the molar mass distribution of the polymer before fractionation. Thus there is considerable overlap of the distribution curves of the individual fractions and this has to be taken into account when constructing the distribution curve for the original polymer.

The procedure just described is rather time-consuming and so attempts have been made to automate fractionation by phase separation using methods of chromatography. One such technique involves depositing the polymer at the top of a column of glass beads and then passing a mixture of solvent and non-solvent through the column which usually has a gradient of temperature along its length (decreasing temperature for UCST behaviour). The concentration of solvent in the eluant is gradually increased so that the low molar mass species dissolve first and are recovered from the initial eluate. Thus fractions are obtained in order of increasing molar mass. More recently, continuous liquid–liquid extraction procedures have been employed whereby the polymer initially is dissolved in one solvent and is extracted continuously from this solution by another solvent which is only partially miscible with the first solvent.

3.17 Gel permeation chromatography

The technique of *gel permeation chromatography* (GPC) was developed during the mid-1960s, and is an extremely powerful method for determining the complete molar mass distribution of a polymer. In GPC a dilute polymer solution is injected into a solvent stream which then flows through a column packed with beads of a porous gel. The porosity of the gel is of critical importance and typically is in the range $50-10^6$Å. The small solvent molecules pass both through and around the beads, carrying the polymer molecules with them where possible. The smallest polymer molecules are able to pass through most of the pores in the beads and so have a relatively long flow-path through the column. However, the largest polymer molecules are excluded from all but the largest of the pores because of their greater molecular size and consequently have a much shorter flow-path. Thus GPC is a form of *size-exclusion chromatography*, in which the polymer molecules elute from the chromatography column in order of decreasing molecular size in solution. The concentration of polymer in the eluate is monitored continuously and the chromatogram obtained is a plot of concentration against elution volume, which provides a qualitative indication of the molar mass distribution. The chromatogram also can reveal the presence of low molar mass additives (e.g. plasticizers), since they appear as separate peaks at large elution volumes.

3.17.1 *Separation by size-exclusion*

The volume of solvent contained in a GPC system from the point of solution injection, through the column to the point of concentration detection can be considered as the sum of a void volume V_0 (i.e. the volume of solvent in the system outside the porous beads) and an internal volume V_i (i.e. the volume of solvent inside the beads). The volume of solvent required to elute a particular polymer species from the point of injection to the detector is known as its *elution volume V_e* and on the basis of *separation by size-exclusion* is given by

$$V_e = V_0 + K_{se}V_i \qquad (3.182)$$

where K_{se} is the fraction of the internal pore volume penetrated by those particular polymer molecules. For small polymer molecules that can penetrate all of the internal volume $K_{se} = 1$ and $V_e = V_0 + V_i$, whereas for very large polymer molecules that are unable to penetrate the pores $K_{se} = 0$ and $V_e = V_0$. Clearly for polymer molecules of intermediate size K_{se} and V_e lie between these limits.

For the flow rates typically used in GPC (about $1\,cm^3\,min^{-1}$) it is found that V_e is independent of flow rate. This means that there is sufficient time for the molecules to diffuse into and out of the pores such that equilibrium concentrations, c_i and c_0, of the molecules are attained inside and outside the pores respectively. Since the separation process takes place under equilibrium conditions, K_{se} can be considered as the *equilibrium constant*

(a)   (b)

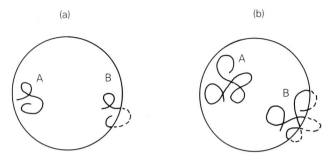

Fig. 3.23 *Schematic illustration of (a) a small and (b) a large polymer molecule inside a cylindrical pore. A, the molecule is positioned with it's centre of mass at it's distance of closest approach to the wall and B, shows a molecular conformation which is excluded when the molecule is positioned very close to the wall.*

defined by $K_{se} = c_i/c_0$ (note that when the polymer molecules penetrate the pores to the same extent as the solvent $c_i = c_0$ and $K_{se} = 1$). From thermodynamics the relation between K_{se} and the Gibbs free energy change ΔG_p^0 for permeation of the pores is given by

$$\Delta G_p^0 = -\mathbf{R}T \ln K_{se} \qquad (3.183)$$

For separation exclusively by size-exclusion, the enthalpy change associated with transfer of solute species into the pores *must be zero*. Thus ΔG_p^0 is controlled by the corresponding entropy change ΔS_p^0 and is given by $\Delta G_p^0 = -T\Delta S_p^0$. Hence from Equation (3.183), K_{se} is given by

$$K_{se} = \exp(\Delta S_p^0/\mathbf{R}) \qquad (3.184)$$

The number of conformations available to a polymer molecule is reduced inside a pore because of its close proximity to the impenetrable walls of the pore, and so ΔS_p^0 is negative. This loss of conformational entropy is equivalent to exclusion of the centre of mass of the molecule from regions closer to the walls of the pores than the hydrodynamic radius of the molecule (see Fig. 3.23). Methods of statistical thermodynamics can be used to obtain expressions for ΔS_p^0 on the basis of simplifying models (e.g. Gaussian coils entering cylindrical pores). The details will not be considered here, but the general result is of the form

$$\Delta S_p^0 = -\mathbf{R}A_s(\overline{L}/2) \qquad (3.185)$$

where A_s is the surface area per unit pore volume and $\overline{L}$ is the mean molecular projection of the molecule when free in solution (e.g. for a spherical molecule, $\overline{L}$ is its mean diameter). Substituting Equations (3.184) and (3.185) into (3.182) gives

$$V_e = V_0 + V_i \exp(-A_s\overline{L}/2) \qquad (3.186)$$

Thus V_e is predicted to vary exponentially with $\overline{L}$ (i.e. with molecular size), and $V_e \rightarrow (V_0 + V_i)$ as $\overline{L} \rightarrow 0$, with $V_e \rightarrow V_0$ as $\overline{L} \rightarrow \infty$. An alternative interpretation of Equation (3.186) is that V_e can be expected to decrease approximately *linearly* with $\log \overline{L}$ (i.e. with the logarithm of molecular size). This interpretation is particularly useful for the purposes of calibration.

3.17.2 *Calibration and evaluation of molar mass distributions*

In order to convert a GPC chromatogram into a molar mass distribution (MMD) curve it is necessary to know the relationship between molar mass (M) and V_e. This relationship results from Equation (3.186) due to the dependence of molecular size upon M. The molecular size of a polymer molecule in solution can be taken as its hydrodynamic volume (Section 3.4.2) which from Equation (3.79) is proportional to $[\eta]M$ where $[\eta]$ is the intrinsic viscosity. Hence it can be predicted that $\log([\eta]M)$ will decrease approximately linearly with V_e. From the Mark–Houwink Equation (3.160)

$$\log([\eta]M) = \log K + (1 + a)\log M$$

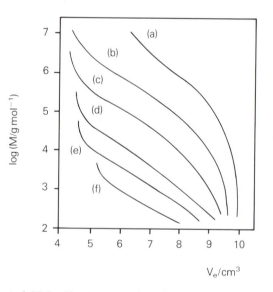

Fig. 3.24 *Some typical GPC calibration curves for polystyrene in tetrahydrofuran. The curves are for six different GPC columns packed with gels of porosity: (a) 10^6 Å, (b) 10^5 Å, (c) 10^4 Å, (d) 10^3 Å, (e) 500 Å, (f) 50 Å (courtesy of Polymer Laboratories Ltd). In each case the curve is approximately linear over the range of resolvable molar masses.*

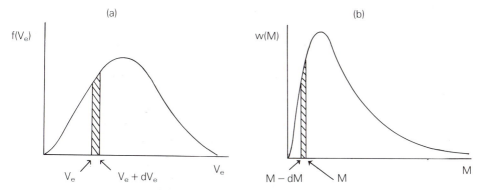

Fig. 3.25 *Schematic illustrations of (a) a GPC chromatogram and (b) the weight-fraction MMD into which it transposes.*

and so $\log([\eta]M)$ increases linearly with $\log M$ as long as K and a remain truly constant. Thus a calibration plot of $\log M$ against V_e should be approximately linear, as is found in practice. The calibration plot is specific to the polymer under study (through the values of K and a) and is obtained by measuring V_e for a series of narrow MMD samples of the polymer with different M (see Fig. 3.24). The value of V_e usually is taken as that at the peak of the chromatogram for this purpose.

Conversion of the GPC chromatogram of detector response $f(V_e)$ against V_e to the *weight-fraction MMD* (i.e. $w(M)$ against M) can now be considered (see Fig. 3.25). Assuming that $f(V_e)$ is directly proportional to the concentration of the polymer in the eluate and that the proportionality constant is independent of M, then the weight fraction dw of polymer which elutes between V_e and $V_e + dV_e$ is given by

$$dw = f(V_e)dV_e \Big/ \int_0^\infty f(V_e)dV_e \qquad (3.187)$$

where the integral in the denominator is simply the area A under the GPC chromatogram and serves to normalize $f(V_e)dV_e$ to give a weight fraction. Thus Equation (3.187) can be re-written as

$$dw/dV_e = f(V_e)/A \qquad (3.188)$$

Corresponding to the elution volume interval V_e to $(V_e + dV_e)$, the same weight fraction dw of polymer exists between M and $(M - dM)$ in the weight-fraction MMD (because M decreases as V_e increases). The distribution function $w(M)$ is normalized and so $dw = -w(M)dM$ from which

$$w(M) = -dw/dM \tag{3.189}$$

Applying the chain rule for derivatives

$$-dw/dM = (dw/dV_e)(-dV_e/d\ln M)(d\ln M/dM) \tag{3.190}$$

Now $-dV_e/d\ln M = 1/2.303|S(V_e)|$ where $|S(V_e)|$ is the magnitude of the (negative) slope of the calibration plot ($\log M$ against V_e) at V_e, and $d\ln M/dM = 1/M(V_e)$ where $M(V_e)$ is the value of M corresponding to V_e. Combining these relationships with Equations (3.190), (3.188) and (3.189) leads to

$$w(M) = (f(V_e)/A)(1/2.303|S(V_e)|)(1/M(V_e)) \tag{3.191}$$

The first term in this equation is obtained from the chromatogram and the remaining two terms from the calibration plot. Thus $w(M)$ can be evaluated point by point to give the weight-fraction MMD.

Due to the dependence of V_e upon the *logarithm* of M, accurate conversion of the GPC chromatogram to the weight-fraction MMD necessitates precise knowledge of the form of the calibration curve. Whilst the curve nominally is linear, in detail it is not, and use of a linear fit to the calibration data can lead to serious errors in $w(M)$. Nowadays it is common practice to fit $\log M$ to a power series in V_e (i.e. $\log M = \sum_{x=1}^{n} k_x V_e^x$, typically with $n = 4$, 5 or 6), the first differential of which enables $S(V_e)$, $= d\log M/dV_e$, to be calculated for each value of V_e. In this way small, but significant, local variations in $S(V_e)$ are taken into account to give $w(M)$ more accurately. Obviously a wide range of calibration standards are required to define accurately the calibration curve over the full range of M.

A further improvement in the accuracy of $w(M)$ can be achieved by first correcting the GPC chromatogram for diffusional broadening of the peak. Such corrections are difficult to perform and are negligible for most polymers analysed using modern gels with small bead diameters. However, they are important in the analysis of narrow MMD polymers which generally give diffusion-broadened peaks that are skewed towards low V_e.

3.17.3 *Evaluation of molar mass averages*

From $w(M)$ any molar mass average can be evaluated. For example, the number-average molar mass $\bar{M}_n$ can be defined in integral form (cf. Equation (1.6)) as

$$\bar{M}_n = 1 \bigg/ \int_0^\infty w(M)(1/M)dM \tag{3.192}$$

and the weight-average molar mass $\bar{M}_w$ (cf. Equation (1.7)) as

$$\overline{M}_w = \int_0^\infty w(M)M\mathrm{d}M \qquad (3.193)$$

Similarly from Equation (3.161) with $w_i = n_i M_i$ (recognizing that $\int_0^\infty w(M)\mathrm{d}M = 1$, i.e. $w(M)$ is normalized), an expression for the viscosity-average molar mass $\overline{M}_v$ is obtained

$$\overline{M}_v = \left[\int_0^\infty w(M)M^a\mathrm{d}M \right]^{1/a} \qquad (3.194)$$

3.17.4 Universal calibration

The need to determine the precise form of the calibration curve for each polymer to be analysed using a given solvent/temperature/GPC column system is a formidable task. Furthermore, with the exception of the more common polymers, standard samples suitable for calibration are not

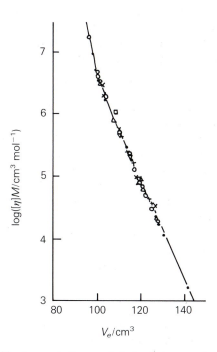

Fig. 3.26 *Universal calibration curve for crosslinked polystyrene gels with tetrahydrofuran as solvent:* ●, *linear polystyrene;* ○, *branched polystyrene (comb type);* +, *branched polystyrene (star type);* △, *branched block copolymer of styrene/methyl methacrylate;* ×, *poly(methyl methacrylate);* ○, *poly(vinyl chloride);* ▽, *graft copolymer of styrene/methyl methacrylate;* □, *polybutadiene (reprinted with permission from Comprehensive Polymer Science, copyright 1989, Pergammon Press plc).*

available for most polymers. Thus the concept of a universal calibration which applies to all polymers analysed using a given system is very attractive.

In Section 3.17.2 it was stated that molecular size in solution can be represented by the *hydrodynamic volume* of the molecule, which is proportional to $[\eta]M$. Thus a plot of $\log([\eta]M)$ against V_e should be approximately linear and *independent of the polymer* under consideration. As shown by the data presented in Fig. 3.26, there is ample experimental evidence confirming that such a plot does indeed provide a *universal calibration* for a given solvent/temperature/GPC column system. From the Mark–Houwink Equation (3.160) $[\eta]M = KM^{1+a}$ and so the molar mass $M(V_e)$ corresponding to V_e for a polymer for which the calibration curve is not known, is given by

$$M(V_e) = [([\eta]M)_{V_e}/K]^{1/(1+a)} \tag{3.195}$$

where $([\eta]M)_{V_e}$ is the value of $[\eta]M$ taken from the universal calibration curve at V_e, and K and a are the Mark–Houwink constants for the polymer under the conditions of the GPC analysis. Thus the system is calibrated using readily-available standards (s), such as narrow MMD polystyrenes, for which K_s and a_s are known. This enables a universal calibration curve of $\log([\eta]_s M_s)$, $= \log(K_s M_s^{1+a_s})$, against V_e to be obtained from which the $\log M$ against V_e calibration curve can be evaluated for any other polymer using Equation (3.195) assuming that K and a are known.

3.17.5 *Porous gels and eluants for GPC*

The GPC column packings most commonly used with organic solvents are rigid porous beads of either crosslinked polystyrene or surface-treated silica gel. For aqueous GPC separations, porous beads of water-swellable crosslinked polymers (e.g. crosslinked polyacrylamide gels), glass or silica are employed.

The ability to resolve the different molar mass species present in a polymer sample depends upon a number of factors. As is evident from Fig. 3.24, an appropriate range of gel porosities is required to obtain the necessary resolution and is obtained by using either a connected series of short columns each of which is packed with a gel of different porosity, or one long column packed with mixed gels. Resolution increases approximately with $l^{1/2}$ where l is the total column length, and also with $(1/d^2)$ where d is the bead diameter. The early GPC gels had d of about $50\,\mu$m and large column lengths were required for the separations. Nowadays, however, most GPC gels have d of either $5\,\mu$m or $10\,\mu$m, and much shorter column lengths can be used. This means that the elution volumes are also much smaller and the separation process is much faster. Thus GPC performed using gels of small bead diameter is termed *high-performance GPC* (HPGPC).

The choice of eluant is of considerable importance because it can have a significant influence upon the contribution of secondary modes of separation. For example, adsorption of polymer molecules to the walls of the pores will retard their movement through the column, causing V_e to increase beyond its value for size-exclusion alone. Such effects are more probable if the solvency conditions are poor, and so a good solvent for the polymer should be used as the eluant. The most common eluants are tetrahydrofuran for polymers soluble at room temperature, 1,2-dichlorobenzene at about 130°C for crystalline polyolefins, and 2-chlorophenol at about 90°C for crystalline polyesters and polyamides.

3.17.6 *Practical aspects of GPC*

The essential components of a GPC apparatus are: (i) a solvent pump, (ii) an injection valve, (iii) a column (or series of columns) packed with beads of porous gel, and (iv) a detector (Fig. 3.27). Since almost all size-exclusion separations are performed by HPGPC nowadays, only HPGPC will be considered here.

High-quality solvent pumps which give pulse-free constant volumetric flow rates (Q) are essential for HPGPC because it is usual to record elution times t_e rather than elution volumes ($=Qt_e$) due to the relatively small

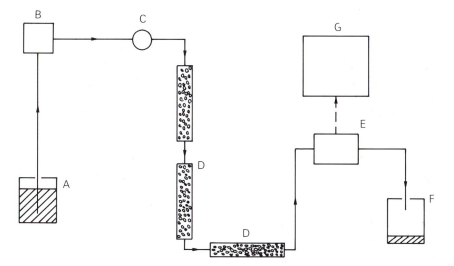

Fig. 3.27 *Schematic illustration of a GPC apparatus: A, solvent reservoir; B, solvent pump; C, injection valve; D, GPC columns; E, detector; F, waste solvent/solution; and G, computer.*

values of $(V_0 + V_i)$, typically $30-40\,cm^3$. Also, the pump must generate high pressures (typically $4-14\,MPa$) to force the solvent through the columns (of tightly-packed small-diameter beads of gel) at the usual flow rate of about $1\,cm^3\,min^{-1}$. About $0.05\,cm^3$ of a dilute solution (e.g. $2\,g\,dm^{-3}$) of the polymer to be analysed is injected into the solvent stream. Since this corresponds to injection of about $0.1\,mg$ of polymer, highly sensitive stable detectors are required. In this respect detectors which measure absorption of ultraviolet (UV) light are best, though they are restricted to systems in which the polymer absorbs UV light at wavelengths at which the solvent does not absorb. Many polymers do not absorb UV light and detectors which measure absorption of infrared (IR) radiation are more appropriate, but much more expensive. The most commonly used detectors are differential refractometers (RI) which measure continuously the difference between the refractive index of the eluate and that of the pure solvent. The responses from UV, IR and RI detectors are proportional to the concentration of polymer in the eluate. More specialized detectors which enable the molar mass of the polymer in the eluate to be measured continuously are also available. For example, on-line viscometric (IV) and low-angle laser light scattering (LALLS) detectors measure respectively the specific viscosity, η_{sp}, and the limiting excess Rayleigh ratio, $\Delta R_{\theta \to 0}$, of the eluate continuously. When used in series with a concentration detector (e.g. RI), the combined signals enable $M(V_e)$ to be determined continuously on the basis that η_{sp} and $\Delta R_{\theta \to 0}$ are equal to their limiting values at $c = 0$ (because c is very small). The attraction for such dual detection systems is that they yield the weight-fraction MMD directly, thus obviating the need for calibration.

In most modern HPGPC systems, the signal from the detector is stored directly in the memory of a computer which then is used to perform the necessary data manipulation. In this way, the MMD and the molar mass averages can be obtained less than an hour after injecting the sample.

It should now be obvious why HPGPC is a very important technique for polymer characterization and why its use is widespread throughout academy and industry. A natural progression from the analytical systems described here, has been the development of much larger volume *preparative* HPGPC systems for fractionation of polymers. Typically the volume of preparative systems is about $10 \times$ that of analytical systems and enables $10-25\,mg$ of polymer to be fractionated in a single run requiring about the same time as an analytical HPGPC analysis. The eluate is collected in successive portions corresponding to specific ranges of molar mass in the MMD, and the different polymer fractions are recovered from these portions by removal of the solvent. If larger quantities of each fraction are required, the procedure can be repeated several times. Thus preparative HPGPC has significant advantages over classical methods for fractionation (Section 3.16), the most important being that it is much faster and can yield fractions with molar masses in pre-defined ranges.

GPC of copolymers is complicated by superposition of the molecular size and copolymer composition distributions. Thus when analysing copolymers it is good practice to use dual detector systems (e.g. UV + RI or IR + RI) that facilitate determination of the average composition of the copolymer molecules eluting at each point in the chromatogram (i.e. in the molecular size distribution). However, a more recent development known as *chromatographic cross-fractionation* is far more elegant for characterization of copolymers and gives both the molecular size distribution and the distribution of copolymer composition. In its most efficient form it involves connecting the outlet from a GPC system directly to another high-performance column chromatography system in which the molecules are separated according to composition via differential solubility and/or adsorption effects. Thus the copolymer molecules are separated first according to their size in solution and then according to their composition.

Chemical Composition and Molecular Microstructure

3.18 **Determination of chemical composition and molecular microstructure**

The previous sections of this chapter have been concerned with methods for determination of average molar masses, molar mass distributions and molecular dimensions. In many instances this information is all that is necessary to characterize a homopolymer when its method of preparation is known. However, for certain homopolymers (e.g. polypropylene, polyisoprene) knowledge of molecular microstructure is of crucial importance. Additionally, for a copolymer it is necessary to determine the chemical composition in terms of the mole or weight fractions of the different repeat units present. It is also desirable to determine the distribution of chemical composition amongst the different copolymer molecules which constitute the copolymer (Section 3.17.6), and to determine the sequence distribution of the different repeat units in these molecules. Furthermore, when characterizing a sample of an unknown polymer the first requirement is to identify the repeat unit(s) present. Thus methods for determination of chemical composition and molecular microstructure are of great importance.

Simple methods of chemical analysis are useful in certain cases. Qualitative chemical tests such as the Beilstein and Lassaigne tests for the presence of particular elements (e.g. halogens, nitrogen, sulphur) are of use in the identification of an unknown polymer. *Combustion analysis* for quantitative determination of elemental composition can be used to confirm the purity of a homopolymer and to determine the average chemical composition of a copolymer for which the repeat units are known and have significantly different elemental compositions (e.g. the acrylonitrile contents of different nitrile rubbers, which are poly(butadiene-*co*-acrylonitrile)s, can be evaluated from their percentage carbon and/or percentage nitrogen contents). The usefulness of these analyses, however, is restricted by their inability to give structural information. *Spectroscopic methods* are by far the most powerful for this purpose, and in addition to giving overall structural information (e.g. identification of repeat units, determination of the average chemical composition of a copolymer), they enable molecular microstructure, repeat unit sequence distributions and structural defects to be determined.

The use of spectroscopic methods to analyse polymers was a natural extension of their initial use for studying the structures of low molar mass compounds. There now are available an enormous number of fully-interpreted spectra of low molar mass compounds and of polymers, and these provide a firm foundation for structural characterization. In the following sections, spectroscopic characterization of polymers is reviewed with particular emphasis upon *infrared spectroscopy* and *nuclear magnetic resonance spectroscopy*, since they are most widely used. Fundamental aspects, and descriptions of the instrumentation and experimental procedures used, will be treated only briefly because a full account would require a disproportionate amount of space and there already exist many excellent texts on spectroscopy which deal with these topics. Also due to the limitations of space, only a few spectra and a brief survey of the vast potential of these and other spectroscopic methods for characterization of polymers are given. The reader is referred to specialist texts on spectroscopy of polymers for a more complete appreciation of their uses.

3.19 Infrared (IR) spectroscopy

The energies associated with vibrations of atoms in a molecule with respect to one another are quantized and absorption of electromagnetic radiation in the infrared region (approximately $1 > \lambda > 50\mu m$ where λ is the wavelength) gives rise to transitions between these different vibrational states. Absorption results from coupling of a vibration with the oscillating electric field of the infrared radiation, and this interaction can occur only when the vibration produces an oscillating dipole moment (cf. Raman spectroscopy, Section 3.21). Since vibrating atoms are linked together by chemical bonds it is usual to refer to the vibrations as bond deformations, of which the simplest types are stretching and bending. For a particular infrared active bond deformation (e.g. $C{=}O$ stretching), absorption will occur when the frequency, ν, of the radiation is that defined by the following rearranged form of *Planck's equation*

$$v = \Delta E/h \qquad\qquad (3.196)$$

where ΔE is the energy difference between the upper and lower vibrational energy levels for the particular bond deformation, and **h** is Planck's constant. Usually, the only significant absorptions correspond to promotion of bond deformations from their ground states to their next highest energy levels. The basis of *infrared (IR) spectroscopy* is that for a particular type of bond deformation ΔE, and hence v, depend upon the atoms involved (e.g. ΔE and v for N—H stretching are higher than for C—H stretching). Thus by measuring the absorption of infrared radiation over a range of v a *spectrum* is obtained which contains a series of absorptions at different v. Each absorption corresponds to a specific type of deformation of a particular bond (or bonds) and is characteristic of that bond deformation with v not greatly affected by the other atoms present in the molecule. Thus the absorption due to a particular bond deformation occurs at approximately the same v for all molecules (including polymers) which contain that bond. The *characteristic absorption regions* for some of

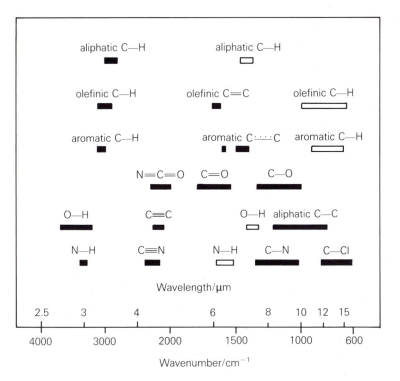

Fig. 3.28 *Characteristic absorption regions for some commonly observed bond stretching (━)
and bending (═) deformations.*

the most commonly-observed bond deformations are shown in Fig. 3.28. Note that infrared radiation is defined either by its wavelength or, more usually nowadays, by its *wavenumber*, $\bar{v}$, which is the reciprocal of its wavelength (i.e. $\bar{v} = 1/\lambda$) and is related to its frequency by $v = \bar{v}c$ where c is the velocity of light. Thus the most common absorptions occur in the wavenumber range $4000–650\,\text{cm}^{-1}$ and for this reason IR spectra usually are recorded over this range.

3.19.1 *IR spectroscopy of polymers*

The simplest application of IR spectroscopy is for *polymer identification*. Comparison of the positions of absorptions in the IR spectrum of a polymer sample with the characteristic absorption regions, leads to identification of the bonds and functional groups present in the polymer. In many cases this information is sufficient to identify the polymer. However, for confirmation the spectrum can be compared in detail with that of an authentic sample since the two spectra should be identical if the identification is correct, i.e. the IR spectrum of a polymer can be considered as a 'fingerprint' for this purpose. Interpretation of the spectra can become more complex when the polymer contains additives which contribute to the absorptions in the spectrum. For example, samples of flexible poly(vinyl chloride), PVC, usually show strong C=O stretching absorptions even though such bonds are not present in the structure of PVC. The absorptions are due to the ester C=O groups in the phthalate ester plasticizers present in the flexible PVC, and the IR spectrum is the sum of the spectra of the plasticizer and PVC. A similar situation obtains in the identification of blends of two or more polymers.

IR spectroscopy also finds application in the characterization of branched polymers, and copolymers. The *degree of branching* in a branched polymer can be determined provided that characteristic absorptions due to the branches can be identified. For example, branching in low density polyethylene is determined using the absorptions due to the terminal CH_3 groups on the branches. *Average compositions of copolymers* can be determined if the different repeat units give characteristic IR absorptions which are not coincident. For example, the composition of a sample of poly[styrene-*co*-(n-butyl acrylate)] can be determined from the out-of-plane C—H bending absorptions of the benzene rings in styrene repeat units and the C=O stretching absorption of the ester groups in the n-butyl acrylate repeat units. For such quantitative determinations the spectra are recorded as *absorbance*, A, against wavenumber (or wavelength), where $A = \log(I_0/I)$ in which I_0 and I are the intensities of the radiation incident to and after transmission through the sample respectively. The difference, ΔA, between the absorbance at the absorption maximum and the background absorbance at that position (obtained by linking the baselines before and after the absorption) is directly related by the *Beer–Lambert equation* to the molar concentration, c, of the bond(s) or group(s) giving rise to the absorption

$$\Delta A = \varepsilon c l \qquad (3.197)$$

where ε is the molar absorptivity of the absorption and l is the path length of the radiation through the sample. Since l is constant for a given sample, by taking the ratio of the ΔA values for the characteristic absorptions the molar ratio of the concentrations of the bonds or groups of interest can be calculated (e.g. $c_1/c_2 = \varepsilon_2 \Delta A_1/\varepsilon_1 \Delta A_2$) as long as the ratio of the molar absorptivities has been determined previously using samples of known composition.

Bond deformations, particularly bending deformations, are not completely insensitive to the presence of other atoms and bonds in the molecular structure or to molecular conformation and interactions with other molecules. Hence, the precise location of an IR absorption carries more information than simply the existence of a particular bond or group. It is for this reason, and also because more complex deformations are specific to a particular structure, that IR spectra can be used as polymer fingerprints and can provide information on *molecular microstructure*. Thus by using IR spectroscopy it is possible to distinguish the different repeat unit structures that can arise from polymerization of 1,3-diene monomers (Table 2.10 of Section 2.8.2). Similarly, the IR spectra of isotactic, syndiotactic and atactic forms of a polymer show characteristic differences (Fig. 3.29), though these differences do not result directly from the differences in configuration but instead arise from the effects of configuration upon local chain conformation (Section 3.3.2). Such differences, for example, have been used to estimate the fractions of isotactic and syndiotactic sequences in samples of polypropylene and poly(methyl methacrylate). The effects of chain conformation upon IR absorptions are more general and also apply to polymers which are not tactic. For example, the IR spectrum of semi-crystalline poly(ethylene terephthalate) shows additional absorptions to that of the purely amorphous polymer due to differences in the conformations of chains in the crystalline and amorphous phases.

Other uses of IR spectroscopy include monitoring reactions through the appearance or loss of an absorption and measurements of orientation by use of polarized IR radiation.

3.19.2 *Practical aspects*

There are two general types of instrument for recording IR spectra, namely *double-beam* and *Fourier transform (FT) IR spectrometers*. Modern

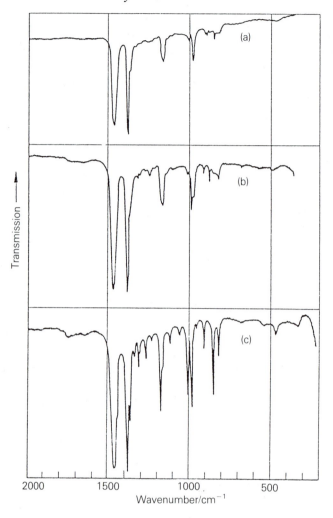

Fig. 3.29 *IR spectra of (a) atactic, (b) syndiotactic, and (c) isotactic polypropylene (reproduced with permission from Klöpffer, 'Introduction to Polymer Spectroscopy'). The absorptions at 970 and 1460 cm^{-1} do not depend upon tacticity, whereas the absorptions at 840, 1000, and 1170 cm^{-1} are characteristic of the 2*3/1 helix of isotactic polypropylene, and the absorption at 870 cm^{-1} is characteristic of the 4*2/1 helix of syndiotactic polypropylene.*

instruments are of the latter type which have a number of advantages over the former, including better signal-to-noise ratios and the ability to record complete spectra in very much shorter timescales (typically a few seconds rather than several minutes). Thus using FTIR spectrometers it also is possible to make *time-resolved measurements* such as monitoring fast reactions, continuous monitoring of the composition of the eluant in gel permeation chromatography and, with more specialized instrumentation, it is even possible to examine molecular dynamics (e.g. segmental motion) in timescales of less than a second.

Sample handling is the same for both types of spectrometer and use is made of salts such as potassium bromide and sodium chloride which do not absorb in the IR region. Solid polymers usually are analysed in the form of either (i) discs pressed from finely powdered dilute (1–2 per cent) dispersions of polymer in potassium bromide, or (ii) melt-pressed or solution-cast thin films. Liquid polymers are analysed as thin films between the polished faces of two blocks (known as plates) of sodium chloride. Analysis of polymers in solution tends to be avoided where possible because a significant proportion of the IR spectrum of the polymer is obscured by the IR absorptions of the solvent. IR spectra of polymer surfaces can be recorded using techniques such as attenuated total reflectance and specular reflectance, and their use has grown with the increasing importance of polymer surface chemistry.

IR spectroscopy is the most widely-used method for characterizing the molecular structures of polymers, principally because it provides a lot of information and is relatively inexpensive and easy to perform. However, it is not simple to interpret absolutely the more subtle features of IR spectra, such as those due to differences in tacticity. Such interpretations are usually made on the basis of information obtained from other techniques, in particular *nuclear magnetic resonance spectroscopy* which is by far the most powerful method for determining the detailed molecular microstructures of polymers.

3.20 Nuclear magnetic resonance (NMR) spectroscopy

All atomic nuclei have a positive charge due to the proton(s) they contain. In the nuclei of isotopes which contain an odd number of protons and/or an odd number of neutrons, this charge spins about the nuclear axis generating a *magnetic moment* along the axis. Such nuclei have *nuclear spin quantum numbers*, I, that are either integral (1, 2, ...) or half integral (1/2, 3/2, ...) and in the presence of an external magnetic field align themselves in $2I + 1$ different ways, each having a characteristic energy. This is the origin of *nuclear magnetic resonance (NMR) spectroscopy*, which is a form of spectroscopy in which absorption of electromagnetic radiation from the radiofrequency range (typically 1–500 MHz) promotes nuclei from low energy to high energy alignments in an external magnetic field. The simplest situation is that for nuclei with $I = 1/2$ (e.g. ^1H, ^{13}C, ^{19}F) since their nuclear magnetic moments align either parallel with (low-energy state) or against (high-energy state) the magnetic field. The energy difference

TABLE 3.10 *Some properties of important nuclei**

Isotope	$a/\%$[†]	I	$\gamma/\mathrm{rad\ T^{-1}\ s^{-1}}$	ν_0/MHz[‡]
$^1\mathrm{H}$	99.98	$\frac{1}{2}$	2.674×10^8	100.0
$^2\mathrm{H}$	0.016	1	4.106×10^7	15.3
$^{13}\mathrm{C}$	1.108	$\frac{1}{2}$	6.724×10^7	25.1
$^{14}\mathrm{N}$	99.63	1	1.932×10^7	7.2
$^{19}\mathrm{F}$	100.00	$\frac{1}{2}$	2.516×10^8	94.1

* $^{12}\mathrm{C}$ and $^{16}\mathrm{O}$ have no spin ($I = 0$) and do not show an NMR effect.
† a is the natural abundance of the isotope.
‡ ν_0 is the fundamental resonance frequency calculated from Equation (3.198) using $B_0 =$ 2.3487 T where B_0 is the strength of the applied magnetic field.

between these two states depends upon the *magnetogyric ratio*, γ, of the nucleus and the *strength*, B, of the *external magnetic field* **at the nucleus**. The frequency, v, of radiation required to flip the nuclear magnetic moment from being aligned with to being aligned against the external magnetic field is known as the *resonance frequency* and is given by

$$v = \gamma B / 2\pi \qquad\qquad (3.198)$$

showing that v is proportional to both γ and B. The properties of some important nuclei are given in Table 3.10. The energy differences corresponding to the values of v are very small (about $0.01-0.1\,\mathrm{J\,mol^{-1}}$) and so even prior to absorption the population of a low-energy state is only very slightly greater than that of the high-energy state.

For each nucleus in a molecule, the effects of its electronic environment and neighbouring nuclei cause B to be slightly different from the strength, B_0, of the applied magnetic field. The largest effects are those due to the electrons surrounding a nucleus. These electrons circulate under the influence of B_0, and in doing so they generate a local magnetic field which opposes B_0, thus *shielding* the nucleus from the external field. As the electron density at a nucleus increases, B (and hence v) decreases. Hence inductive effects are important in determining the shielding: when a group bonded to a nucleus donates electrons it is said to *shield* the nucleus. Alternatively, if it withdraws electrons it is said to *deshield* the nucleus. Additionally, circulation of π-electrons can have very significant effects upon B at adjacent nuclei. For example, $^1\mathrm{H}$ nuclei of benzene rings (which have C—C π-bonds) and of aldehyde groups (which have a C—O π-bond) are strongly deshielded by the magnetic fields arising from circulation of the π-electrons. It should now be evident that if nuclei of the same type (e.g. $^1\mathrm{H}$) in the same molecule do not have exactly identical local environments in the molecule, then they will have different B and will absorb (i.e. *magnetic resonance* will occur) at different frequencies. Although these differences in B are only of the order of parts per million of B_0, they are measurable and form the basis of NMR spectroscopy.

NMR spectra can be obtained either (i) by fixing B_0 and measuring the absorption of radiation over a range of v close to the fundamental resonance frequency, v_0, $(=\gamma B_0/2\pi)$, of the type of nuclei under study, or (ii) by fixing v at v_0 and measuring the absorption of this radiation as B_0 is varied over a small range. These methods are equivalent, though modern spectrometers invariably use method (i). A strongly shielded nucleus will absorb at low v (corresponding to low B) using method (i), or equivalently, at high B_0 using method (ii). In both cases the absorption is said to occur *upfield*. Similarly an absorption which due to deshielding effects occurs at relatively high v, or equivalently, at relatively low B_0, is said to be *downfield*. For both methods, the absorptions of the sample are recorded as chemical shifts from an absorption of a reference compound. The δ-scale of chemical shifts is universally used nowadays and does not depend upon the method used to record the spectrum. It is defined by

$$\delta = 10^6(v^{\text{sample}} - v^{\text{ref}})/v^{\text{ref}} \qquad (3.199a)$$

for measurements made using method (i), or by

$$\delta = 10^6(B_0^{\text{ref}} - B_0^{\text{sample}})/B_0^{\text{ref}} \qquad (3.199b)$$

for those made using method (ii) (N.B. the older τ-scale for ^{1}H NMR spectroscopy is related to the δ-scale by $\tau = 10 - \delta$). The (dimensionless) δ-scale is used because it is independent of B_0 and v_0, whereas the differences $v^{\text{sample}} - v^{\text{ref}}$ and $B_0^{\text{ref}} - B_0^{\text{sample}}$ increase as B_0 and v_0 increase. Thus NMR spectra with greatly improved resolution (i.e. separation) of the absorptions, recorded by using much stronger applied magnetic fields, can easily be compared with other spectra because the δ-scale is unaffected.

In NMR studies of polymers, ^{1}H and ^{13}C NMR spectroscopy are most widely used. Both employ tetramethylsilane (TMS), $Si(CH_3)_4$, as the reference compound. Since silicon is more electropositive than carbon, the hydrogen and carbon atoms in TMS give single ^{1}H and ^{13}C absorptions (corresponding to $\delta = 0$) that almost invariably are upfield of ^{1}H and ^{13}C absorptions of organic compounds. Thus ^{1}H and ^{13}C δ-values are normally positive and increase in the downfield (i.e. deshielding) direction. Some generalized correlations between chemical environment and chemical shift for ^{1}H and ^{13}C nuclei are shown in Fig. 3.30 and 3.31. These correlations illustrate the potential of NMR spectroscopy for evaluation of molecular structure and show that whilst most ^{1}H absorptions occur in the range $0 < \delta < 13$, the corresponding ^{13}C absorptions occur over the much wider range $0 < \delta < 220$. Thus ^{13}C NMR spectroscopy has the distinct advantage of having far greater resolution than ^{1}H NMR spectroscopy.

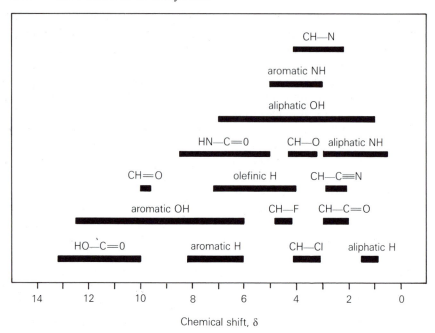

Fig. 3.30 *Characteristic ^1H NMR absorption regions.*

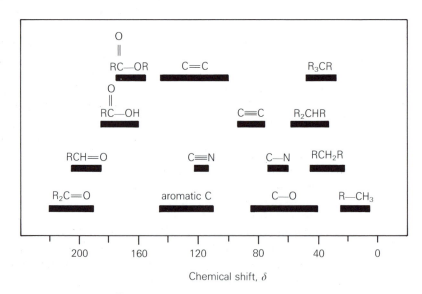

Fig. 3.31 *Characteristic ^{13}C NMR absorption regions.*

The effects, intimated earlier, of neighbouring nuclei upon NMR absorptions arise because nuclear spins interact through the intervening bonding electrons. In this interaction, known as *scalar spin–spin coupling* (referred to here as *J-coupling*), the magnetic moment of one nucleus is sensed by other nuclei in the same molecule and vice versa (though the interaction usually is significant only when the coupled nuclei are separated by no more than three bonds). This *J*-coupling causes an NMR absorption to be *split* into essentially symmetric multiple absorptions, the number of absorptions in most cases of interest being one greater than the number of neighbouring nuclei to which the nucleus is coupled and their separation being determined by the strength of the coupling. Hence, the splitting of an NMR absorption due to *J*-coupling provides valuable structural information. In ^{1}H NMR, *J*-coupling of atoms on adjacent carbon atoms (i.e. in the molecular fragment H—C—C—H) is of great importance for the elucidation of molecular structure. In most cases, ^{1}H—^{1}H *J*-splitting is small compared with the chemical shift differences between the different ^{1}H nuclei, thus enabling ^{1}H NMR spectra to be interpreted without difficulty. The low natural abundance of ^{13}C means that in ^{1}H NMR the effects of ^{1}H—^{13}C *J*-coupling are insignificant, as are the effects of ^{13}C—^{13}C *J*-coupling in ^{13}C NMR. However, ^{1}H—^{13}C *J*-coupling is significant in ^{13}C NMR but makes interpretation of ^{13}C spectra difficult because the splitting is strong and in many cases is greater than the chemical shift differences between the absorptions of the different ^{13}C nuclei. Thus when recording ^{13}C NMR spectra it is usual to employ techniques which *decouple* ^{13}C spins from ^{1}H spins so that the different ^{13}C nuclei each give rise to only a single absorption in the spectrum. Although this facilitates interpretation of the spectrum, the structural information available from splitting patterns is lost. However, this information can now be obtained without prejudicing interpretation due to recent advances in NMR techniques which enable *two-dimensional (2D NMR) spectra* to be recorded in which the chemical shifts of the different ^{13}C absorptions are shown on one axis and their splitting by ^{1}H—^{13}C *J*-coupling is shown on a second axis perpendicular to the first. In addition, by using other specialized techniques (e.g. 'DEPT') it is possible to obtain a ^{1}H–decoupled ^{13}C NMR spectrum from which it is easy to identify the numbers of hydrogen atoms bonded directly to each carbon atom giving rise to an absorption in the spectrum.

For quantitative studies, the areas under each of the absorptions are required. These areas are evaluated by an integrator and so are called *integrations*. Provided that the procedures for recording an NMR spectrum are chosen properly (Section 3.20.2), the ratios of the integrations for the different absorptions are equal to the ratios of the numbers of the respective nuclei present in the molecules, and can be used, for example, to evaluate relative molar compositions.

In this relatively brief introduction it has only been possible to give a general indication of the immense power of NMR spectroscopy for elucidation of molecular structures. In order to gain a deeper appreciation of its utility, up-to-date specialist texts on NMR spectroscopy should be consulted.

3.20.1 *NMR spectroscopy of polymers*

Many of the applications of NMR spectroscopy are the same as those of IR spectroscopy (Section 3.19.1). Thus both ^1H and ^{13}C NMR are widely used

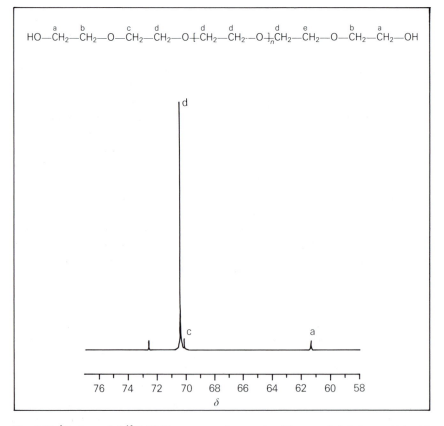

Fig. 3.32 *^1H–decoupled ^{13}C NMR spectrum of a sample of linear poly(ethylene oxide) with hydroxyl end-groups. The (weak) absorptions at δ 72.5, 70.1 and 61.2 are due to the ^{13}C atoms nearest the chain-ends, and are shifted from that of the main-chain ^{13}C atoms at δ 70.3 because of effects arising from the hydroxyl end-groups. From the ratio of the integrations for the 'end-group' absorptions to that for the main-chain absorption, the number-average molar mass of the poly(ethylene oxide) sample is calculated to be 1500 g mol^{-1} (courtesy of F. Heatley).*

for routine purposes such as *polymer identification*, confirmation of *molecular structure* and evaluation of *average copolymer composition*. NMR spectroscopy, however, offers much greater scope than IR spectroscopy for elucidating detailed features of molecular microstructure. This is especially true of ^{13}C NMR spectroscopy which nowadays is by far the most important technique for structural characterization of polymers.

NMR spectra of linear polymers of low molar mass often show unique absorptions due to the end-groups (Fig. 3.32). By referencing these absorptions to those of nuclei in the repeat units, it is possible to obtain the ratio of the number of end-groups to the number of repeat units and thereby to evaluate the *number-average molar masses* of such polymers.

For *branched polymers*, NMR absorptions due to the branch points can be identified and reveal the chemical structure of those branch points, thus leading to a better understanding of the mechanism(s) by which the branches form. It also is possible to calculate the degree of branching by referencing these absorptions to those of nuclei in the repeat units.

The power of NMR spectroscopy (especially ^{13}C NMR) is most clearly demonstrated by its ability to yield quantitative information on features of *molecular microstructure* not accessible by other techniques. These features include: (i) head-to-tail and head-to-head repeat unit linkages, (ii) the different types of repeat unit structures arising from 1,3-diene monomers, (iii) isotactic and syndiotactic sequences of repeat units (Fig. 3.33), and (iv) in copolymers, the sequence distributions of the different repeat units present along the copolymer chains.

Another important application of NMR spectroscopy is the study of *chain dynamics* by monitoring the magnetic relaxation behaviour of the excited nuclei. It is not possible here to give details of such measurements, but it should be noted that the motion of side groups and chain segments can be characterized, and that diffusion coefficients can be determined.

3.20.2 *Practical aspects*

All NMR spectrometers have large magnets which give highly homogeneous magnetic fields and commonly are specified in terms of the fundamental ^1H resonance frequency corresponding to the strength of the magnet, e.g. 100 MHz if $B_0 = 2.3\,T$ (Table 3.10). Early NMR spectrometers used either permanent magnets or electromagnets (e.g. 60 MHz and 100 MHz) and were of the *continuous-wave (CW) type* in which a single radiowave frequency was applied continuously whilst a small sweep coil slowly swept B_0 over an appropriate range. Modern NMR spectrometers commonly have superconducting magnets (e.g. 300 MHz and 500 MHz) and are of the *pulsed Fourier transform (PFT) type*. In PFT NMR, B_0 is fixed and a short pulse (e.g. a few microseconds) of radiowave frequencies covering the range of interest is applied to promote the nuclei to their

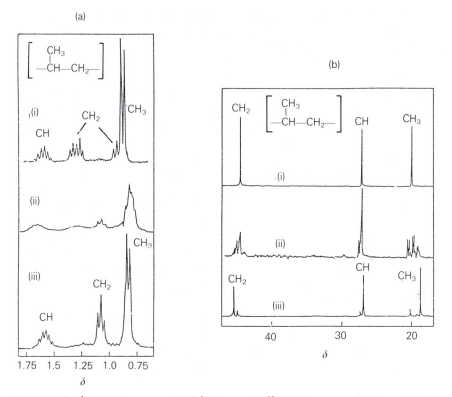

Fig. 3.33 *(a)* ^1H *NMR spectra and (b)* ^1H*–decoupled* ^{13}C *NMR spectra of (i) isotactic, (ii) atactic, and (iii) syndiotactic polypropylene (reproduced with permission from Tonelli, 'NMR Spectroscopy and Polymer Microstructure', VCH). The* ^1H *NMR spectra show multiplet absorptions due to* ^1H–^1H *spin–spin coupling (e.g. the* CH_3 ^1H *absorption is split by spin–spin coupling with the CH* ^1H *atoms and is observed as a doublet at* $0.9 > \delta > 0.8$*). The* ^{13}C *absorptions appear only as singlets because the* ^{13}C *NMR spectra were recorded under conditions of* ^1H *decoupling. The two sets of spectra clearly reveal the sensitivity of NMR absorptions to molecular microstructure and show that for polypropylene the* CH_2*, CH and* CH_3 *absorptions are sensitive to tacticity (in particular the* CH_2 ^1H *absorption, and the* CH_2 *and* CH_3 ^{13}C *absorptions).*

high-energy states. The simultaneous relaxation of these nuclei back to their low-energy states is detected as an interferogram, known as a *free induction decay* (FID), which is converted to the NMR spectrum by Fourier transformation. Usually this procedure is repeated many times allowing a suitable time interval (e.g. a few seconds) between successive pulses for acquisition of the FIDs. The data from each of these scans is accumulated to give the final NMR spectrum. The number of scans required to obtain a good spectrum depends upon the sensitivity of the

NMR effect. For example, a single scan may be adequate for ^{1}H NMR whereas several thousand scans often are required for ^{13}C NMR due to the low natural abundance of ^{13}C. In quantitative studies it is essential to ensure that the time interval between successive pulses is sufficient to ensure that all of the excited nuclei have relaxed back to their low-energy states before each subsequent pulse is applied. For ^{13}C NMR spectroscopy, ^{1}H-decoupling enhances the intensity of the ^{13}C absorptions due to an effect known as the *nuclear Overhauser effect (NOE)* and so in quantitative ^{13}C NMR studies the NOE enhancement must either be taken into account or suppressed by use of specialized techniques.

Polymer samples can be analysed in the form of dilute, semi-dilute or concentrated solutions, as solvent-swollen gels and also as solids. Polymer solutions and gels for ^{1}H and ^{13}C NMR studies are prepared using deuterated solvents such as deutero-chloroform ($CDCl_3$) and deutero-benzene (C_6D_6). With the general exception of dilute solutions, as the polymer concentration increases the absorptions become broader leading ultimately to loss of information such as that available from the splitting due to *J*-coupling. This can be particularly severe in spectra recorded from solid polymers. However, techniques have been developed which greatly suppress this broadening (e.g. dipolar decoupling with magic-angle spinning) and enable useful spectra to be recorded from solid polymers. Thus solid-state NMR spectroscopy is finding increasing use in the study of bulk polymers.

3.21 Other spectroscopic methods

In addition to IR and NMR spectroscopy, other spectroscopic methods which were developed for studies of low molar mass substances can be used to characterize polymers.

Ultraviolet and visible light absorption spectroscopy can be used to identify chromophores (e.g. benzene rings and carbonyl groups) and to determine the lengths of sequences of conjugated multiple bonds in polymers. It also can be used to analyse polymers for the presence of additives such as antioxidants or for detection of residual monomer(s). Additionally, fluorescence and phosphorescence techniques are important in studies of polymer photophysics.

Raman spectroscopy (Section 3.11) is another form of vibrational spectroscopy, but it complements IR spectroscopy in that bond vibration modes which are Raman-active tend to be IR-inactive, and vice versa. This is because the bond vibration must produce a change in polarizability for Raman absorption, but a change in dipole moment for IR absorption. Raman spectroscopy can be used to identify particular bonds and functional groups in the structure of a polymer in much the same way as IR spectroscopy. It can also be used to examine morphological features of bulk polymers and is especially useful for such studies of crystalline polymers. More recently it has been applied with great success to studies of the micromechanics of deformation of polymer fibres and composites.

Electron spin resonance (ESR) spectroscopy is similar to NMR spectroscopy except that the absorptions arise from the different alignments, again in an applied magnetic field, of the magnetic moments resulting from the spins of *unpaired electrons*. Thus ESR spectroscopy can be used to study species such as radicals, radical-anions and radical-cations. The splitting of an ESR absorption due to spin–spin coupling with magnetic nuclei (e.g. ^1H) in the molecular species containing the unpaired electron provides details of the molecular structure of that species. ESR spectroscopy finds use in the identification and study of reactive intermediates formed in polymerizations and during oxidation, high-energy irradiation, and thermal and mechanical degradation of polymers. Additionally, it can be used to study molecular dynamics.

Further reading

Abraham, R.J. and Loftus, P. (1979), *Proton and Carbon-13 NMR Spectroscopy*, Heyden, London.

Allen, G. and Bevington, J.C. (1989), *Comprehensive Polymer Science*, Vols 1 and 2, Pergamon Press, Oxford.

Billingham, N.C. (1977), *Molar Mass Measurements in Polymer Science*, Kogan Page, London.

Bovey, F.A. (1988), *Nuclear Magnetic Resonance Spectroscopy*, 2nd edn, Academic Press, New York.

Bower, D.I. and Maddams, W.F. (1989), *The Vibrational Spectroscopy of Polymers*, Cambridge University Press, Cambridge.

Colthrup, N.B., Daly, L.H. and Wiberley, S.E. (1975), *Introduction to Infrared and Raman Spectroscopy*, 2nd edn, Academic Press, New York.

Elias, H-G. (1984), *Macromolecules*, Vol. 1, 2nd edn, John Wiley, New York.

Flory, P.J. (1953), *Principles of Polymer Chemistry*, Cornell University Press, Ithaca.

Klöpffer, W. (1984), *Introduction to Polymer Spectroscopy*, Springer-Verlag, Berlin.

Lambert, J.B., Shurvell, H.F., Lightner, D.A. and Cooks, R.G. (1987), *Introduction to Organic Spectroscopy*, Macmillan, New York.

Mark, H.F., Bikales, N.M., Overberger, C.G. and Menges, G. (eds) (1985–89), *Encyclopedia of Polymer Science and Engineering*, Wiley-Interscience, New York.

Morawetz, H. (1975), *Macromolecules in Solution*, 2nd edn, John Wiley, New York.

Randall, J.C. (1977), *Polymer Sequence Determination: Carbon-13 NMR Method*, Academic Press, New York.

Tanford, C. (1961), *Physical Chemistry of Macromolecules*, John Wiley, New York.

Tonelli, A.E. (1989), *NMR Spectroscopy and Polymer Microstructure: the Conformational Connection*, VCH Publishers, New York.

Wehrli, F.W., Marchand, A.P. and Wehrli S. (1988), *Interpretation of Carbon-13 NMR Spectra*, 2nd edn, John Wiley, New York.

Problems

3.1 Three mixtures were prepared from three monodisperse polystyrene samples (with molar masses of $10\,000$, $30\,000$, and $100\,000\,g\,mol^{-1}$) as indicated below:

(a) equal numbers of molecules of each sample,
(b) equal masses of each sample, and
(c) by mixing in the mass ratio $0.145 : 0.855$ the two samples with molar masses of $10\,000$ and $100\,000\,g\,mol^{-1}$.

For each of the mixtures calculate the number-average, $\overline{M}_n$, and weight-average, $\overline{M}_w$, molar mass and comment upon the meaning of the $\overline{M}_w/\overline{M}_n$ values.

3.2 The hydroxyl end-groups of a sample ($2.00\,g$) of linear poly(ethylene oxide) were acetylated by reaction with an excess of acetic anhydride ($2.50 \times 10^{-3}\,mol$) in pyridine:

$$R-OH + CH_3-\overset{\overset{O}{\|}}{C}-O-\overset{\overset{O}{\|}}{C}-CH_3 \rightarrow R-O-\overset{\overset{O}{\|}}{C}-CH_3 + CH_3-\overset{\overset{O}{\|}}{C}-OH$$

Water was then added to convert the excess acetic anhydride to acetic acid, which together with the acetic acid produced in the acetylation reaction was neutralized by addition of $23.30\,cm^3$ of a $0.100\,mol\,dm^{-3}$ solution of sodium hydroxide. Calculate $\overline{M}_n$ for the sample of poly(ethylene oxide) given that each molecule has two hydroxyl end-groups.

3.3 Starting from the Flory–Huggins Equation (3.22), show that for the formation of a solution from a monodisperse polymer the partial molar Gibbs free energy of mixing, $\overline{\Delta G}_1$ for the solvent is given by

$$\overline{\Delta G}_1 = \mathbf{R}T\left[\ln(1 - \phi_2) + \left(1 - \frac{1}{x}\right)\phi_2 + \chi\phi_2^2\right]$$

where ϕ_2 is the volume fraction of polymer, x is the ratio of the molar volume of the polymer to that of the solvent, and χ is the polymer–solvent interaction parameter.

Use this equation to calculate an estimate of χ for solutions of natural rubber ($\overline{M}_n = 2.5 \times 10^5\,g\,mol^{-1}$) in benzene given that vapour pressure measurements showed that the activity of the solvent in a solution with $\phi_2 = 0.250$ was 0.989.

3.4 The radial distribution function $W(r)$ of end-to-end distances r for an isolated flexible polymer chain is given by

$$W(r) = 4\pi(\beta/\pi^{1/2})^3 r^2 \exp(-\beta^2 r^2)$$

where $\beta = [3/(2nl^2)]^{1/2}$ in which n is the number of links of length l forming the chain.

Show by integration that this distribution function normalizes to unity, and derive equations for (a) the most probable value of r, (b) the root-mean-square value of r, and (c) the mean value of r.

Note that for $I(b) = \int_0^\infty x^b \exp(-ax^2)\mathrm{d}x$

$$I(0) = \left(\frac{1}{2}\right)\pi^{1/2}a^{-1/2}$$

$$I(1) = \left(\frac{1}{2}\right)a^{-1}$$

$$I(2) = \left(\frac{1}{4}\right)\pi^{1/2}a^{-3/2}$$

$$I(3) = \left(\frac{1}{2}\right)a^{-2}$$

$$I(4) = \left(\frac{3}{8}\right)\pi^{1/2}a^{-5/2}$$

3.5 For a linear molecule of polyethylene of molar mass $119\,980$ g mol^{-1} calculate:

(a) the contour length of the molecule,
(b) the end-to-end distance in the fully-extended molecule, and
(c) the root-mean-square end-to-end distance according to the valence angle model.

In the calculations, end-groups can be neglected and it may be assumed that the C—C bonds are of length 0.154 nm and that the valence angles are $109.5°$. Comment upon the values obtained. Indicate, giving your reasoning, which of the very large number of possible conformations of the molecule is the most stable.

3.6 Equilibrium osmotic pressure heads (h) measured at 22°C for a series of solutions of a polystyrene sample in toluene are tabulated below.

Polystyrene concentration/g dm^{-3}	2.6	5.2	8.7	11.8	15.1
h/mm toluene	5.85	12.93	24.37	36.34	50.99

Assuming that the densities of toluene and polystyrene at 22°C are respectively 0.867 g cm^{-3} and 1.05 g cm^{-3}, calculate $\bar{M}_n$ for the sample of polystyrene and an estimate of the polystyrene–toluene interaction parameter.

3.7 Measurements of the equilibrium resistance difference (ΔR) between the solution and solvent thermistor beads in a vapour pressure osmometer were made under identical conditions for solutions in chloroform of (i) benzil and (ii) a sample of poly(propylene oxide). The results are given in the table below, in which c_B is the concentration of benzil and c_P is the concentration of poly(propylene oxide). Calculate $\overline{M}_n$ for the sample of poly(propylene oxide).

c_B/mol dm^{-3}	0.030	0.070	0.100	0.200
ΔR/ohm	11.541	26.819	38.183	75.481
c_P/g dm^{-3}	6.12	17.52	25.06	50.54
ΔR/ohm	1.246	4.135	5.920	15.130

3.8 Rayleigh ratios (R_θ) were obtained at 25°C for a series of solutions of a polystyrene sample in benzene, with the detector situated at various angles (θ) to the incident beam of unpolarized monochromatic light of wavelength 546.1 nm. The results are tabulated below.

Polystyrene concentration (g dm^{-3})	$10^4 \times R_\theta$/m^{-1} measured at $\theta =$			
	30°	60°	90°	120°
0.50	72.3	69.4	66.2	64.3
1.00	89.8	85.7	81.3	78.2
1.50	100.8	97.1	92.0	88.1
2.00	108.7	103.8	99.7	95.9

Under the conditions of these measurements the Rayleigh ratio and refractive index of benzene are 46.5×10^{-4} m^{-1} and 1.502 respectively, and the refractive index increment for the polystyrene solutions is 1.08×10^{-4} dm^3 g^{-1}.

Using a Zimm plot, determine: (a) the weight-average molar mass ($\overline{M}_w$) of the polystyrene sample, (b) the z-average radius of gyration ($\langle s^2 \rangle_z^{1/2}$) of the polystyrene molecules in benzene at 25°C, and (c) the second virial coefficient (A_2) for the polystyrene–benzene interaction at 25°C.

3.9 The tables below give the mean flow times (t) in a suspended-level viscometer recorded for solutions of two of five monodisperse samples of polystyrene at various concentrations (c) in cyclohexane at 34°C. Under these conditions, the mean flow time (t_0) for cyclohexane is 151.8 s.

	Sample B		Sample E	
$10^3 \times c$/g cm^{-3}	t/s	$10^3 \times c$/g cm^{-3}	t/s
1.586	158.5	1.040	176.1
5.553	178.6	1.691	195.4
7.403	189.6	2.255	214.8
8.884	199.5	2.706	232.4

Determine the intrinsic viscosities, $[\eta]$, of these samples. The intrinsic viscosities of the other three polystyrene samples were evaluated under the same conditions and are given in the following table together with $\bar{M}_w$ values determined by light scattering.

Polystyrene sample	$\bar{M}_w/\text{g mol}^{-1}$	$[\eta]/\text{cm}^3\,\text{g}^{-1}$
A	37 000	15.77
B	102 000	–
C	269 000	42.56
D	690 000	68.12
E	2 402 000	–

Using these data together with the calculated values of $[\eta]$ for samples B and E, evaluate the constants of the Mark–Houwink equation for polystyrene in cyclohexane at 34°C.

What can you deduce about the conformation of the polystyrene chains under the conditions of the viscosity determinations?

3.10 A polymer solution was cooled very slowly until phase-separation took place to give two phases in equilibrium. Subsequent analysis of the phases showed that the volume fractions of polymer in the phases were $\phi_2'' = 0.89$ and $\phi_2' = 0.01$ respectively. Using the equation for ΔG_1 given in Problem 3.3, together with the equilibrium condition $\Delta G_1'' = \Delta G_1'$, calculate an estimate (2 significant figures) of the polymer–solvent interaction parameter for the conditions of phase separation.

3.11 Polystyrene and poly(methyl methacrylate) samples of narrow molar mass distribution were analysed under identical conditions using a high-performance gel permeation chromatography system. The elution time for one of the poly(methyl methacrylate) samples was found to be the same as that for the polystyrene sample of molar mass $97\,000\,\text{g mol}^{-1}$. Evaluate the molar mass of this poly(methyl methacrylate) sample given that for the solvent and temperature used the Mark–Houwink constants are $K = 1.03 \times 10^{-2}\,\text{cm}^3\,\text{g}^{-1}\,(\text{g mol}^{-1})^{-a}$ with $a = 0.74$ for polystyrene and $K = 5.7 \times 10^{-3}\,\text{cm}^3\,\text{g}^{-1}\,(\text{g mol}^{-1})^{-a}$ with $a = 0.76$ for poly(methyl methacrylate). State clearly any assumptions you make in your calculation.

3.12 State the methods you would use to determine:

(a) $\bar{M}_n$ for samples of poly(ethylene glycol) with values in the range 4×10^2 to $5 \times 10^3\,\text{g mol}^{-1}$;

(b) $\bar{M}_n$ for samples of polyacrylonitrile with values in the range 5×10^4 to $2 \times 10^5\,\text{g mol}^{-1}$, and

(c) the polydispersity indices of samples of polystyrene with molar masses in the range 10^4 to $10^6\,\text{g mol}^{-1}$.

For each method, give the reasons for your choice, name a solvent that would be suitable for the measurements, and discuss briefly possible errors in the determinations.

4 Structure

4.1 **Polymer crystals**

It has been recognized for many years that polymer molecules possess the ability to crystallize. The extent to which this occurs varies with the type of polymer and its molecular microstructure. The main characteristic of crystalline polymers that distinguishes them from most other crystalline solids is that they are normally only *semi-crystalline*. This is self-evident from the fact that the density of a crystalline polymer is normally between that expected for fully crystalline polymer and that of amorphous polymer. Also X-ray diffraction patterns from melt-crystallized polymers are usually in the form of rings superimposed on a diffuse background as shown in Fig. 4.1. This background indicates the presence of a non-crystalline (amorphous) phase and the rings indicate that there is a second phase consisting of small randomly oriented crystallites also present. The degree of crystallinity and the size and arrangement of the crystallites in a semi-crystalline polymer have a profound effect upon the physical and mechanical properties of the polymer and in order to explain these properties fully it is essential to have a clear understanding of the nature of these crystalline regions.

4.1.1 *Crystallinity in polymers*

The crystallization of polymers is of enormous technological importance. Many thermoplastic polymers will crystallize to some extent when the molten polymer is cooled below the melting point of the crystalline phase. This is a procedure which is done repeatedly during polymer processing and the presence of the crystals has an important effect upon polymer properties. There are many factors which can affect the rate and extent to which crystallization occurs for a particular polymer. They can be processing variables such as the rate of cooling, the presence of orientation in the melt and the melt temperature. Other factors include the tacticity and molar mass of the polymer, the amount of chain branching and the presence of any additives such as nucleating agents.

The obvious questions to ask concerning crystallinity in polymers are why and how do polymer molecules crystallize? The answer to the question why is given by consideration of the thermodynamics of the crystallization

Fig. 4.1 *Flat-plate X-ray diffraction pattern obtained from an isotropic sample of melt-crystallized polypropylene.*

process. The Gibbs free energy G of any system is related to the enthalpy H and the entropy S by the equation

$$G = H - TS \qquad (4.1)$$

where T is the thermodynamic temperature. The system is in equilibrium when G is a minimum. A polymer melt consists of randomly coiled and entangled chains. This gives a much higher entropy than if the molecules are in the form of extended chains since there are many more conformations available to a coil than for a fully extended chain. The higher value of S leads to a lower value of G. Now if the melt is cooled to a temperature below the melting point of the polymer, T_m, crystallization may occur. There is clearly a high degree of order in polymer crystals as in any other crystalline materials and this ordering leads to a considerable reduction in entropy, S. This entropy penalty is, however, more than offset by the large reduction in enthalpy that occurs during crystallization. If the magnitude of the enthalpy change ΔH_m (latent heat) is greater than that of the product of the melting temperature and the entropy change $(T_m \Delta S_m)$ crystallization will be favoured thermodynamically since a lower value of G will result.

As with all applications of thermodynamics, it can only be strictly applied to processes which occur quasi-statically, that is very slowly. Polymers are often cooled rapidly from the melt, particularly when they are being processed industrially. In this situation crystallization is controlled by kinetics and the rate at which the crystals nucleate and grow becomes important. With many crystallizable polymers it is possible to cool the melt so rapidly that crystallization is completely absent and an amorphous glassy polymer results. With these systems crystallization can normally be induced by annealing the amorphous polymer at a temperature between the glass transition temperature and the melting-point, T_m, of the crystals.

The second question that was raised is how do polymer molecules crystallize? A useful way of imagining a molten polymer is as an enormous bowl of tangled wriggling spaghetti. The actual strands of polymer spaghetti would be very much longer than those of the edible pasta. They would have a length to diameter ratio typically of 10 000 to 1 rather than about 100 to 1 for spaghetti. It is quite remarkable that such a random mass could crystallize to give an ordered structure and clearly significant molecular rearrangement and cooperative movement is required for crystallization to occur.

Well-defined examples of polymer crystals can be found on crystallization from dilute polymer solutions. Small isolated lamellar crystals are obtained although detailed measurements have shown that they are still only semi-crystalline. It is thought that in this case the molecules are folded and the fold regions give rise to the non-crystalline component in the crystals. There is a higher degree of perfection in solution-grown polymer crystals than in the melt-crystallized counterparts. The polymer molecules are thought to be in the form of isolated coils in dilute solution and crystallization is not hindered by factors such as entanglements.

Single crystals of many materials occur naturally (e.g. quartz and diamond). For other materials such as metals and semi-conductors the growth of single crystals from the melt is now a routine matter. With polymers it is only recently that true single crystals have been prepared. This can only be done by using the process of solid-state polymerization outlined in Section 2.11. The most perfect crystals are obtained with certain substituted diacetylene monomers which polymerize topochemically to give polymer crystals which can be 100% crystalline and are of macroscopic dimensions.

4.1.2 *Determination of crystal structure*

Crystalline solids consist of regular three-dimensional arrays of atoms. In polymers, the atoms are joined together by covalent bonds along the macromolecular chains. These chains pack together side-by-side and lie along one particular direction in the crystals. It is possible to specify the structure of any crystalline solid by defining a regular pattern of atoms which is repeated in the structure. This repeating unit is known as the *unit cell* and the crystals are made up of stacks of the cells. In atomic solids such as metals the atoms lie in simple close-packed arrays (like billiard balls) and the structure can be specified by defining simple cubic or hexagonal unit cells containing only a few atoms. The packing units in molecular solids such as polymer crystals are more complicated than in atomic solids. In this case the unit cells are made up of the repeating segments of the polymer chains packed together, often with several segments in each unit cell. Depending upon the complexity of the polymer chain there can be tens or even hundreds of atoms in the unit cell. The spatial arrangement of the atoms is controlled by the covalent bonding within a particular molecular segment, with the polymer segments held together in the crystals by secondary van der Waals or hydrogen bonding. Since the polymer chains lie along one particular direction in the cell and there is only relatively weak secondary bonding between the molecules, the crystals have very anisotropic physical properties.

When the crystal structure of a molecular compound is analysed both the relative positions of the atoms on the molecular repeat units and the arrangement of these segments in the unit cell must be determined. This three-dimensional structure is normally determined using X-ray diffraction, involving measurement of the positions and intensities of all the X-ray maxima from a single crystal sample. Computer-controlled, four-circle diffractometers are now available which allow the relative positions of all the atoms in crystals with quite complex structures to be determined as a matter of routine in a period of a few days, but such machines required good, relatively large crystals.

For most conventional polymers, large single crystals are not available, but the complete crystal structures of several polydiacetylenes polymerized as macroscopic single crystals in the solid state (Section 2.11) have been determined recently. The degree of perfection shown by these polydiacetylene single crystals can be seen in Fig. 4.2. This is a single crystal rotation photograph obtained by using a cylindrical film with the chain direction as the oscillation axis. The diffraction pattern is obtained by rotating the crystal about one particular axis and is in the form of spots lying along parallel layer lines. The lattice parameter in the direction of the rotation axis of the crystal can be determined directly from the layer line spacing. Since the rotation axis coincides with the chain direction then this is, of course, the chain direction repeat conventionally indexed for polymers as c in the unit cell. If the layer lines are indexed $l = 0$ for the equatorial line and $l = 1$ for the first layer etc., then c is simply related to the height, h_l, of the lth layer line above the equatorial line. The crystal lattice can be considered as acting as a diffraction grating as shown in Fig. 4.3(a). The layer lines are then due to discrete differences in path length, $l\lambda$, where λ is the wavelength of the radiation. The angle of deviation of the beam, ϕ, is therefore given

$$\sin \phi = l\lambda/c \qquad (4.2)$$

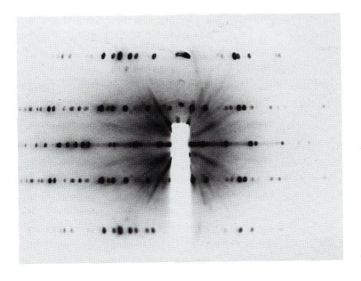

Fig. 4.2 *Single-crystal X-ray oscillation photograph obtained from a polydiacetylene single crystal. The chain direction is vertical.*

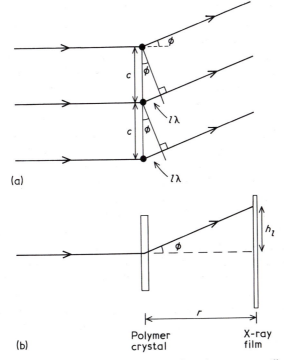

(a)

(b)   Polymer   X-ray
             crystal   film

Fig. 4.3 *Schematic illustration of the formation of the layer lines on an oscillation photograph.*
(a) *Diffraction by lattice in one dimension.* (b) *Experimental arrangement.*

The angle ϕ is related to h_l and the radius of the film, r, through the simple geometrical construction shown in Fig. 4.3(b). This leads to the relation

$$\tan \phi = h_l/r \tag{4.3}$$

Combining Equations (4.2) and (4.3) gives the lattice parameter c as

$$c = \frac{l\lambda}{\sin \tan^{-1}(h_l/r)} \tag{4.4}$$

Macroscopic single crystals are a special case and large single crystals of conventional polymers are not available. Crystal-structure determinations for semi-crystalline polymers are less precise and the sophisticated methods of crystal-structure determination cannot be used without modification. Instead of using single crystals, samples are normally prepared in the form of highly oriented, drawn or rolled fibres. They can be analysed by obtaining rotation photographs, but flat film X-ray patterns are more usually obtained. A flat film pattern of oriented polypropylene is shown in Fig. 4.4. This type of X-ray photograph is often called a fibre pattern and they are characteristic of oriented fibres of semicrystalline polymers. They are analogous to rotation photographs, but the layer lines are in the form of hyperbolae because the film is flat rather than cylindrical. It is not necessary to rotate the polymer fibres because they are polycrystalline with a spread of crystal orientation about the fibre axis. In polymer fibres the polymer molecules are oriented approximately parallel to the fibre axis and since this is parallel to a crystal axis (normally defined as c), the fibre pattern is equivalent to a rotation photograph about this crystallographic axis for a single crystal sample. It is possible to analyse the layer lines in fibre patterns in a similar way to rotation photographs if certain precautions are taken. Equation (4.4) can only be applied when h_l is measured directly above and below the central spot and if r is taken as the specimen-to-film distance rather than the film radius. Since the chain direction normally coincides with the fibre axis in oriented polymer samples the dimensions of the c axis in the unit cell can be found directly from the height of the lth layer line.

In order to determine other unit cell dimensions it is necessary to assign the crystallographic indices (hkl) to each spot in the fibre pattern. This is often difficult to do for single crystal patterns and there are even greater problems with semi-crystalline polymers. The reflections tend to be in the form of arcs rather than spots because the orientation of the crystals is somewhat imperfect. Also they tend to be rather diffuse because the dimensions of the crystallites are usually small (a few hundred Å) and they contain many lattice imperfections. It is sometimes possible to improve the resolution of the diffraction patterns by using doubly-oriented samples. These are tapes which have been rolled as well as stretched. They can have

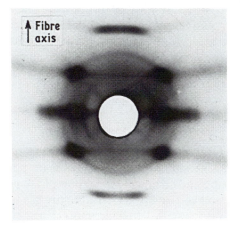

Fibre
axis

Fig. 4.4 *Flat-plate X-ray diffraction pattern obtained from a sample of oriented polypropylene.*

a three-dimensional texture rather than the normal cylindrical symmetry possessed by fibres.

Once the position and intensities of all the (*hkl*) reflections have been measured as accurately as sample perfection allows, the polymer crystallographer does not normally have sufficient information to determine the crystal structure of the polymer and so he has to resort to a series of educated guesses as to what this structure may be. There is now a considerable amount of experience in this area and some empirical rules have been noted to help with structure determination. The polymer chains are normally in their lowest energy conformations within the crystal as the bonding in the crystal is not sufficiently strong to perturb the nature of the chain. The chains pack into the crystal so as to fill space as efficiently as possible. The van der Waals' radii of the pendant atoms on the chain limit the distance of closest approach of each chain.

It is also possible to call upon other information which is generally available. The structures of chemically similar polymers are often known and these are often useful as a starting-point for structure determination. The stereochemical nature of the polymer chains depends upon the method of synthesis (Section 2.8) and knowledge of this is important. Spectroscopic techniques (Sections 3.19–3.20) give the detailed microstructure of the polymer molecules and information on the conformation and packing of the chains within the crystals. Also it is possible to determine the lowest-energy crystal structure by taking into account inter- and intramolecular interactions for different structural models.

Once a promising structure has been found it is necessary to compare the measured positions and intensities of the (*hkl*) reflections with those predicted for the structure that is proposed. The agreement that is obtained is never perfect and normally one has to refine the proposed structure until a structure with the best fit to the measured data is obtained. This means that the crystal structures that are quoted below tend to be somewhat idealized and in many cases better fits are almost certainly possible.

4.1.3 *Polymer crystal structures*

The crystal structures of several hundred polymer molecules have now been determined. The structure of a few common polymers are listed in Table 4.1. The name and chemical repeat unit of the polymer are given in column one. The unit cell of any crystal structure can be assigned to one of the seven basic crystal systems (triclinic, monoclinic, etc.) and this is given as the first entry in column two. The packing and symmetry of the polymer segments in the unit cell allows the structure to be assigned to one of the 230 space groups* which are given as the second entry in column two. Polymer molecules pack into crystals either in the form of zig-zags or helices and this is also indicated in column two. The nomenclature used to describe the zig-zag or helix is $A*u/t$, where A represents the class of the helix given by the number of skeletal atoms in the asymmetric unit of the chain. The parameter u represents the number of these units on the helix corresponding to the crystallographic (chain direction) repeat and t is the number of turns of the helix in this crystallographic repeat. This nomenclature can be used to describe both helices and zig-zags. For example, if polyethylene is considered as polymethylene $(-CH_2-)_n$ the $1*2/1$ helix represents a planar zig-zag. Alternatively it could be considered as $(-CH_2-CH_2-)_n$ and the zig-zag chain conformation would be designated $2*1/1$. The unit cell dimensions (Å) are listed in column three in the order a, b, c, and the chain axis is indicated by an asterisk. The angles between the axes of the unit cell α, β and γ are given in order in column four. The number of chain repeat units per unit cell are given in column five and the densities of the crystals determined from the crystal structure are given in the final column.

4.1.4 *Factors determining crystal structure*

There are certain structural requirements that are essential before a polymer molecule can crystallize. It is necessary that the polymer chains must be linear although a limited number of branches on a polymer chain tend to limit the extent of crystallization but do not stop it completely. In a

---

*See International Tables for X-ray crystallography, Vol. 1.

TABLE 4.1 *Details of the crystal structures of various common polymers (after Wunderlich, reproduced with permission)*

Macromolecule	Crystal System Space group Mol. Helix	Unit cell axes	Unit cell angles	No. units	ρ_c $(g\,cm^{-3})$
Polyethylene I —CH$_2$—	Orthorhombic Pnam 1*2/1	7.418 4.946 2.546*	90° 90° 90°	4	0.9972
Polyethylene II —CH$_2$—	Monoclinic C2/m 1*2/1	8.09 2.53* 4.79	90° 107.9° 90°	4	0.998
Polytetrafluoroethylene I —CF$_2$—	Triclinic P1 1*13/6	5.59 5.59 16.88*	90° 90° 119.3°	13	2.347
Polytetrafluoroethylene II —CF$_2$—	Trigonal P3$_1$ or P3$_2$ 1*15/7	5.66 5.66 19.50*	90° 90° 120°	15	2.302
Polypropylene (iso) —CH$_2$—CHCH$_3$—	Monoclinic P2$_1$/c 2*3/1	6.66 20.78 6.495*	90° 99.62° 90°	12	0.946
Polystyrene (iso) —CH$_2$—CHC$_6$H$_5$—	Trigonal R3c 2*3/1	21.9 21.9 6.65*	90° 90° 120°	18	1.127
Polypropylene (Syndio) —CH$_2$—CHCH$_3$—	Orthorhombic C222$_1$ 4*2/1	14.50 5.60 7.40*	90° 90° 90°	8	0.930
Poly(vinyl chloride) (Syndio) —CH$_2$—CHCl—	Orthorhombic Pbcm 4*1/1	10.40 5.30 5.10*	90° 90° 90°	4	1.477
Poly(vinyl alcohol) (atac) —CH$_2$—CHOH—	Monoclinic P2/m 2*1/1	7.81 2.51* 5.51	90° 91.7° 90°	2	1.350
Poly(vinyl fluoride) (atac) —CH$_2$—CHF—	Orthorhombic Cm2m 2*1/1	8.57 4.95 2.52*	90° 90° 90°	2	1.430
Poly(4-methyl-1-pentene) (iso) —CH$_2$—CH— | CH$_2$—CH(CH$_3$)$_2$	Tetragonal P4̄ 2*7/2	20.3 20.3 13.8*	90° 90° 90°	28	0.822
Poly(vinylidene chloride) —CH$_2$—CCl$_2$—	Monoclinic P2$_1$ 4*1/1	6.73 4.68* 12.54	90° 123.6° 90°	4	1.957
1,4-Polyisoprene (*cis*) —CH$_2$—CCH$_3$=CH—CH$_2$—	Orthorhombic Pbac 8*1/1	12.46 8.86 8.1*	90° 90° 90°	8	1.009
1,4-Polyisoprene (*trans*) —CH$_2$—CCH$_3$=CH—CH$_2$—	Orthorhombic P2$_1$2$_1$2$_1$ 4*1/1	7.83 11.87 4.75*	90° 90° 90°	4	1.025

Macromolecule	Crystal System Space group Mol. Helix	Unit cell axes	Unit cell angles	No. units	ρ_c (g cm^{-3})
Polyoxymethylene I	Trigonal	4.471	90°	9	1.491
—CH$_2$—O—	P3$_1$ or P3$_2$	4.471	90°		
	2*9/5	17.39*	120°		
Polyoxymethylene II	Orthorhombic	4.767	90°	4	1.533
—CH$_2$—O—	P2$_1$2$_1$2$_1$	7.660	90°		
	2*2/1	3.563*	90°		
Poly(ethylene terephthalate)	Triclinic	4.56	98.5°	1	1.457
—(CH$_2$)$_2$—O—CO—C$_6$H$_4$—	P$\bar{1}$	5.96	118°		
—CO—O—	12*1/1	10.75*	112°		
Nylon 6, α	Monoclinic	9.56	90°	8	1.235
—(CH$_2$)$_5$—CO—NH—	P2$_1$	17.24*	67.5°		
	7*2/1	8.01	90°		
Nylon 6, γ	Monoclinic	9.33	90°	4	1.163
—(CH$_2$)$_5$—CO—NH—	P2$_1$/a	16.88*	121°		
	7*2/1	4.78	90°		
Nylon 6.6, α	Triclinic	4.9	48.5°	1	1.24
—(CH$_2$)$_6$NH—CO—(CH$_2$)$_4$—	P$\bar{1}$	5.4	77°		
—CO—NH—	14*1/1	17.2*	63.5°		
Nylon 6.6, β	Triclinic	4.9	90°	2	1.25
—(CH$_2$)$_6$NH—CO—(CH$_2$)$_4$—	P$\bar{1}$	8.0	77°		
—CO—NH—	14*1/1	17.2*	67°		

similar way copolymerization affects the perfection of a polymer chain, but it can be tolerated to a limited extent. One of the most important factors which determines the extent of crystallization and crystal structure is tacticity. In general, isotactic and syndiotactic polymers will crystallize whereas atactic polymers are non-crystalline, although there are notable exceptions to this rule. The most useful way of considering the factors controlling the structure of crystalline polymers is to look at specific examples.

The most widely studied crystalline polymer is *polyethylene*. The lowest-energy chain conformation of this macromolecule can be evaluated from considerations of the conformations of the n-butane molecule (Fig. 3.5). A series of all *trans* bonds produce the lowest-energy conformation in n-butane because of steric interactions and this is reflected in polyethylene where the most stable molecular conformation is the crystal with all *trans* bonds, i.e. the planar zig-zag. These molecules then pack into the orthorhombic unit cell given in Table 4.1. The arrangement of the molecules in the polyethylene crystal structure is illustrated in Fig. 4.5 with the polymer molecules lying parallel to the *c* axis. The molecules are held

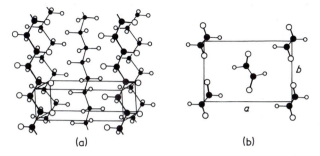

Fig. 4.5 *Crystal structure of orthorhombic polyethylene.* (a) *General view of unit cell.* (b) *Projection of unit cell parallel to the chain direction, c.* ($\bullet$ . . . *carbon atoms*, $\bigcirc$ . . . *hydrogen atoms*).

in position in the cell by the secondary van der Waals bonds between the chain segments. The interactions between the H atoms on the polymer chain determine the setting angle of the molecules in the cell. This is the angle that the molecular zig-zags make with the *a* or *b* axes. Orthorhombic polyethylene (I) is the most stable crystal structure but a monoclinic modification (II) can be formed by mechanical deformation of orthorhombic polyethylene. The molecules are still in the form of a planar zig-zag and the chain direction repeat and crystal density are virtually unchanged. The two structures differ just in the way the molecules are packed in the unit cells. Polymorphism, due to deformation or caused by different crystallization conditions, is not uncommon in polymers and other examples will be discussed below.

The copolymerization of ethylene with small amounts of propylene produces a branched form of polyethylene which is still capable of crystallization. The orthorhombic crystal structure is maintained and the *b* and *c* axes are almost unchanged. On the other hand, the *a* crystal axis undergoes an expansion which depends upon the proportion of —CH_3 groups on the polymer chain.

Polytetrafluoroethylene exists in at least two crystal modifications. The low-temperature form I is stable below 19°C and above this temperature the trigonal II form is found. The polymer can be considered as an analogue of polyethylene and it would undoubtedly have a similar crystal structure if the F atoms were not just too large to allow a planar zig-zag to be stable. The F atoms are significantly larger than H atoms and there is a considerable amount of steric repulsion when (—CF_2—)$_n$ is in the form of a planar zig-zag. Because of this the molecule adopts a helical conformation which allows the larger F atoms to be accommodated. Below 19°C the molecules are in the form of a 13/6 helix and at higher temperatures they untwist slightly into a 15/7 helix. The two types of polytetrafluoroethylene helices have very smooth molecular profiles and can be considered as cylinders, as building molecular models will confirm. This smooth profile has been suggested as being responsible for the good low-friction properties of this polymer. The molecular cylinders pack very efficiently into the crystal structures in a hexagonal array, rather like a stack of pencils, and give rise to unit cells which have hexagonal dimensions (Table 4.1). But they are not assigned hexagonal crystal structures because the twisted molecules do not have a sufficiently high degree of symmetry. *Poly(vinyl fluoride)* is a related polymer which only has 25% of the F atoms possessed by polytetrafluoroethylene. Consequently it does not suffer from the steric repulsion of the pendant atoms and even the atactic form adopts a crystal structure similar to polyethylene I (Table 4.1). In fact the *b* and *c* unit cell dimensions are virtually identical to those of polyethylene, but the *a* repeat is considerably larger, to accommodate the bigger F atoms, as for the ethylene–propylene copolymers.

It is found that atactic vinyl polymers ($-CH_2-CHX-$)$_n$ will not crystallize when X is large. We have just seen that when X = F crystallization can still take place. This is also the case for atactic *poly(vinyl alcohol)*. Since the $-OH$ group is relatively small the molecules crystallize in the form of a planar zig-zag into a distorted monoclinic version of the polyethylene structure. In general it is found that vinyl polymers must be either isotactic or syndiotactic before crystallization can occur. Isotactic vinyl polymers invariably crystallize with the polymer molecules in a helical conformation to accommodate the relatively large side groups. For example, the molecules in the isotactic forms of both *polystyrene* and *polypropylene* are in the form of a 3/1 helix as illustrated schematically in Fig. 4.6. This helix is readily formed by having alternating *trans* and *gauche* positions along the polymer chain. If the side groups are large and bulky they require more space than is available in the 3/1 helix and the molecules form looser helices.

The packing of the helices in isotactic vinyl polymers depends upon the type of helix present and the nature and size of the side groups. The packing is dictated by the intermeshing of the side groups, but it is also complicated by the possibility of having two different kinds of helix both of which can be either right- or left-handed. The trigonal crystal structure of isotactic polystyrene (Table 4.1) is typical of many 2*3/1 helices whereas the monoclinic structure of polypropylene is somewhat exceptional. In *syndiotactic* vinyl polymers the conformation of the molecules is again controlled by the size of the side groups. For example, the Cl atoms in syndiotactic *poly(vinyl chloride)* are sufficiently small to allow the backbone to remain in a planar zig-zag.

The final types of crystal structure that will be considered in detail are those of *nylons* (polyamides) where hydrogen bonding, which can be many times stronger than van der Waals bonding, controls chain packing. In both

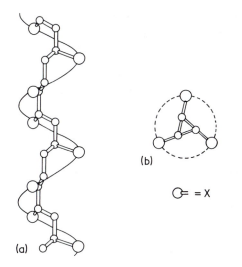

Fig. 4.6 *Schematic representation of the 3/1 helix found for isotactic vinyl polymers of the type* (—CH₂—CHX—)ₙ. (a) *Side view of helix.* (b) *View along helix. The X groups are* —CH₃, —⬡ , —CH=CH₂ *etc.*

forms of *nylon 6* and *nylon 6.6* the molecules are in the form of extended planar zig-zags joined together in hydrogen-bonded sheets. An example of this is given for nylon 6.6 in Fig. 4.7(a) where the hydrogen bonds between oxygen atoms and >NH groups in adjacent molecules are indicated by dashed lines. The α and β forms of this polymer differ in the way in which the hydrogen-bonded sheets are stacked as illustrated in Fig. 4.7(b) and (c). In the α form successive planes are displaced in the *c* direction whereas in the β form the hydrogen-bonded sheets have alternating up and down displacements. The situation is slightly different in nylon 6 where the α and γ forms differ in the orientation of the molecules in the hydrogen-bonded sheets. They are antiparallel in the α form and parallel to each other in the γ form. This difference in orientation leads to a slightly lower chain direction repeat in the γ form.

4.1.5 *Solution-grown single crystals*

It has been known for many years that crystallization can occur in polymers, but it was not until 1957 that the preparation of isolated polymer single crystals was first reported. They were grown by precipitation from dilute solution. The first crystals that were prepared were of polyethylene but, in the years since, single crystals of nearly every known soluble crystalline polymer have been prepared. Polyethylene crystals are by far

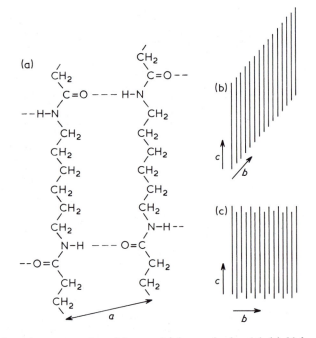

Fig. 4.7 *Schematic representation of the α and β forms of nylon 6.6. (a) Molecular arrangement of hydrogen-bonded sheets common to both forms. (b) Stacking of sheets in the α form, viewed parallel to the sheets. (c) Stacking of the sheets in the β form (after Wunderlich).*

the most widely studied examples of polymer single crystals and most of the discussion here will be concerned with polyethylene crystals.

A characteristic feature of solution-grown polymer crystals is that they are small. They can usually just about be resolved in an optical microscope, but they are most readily examined in an electron microscope. Fig. 4.8(a) is an electron micrograph of a polyethylene single crystal which has been grown by precipitation from dilute solution. The crystals are normally precipitated either by cooling a hot solution, which is the most widely used method, or by the addition of a non-solvent. The polyethylene crystal in Fig. 4.8(a) is plate-like (lamellar) and of the order of 10 μm across. The crystal thickness can be measured by shadowing the crystal under vacuum with a heavy metal such as gold, at a known angle. A simple geometrical construction allows the crystal thickness to be calculated from the length of the shadow. For most solution-grown crystals the thickness is found to be of the order of 100 Å. It can also be seen from Fig. 4.8(a) that the lamellar crystals have straight edges which make angles which are characteristic for a given polymer. This regular shape reflects the underlying crystalline nature of the lamellae.

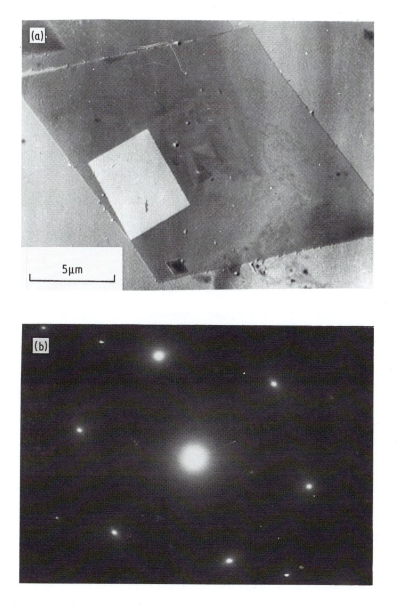

Fig. 4.8 *Solution-grown polyethylene single crystal. (a) Electron micrograph of a lamellar crystal (courtesy of Dr P. Allan). (b) Selected-area diffraction pattern (courtesy of Dr. I.G. Voigt–Martin).*

The individual single crystals are too small to be studied with X-rays and the relative orientation of the polymer molecules within the lamellar crystals can only be determined by electron diffraction. This can conveniently be done in an electron microscope set up for selected-area electron diffraction. An electron diffraction pattern from a polyethylene single crystal with the electron beam perpendicular to the crystal surface is shown in Fig. 4.8(b). The diffraction pattern consists of a regular array of discrete spots and there is no evidence of an amorphous halo. This means that the entity in Fig. 4.8(a) diffracts as good single crystals of non-polymeric solids such as metals. Because of this, the diffraction pattern can be analysed in the standard way and so it can be considered as corresponding to a section of the reciprocal lattice perpendicular to the crystallographic direction parallel to the electron beam. The diffraction pattern in Fig. 4.8(b) can be indexed as that expected for conventional orthorhombic polyethylene (Table 4.1) and it corresponds to a beam direction of [001]. Since the chain direction in orthorhombic polyethylene is c it means that the molecules in polyethylene single crystals are *perpendicular to the crystal surface*. This observation is somewhat surprising since, as the polymer molecules are many thousands of Ångstroms long and the crystals only ∼100 Å thick, it must mean that the polymer molecules *fold back and forth* between the top and bottom surfaces of the lamellae. Although the example given in Fig. 4.8 is for polyethylene it has been found that solution-grown crystals of polymers are normally lamellar and that the molecules are always folded such that they are perpendicular (or nearly perpendicular) to the crystal surface.

A distinctive feature of polymer single crystals is that the crystal edges are normally straight and make well-defined angles with each other. It is also found from electron diffraction that the edge faces correspond to particular crystallographic planes. The crystal shown in Fig. 4.8(a) has four {110} type edge faces. Sometimes polyethylene crystals are truncated and can have {100} edges as well. The different types of polyethylene crystals are shown schematically in Fig. 4.9. It is also found that the crystals are divided into sectors. The crystal illustrated in Fig. 4.9(a) shows four {110}

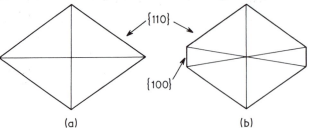

(a)   (b)

Fig. 4.9 *Schematic diagrams of two forms of lamellar single crystals of polyethylene.* (a) *Crystal with* {110} *sectors.* (b) *Crystal with* {100} *sectors as well.*

sectors and that in Fig. 4.9(b) has in addition two extra {100} sectors. This sectorization gives a clue concerning the nature of the folding which will be discussed later.

Measurement of the density and other properties of solution-grown polymer crystals has shown that they are not perfect single crystals. The density of the crystals is always less than the theoretical density which means that non-crystalline material must also be present in the crystal. It is generally thought that the bulk of this non-crystalline component resides in the *fold surfaces* of the crystals. Examination of molecular models of the folding molecule shows that if bonds are to remain unstrained, the folding can be achieved in polyethylene within the space of not less than five bonds with three of them in *gauche* conformations and it is obvious that the packing of folded molecules must be less efficient than that of extended molecules. The non-crystalline nature of the fold surface is also reflected in the observation that single crystals of polyethylene and other polymers are not always flat. Such crystals can often be seen in a collapsed form in the electron microscope containing pleats or corrugations. The true form of the crystals was only found by direct observation of single crystals suspended in a liquid using optical microscopy. It was shown that some polyethylene crystals were in the form of a hollow pyramid as shown schematically in Fig. 4.10. The fold surface in pyramidal polyethylene single crystals has been shown to be close to the {312} plane of the polyethylene unit cell. The exact form of the single crystals depends upon several factors such as the temperature at which the crystals were grown. Pyramidal crystals tend to grow at high temperatures whereas corrugated crystals form at lower crystallization temperatures.

There has been considerable argument over the years concerning the way in which folding occurs and the nature of the fold plane. The different models that have been suggested to account for folding in polymer crystals are illustrated schematically in Fig. 4.11. The models range from random re-entry ones, where a molecule leaves and re-enters a crystal randomly, to adjacent re-entry models, whereby molecules leave and re-enter the crystals in adjacent positions. Two particular adjacent re-entry models

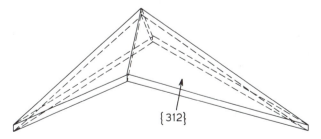

{312}

Fig. 4.10 *Schematic representation of a pyramidal polyethylene single crystal. (After Schultz).*

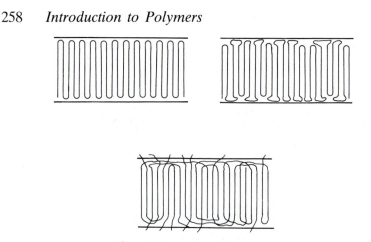

Fig. 4.11 *Schematic illustrations of the different types of folding suggested for polymer single crystals. (a) Adjacent re-entry with sharp folds. (b) Adjacent re-entry with loose folds. (c) Random re-entry or 'switch-board' model.*

have been suggested where the folds are either regular and tight or irregular and of variable length.

The main consensus of opinion appears to be that in single crystals grown from dilute solution the molecules undergo adjacent re-entry. This leads to there being a crystallographic plane on which the molecules fold, normally known as a *fold plane*. In the polyethylene crystals drawn in Fig. 4.9(a) it is thought that the fold planes are parallel to the crystal edges. This means that the fold planes must be of the type {110} and in truncated crystals (Fig. 4.9 (b)) there are thought to be sectors of {100} folding as well. The fold surface itself is thought to be fairly irregular, consisting of a mixture of loose and tight folds and chain ends.

4.1.6 *Solid-state polymerized single crystals*

It has recently become possible to prepare true 100% crystalline polymer crystals by the process of solid-state polymerization described in Section 2.11. Examples of solid-state polymerized polydiacetylene single crystals in the form of both lozenges and fibres are shown in Fig. 4.12. The crystals have regular facets and are of macroscopic dimensions and so similar to single crystals that can be prepared for non-polymeric materials such as metals or ceramics. The polymer molecules are parallel to the fibre axis in the fibrous crystals. A typical X-ray rotation photograph of a similar polydiacetylene crystal has been given in Fig. 4.2 and it indicates that in such crystals the molecules are in a chain-extended conformation. Density measurements have shown that the crystals have their theoretical densities calculated from a knowledge of the crystal structure and so it is concluded that the crystals are virtually defect-free with chain-folding being

Fig. 4.12 *Examples of polydiacetylene single crystals in the form of* (a) *lozenges and* (b) *fibres. The crystals are placed on graph paper with mm spacing (crystals supplied by Dr D. Bloor).*

completely absent. Such crystals have allowed considerable advances to be made in the understanding of the fundamental physical properties of crystalline polymers. In particular, important new aspects of the deformation and fracture of polymer crystals have been revealed.

4.2 Semi-crystalline polymers

The isolated chain-folded lamellar single crystals described in the previous section are only obtained by crystallization from dilute polymer solutions. As the solutions are made more concentrated the crystal morphologies change and more complex crystal forms are obtained. In dilute solution the polymer coils are isolated from each other, but as it becomes more concentrated the molecules become entangled. Particular chains, therefore, have the probability of being incorporated in more than one crystal giving rise to inter-crystalline links. Crystals grown from concentrated solutions are often found either as lamellae with spiral overgrowth as shown in Fig. 4.13 or as aggregates of lamellar crystals. Under certain conditions the crystals can be even more complex showing twinned and/or dendritic structures similar to the morphologies found in non-polymeric crystalline solids.

This particular section is concerned with the structures that are formed principally by crystallization from the melt. This is not fundamentally

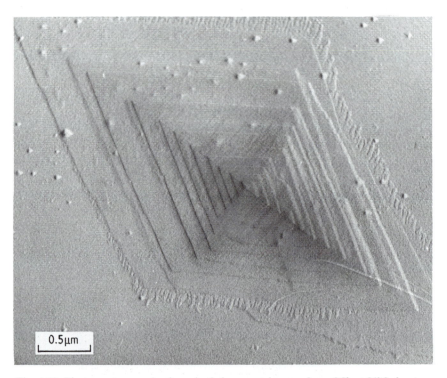

0.5 μm

Fig. 4.13 *Electron micrograph of a polyethylene crystal grown from CCl₄ at 80°C showing a large growth spiral. The micrograph was obtained by replication and the replica has been shadowed with a heavy metal (courtesy of Dr G. Lieser).*

different from solution crystallization especially when the solutions are concentrated. In the melt, chain entanglements are of extreme importance and consequently the crystals that form are more irregular than those obtained from dilute solution. A feature common to both melt and solution crystallization is that the solid polymer is only *semi-crystalline* and contains both crystalline and amorphous components.

4.2.1 *Spherulites*

If a melt-crystallized polymer is prepared in the form of a thin film, either by sectioning a bulk sample or casting the film directly, and then viewed in an optical microscope between crossed polars a characteristic structure is normally obtained. It consists of entities known as *spherulites* which show a characteristic Maltese cross pattern in cross-polarized light. A typical spherulite is shown in Fig. 4.14. They form by nucleation at different points in the sample and grow as spherical entities. The growth of the spherulites stops when impingement of adjacent spherulites occurs. At first sight they look very similar to grains in a metal being roughly of the same dimensions and growing in a similar way. However, the grains in a metal are individual single crystals whereas detailed investigation of polymeric spherulites shows that they consist of numerous crystals radiating from a central nucleus.

There are many examples of spherulites in other crystalline solids,

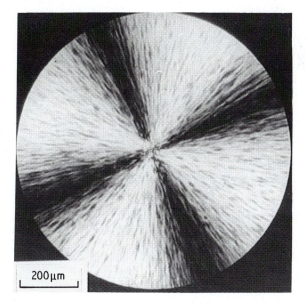

200 µm

Fig. 4.14 *A single spherulite growing in isotactic polystyrene. Optical micrograph taken in polarized light (courtesy of Dr D. Tod).*

particularly in naturally occurring mineral rocks. The typical extinction patterns seen in polarized light are due to the orientation of the crystals, within the spherulites. Analysis of the Maltese cross patterns has indicated that the molecules are normally aligned tangentially in polymer spherulites. It has been shown that the *b* crystals axis is radial in polyethylene spherulites and the *a* and *c* crystallographic directions are tangential. In some polymer samples the spherulites are sometimes seen to be ringed. This has been attributed to a regular twist in the crystals.

The detailed structure of polymer spherulites can best be studied by electron microscopy because of the small size of the individual crystals. Fig. 4.15 shows a scanning electron micrograph of an etched surface of polypropylene crystallized at 128°C. The surface was microtomed and then etched with a strongly-oxidizing solution which preferentially attacks the amorphous regions of the polymer. It exposes the crystals radiating from the central nucleus and terminating at the spherulite boundaries. Close examination of the micrograph shows that the crystals are lamellar. When viewed edge-on they appear as fine lines of the order of 100 Å thick and as flat areas when they are parallel to the sample surface. The dimensions of these crystal lamellae are, of course, similar to those of solution-grown lamellar single crystals. Further evidence for the lamellar nature of

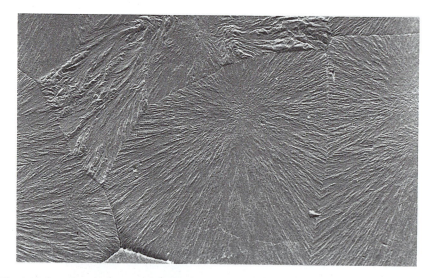

Fig. 4.15 *Scanning electron micrograph of an etched surface of polypropylene crystallized at 128°C showing the structure of the individual spherulites and the spherulite boundaries at a magnification of 400 × (courtesy of M. Burke).*

the crystals in polymer spherulites has been obtained by examination of the debris left after the degradation of spherulitic polyethylene and polypropylene in nitric acid. The acid selectively oxidizes the inter-crystalline material and the crystal fragments that are left are found by electron microscopy to be lamellar. Selected-area electron diffraction has shown that the polymer molecules are oriented approximately normal to the lamellar surfaces. This is further evidence of the similarity between melt-crystallized and solution-grown lamellae.

Electron microscopy has also revealed the twist of the crystals in the spherulites that was suspected from optical microscopy. Surface replicas taken from spherulitic polymers have clearly shown that the lamellae twist but the reason for this twisting is not yet understood. A possible structure of the twisted lamellae in polyethylene spherulites is illustrated schematically in Fig. 4.16.

An obvious question which arises is that of the nature of the conformation of the molecules within the crystals. There is accumulated evidence that inter-crystalline links must exist between the lamellae in spherulitic polymers. The similarity between the morphologies of and molecular orientation in these lamellae and solution-grown crystals clearly implies that a certain degree of chain folding must also take place. However, definitive experiments in this area have been extremely difficult to perform. The best evidence of the conformation of polymer molecules in melt-crystallized polymers has come from recent neutron-scattering studies of mixtures of a homopolymer and deuterated version of the same polymer. The results of such experiments are open to different interpretations but the consensus of opinion appears to be that the structure is similar to that shown schematically in Fig. 4.17. There are thought to be small segments of fairly sharp chain-folding and a large number of inter-crystalline links with a given molecule shared between at least two crystals.

4.2.2 *Degree of crystallinity*

Melt-crystallized polymers are never completely crystalline. This is because there are an enormous number of chain entanglements in the melt and it is impossible for the amount of organization required to form a 100% crystalline polymer to take place during crystallization. The degree of crystallinity is of great technological and practical importance and several methods, which do not always produce precisely the same results, have been devised to measure it.

The crystallization of a polymer from the melt is accompanied by a reduction in specimen volume due to an increase in density. This is because the crystals have a higher density than the molten or non-crystalline

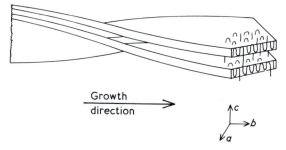

Growth direction →

Fig. 4.16 *Schematic representation of a possible model for twisted lamellae in spherulitic polyethylene showing chain-folds and intercrystalline links.*

polymer and this effect provides the basis of the *density* method for the determination of the degree of crystallinity. The technique relies upon the observation that there is a relatively large and measurable difference (up to 20%) between the densities of the crystalline and amorphous regions of the polymer. This method can yield both the volume fraction of crystals ϕ_c and the mass fraction x_c from measurement of sample density ρ.

If V_c is the volume of crystals and V_a the volume of amorphous material then the total specimen volume, V, is given by

$$V = V_c + V_a \tag{4.5}$$

Similarly the mass of the specimen W is given by

$$W = W_c + W_a \tag{4.6}$$

where W_c and W_a are the masses of crystalline and amorphous material in the sample respectively. Since density ρ is mass per volume then it follows from Equation (4.6) that

$$\rho V = \rho_c V_c + \rho_a V_a \tag{4.7}$$

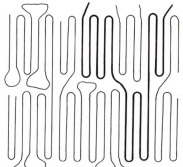

Fig. 4.17 *A schematic representation of one possible model for the conformation of a deuterated polymer molecule (heavy line) in a matrix of protonated molecules (light lines).*

substituting for V_a from Equation (4.5) into Equation (4.7) and rearranging leads to

$$\frac{V_c}{V} = \left(\frac{\rho - \rho_a}{\rho_c - \rho_a}\right) = \phi_c \tag{4.8}$$

since ϕ_c is equal to the volume of crystals divided by the total specimen volume. The mass fraction x_c of crystals is similarly defined as

$$x_c = W_c/W = \rho_c V_c/\rho V \tag{4.9}$$

and combining Equations (4.8) and (4.9) gives

$$x_c = \frac{\rho_c}{\rho}\left(\frac{\rho - \rho_a}{\rho_c - \rho_a}\right) \tag{4.10}$$

This equation relates the degree of crystallinity x_c to the specimen density ρ and the densities of the crystalline and amorphous components.

The density of the polymer sample can be readily determined by *flotation* in a density-gradient column. This is a long vertical tube containing a mixture of liquids with different densities. The column is set up with a light liquid at the top and a steady increase in density towards the bottom. It is calibrated with a series of floats of known density. The density of a small piece of the polymer is determined from the position it adopts when it is dropped into the column. The density of the crystalline regions ρ_c can be calculated from knowledge of the crystal structure (Table 4.1). The term ρ_a can sometimes be measured directly if the polymer can be obtained in a completely amorphous form, for example by rapid cooling of a polymer melt. Otherwise it can be determined by extrapolating either the density of the melt to the temperature of interest or that of a series of semi-crystalline samples to zero crystallinity. Equations (4.8) and (4.10) are only valid if the sample contains no holes or voids, which are often present in moulded samples, and if ρ_a remains constant. In practice, since the packing of the molecules in the amorphous areas is random it is likely that ρ_a will be different in specimens which have had different thermal treatments.

A powerful method of determining the degree of crystallinity is *wide-angle X-ray scattering* (WAXS). A typical WAXS curve for a semi-crystalline polymer is given in Fig. 4.18 where the intensity of X-ray scattering is plotted against diffraction angle, 2Θ. The relatively sharp peaks are due to scattering from the crystalline regions and the broad underlying 'hump' is due to scattering from non-crystalline areas. In principle, it should be possible to determine the degree of crystallinity from the relative areas under the crystalline peaks and the amorphous hump. In practice it is often difficult to resolve the curve into areas due to each phase. The shape of the amorphous hump can be determined from the

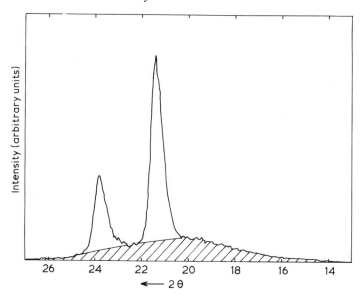

Fig. 4.18 *WAXS curves for a medium-density polyethylene. The intensity of scattering is plotted as a function of 2Θ. The amorphous hump is shaded.*

WAXS curve for a completely amorphous sample, obtained by rapidly cooling a molten sample. For certain polymers such as polyethylene this can be difficult or even impossible to do and the amorphous scattering can only be estimated. Also corrections should be made for disorder in the crystalline regions which can give rise to a reduction in area of the sharp peaks. Nevertheless, an approximate idea of the crystallinity can be obtained from the simple construction shown in Fig. 4.18. A horizontal base-line is drawn between the extremities of the scattering curve to remove the background scattering. The amorphous hump is then traced in either from a knowledge of the scattering from a purely amorphous sample or by estimating from experience with other polymers. The mass fraction of crystals x_c is then given to a first approximation by

$$x_c = A_c/(A_a + A_c) \tag{4.11}$$

where A_a is the area under the amorphous hump and A_c is the area remaining under the crystalline peaks. A more sophisticated analysis involving resolution of the curve in Fig. 4.18 into three distinct components, two crystalline peaks and one amorphous hump, lying on the same baseline can produce slightly more accurate results.

Although the density and WAXS methods of determining the degree of crystallinity of semi-crystalline polymers are widely used techniques several others have been employed with varying degrees of success. One technique often used is differential scanning calorimetry (DSC). This involves determining the change in enthalpy (ΔH_m) during the melting of a semi-crystalline polymer. The degree of crystallinity can then be calculated by comparing with ΔH_m for a 100% crystalline sample. Spectroscopic methods such as NMR and infra-red spectroscopy (Sections 3.18 and 3.19) have also been employed. In general there is found to be an approximate correlation between the different methods of measurement used although the results often differ in detail.

4.2.3 *Crystal thickness and chain extension*

It is known that lamellar crystalline entities are obtained by the crystallization of polymers from both the melt and dilute solution. The most characteristic dimension of these lamellae is their thickness and this is known to vary with the crystallization conditions. The most important variable which controls the crystal thickness is the *crystallization temperature* and the most detailed measurements of this have been made for solution-crystallized lamellae. Several methods of crystal thickness determination have been used. They include shadowing lamellae deposited upon a substrate and measuring the length of the shadow by electron microscopy, measuring the thickness by interference microscopy and determining it by using small-angle X-ray scattering (SAXS) from a stack of deposited crystals. When the lamella single crystals are allowed to settle out from solution they form a solid mat of parallel crystals. The periodicity of the stack in the mat is related to both the crystal thickness and the degree of crystallinity. The stack scatters X-rays in a similar way to the atoms in a crystal lattice and Bragg's law is obeyed

$$n\lambda = 2d \sin \theta \qquad (4.12)$$

where n is an integer, λ the wavelength of the radiation, d is the periodicity of the array and θ the diffraction angle. Since d is of the order of 100 Å then by using X-rays ($\lambda \sim 1$ Å) θ will be very small and the diffraction maxima are very close to the main beam.

The variation of lamellar thickness with crystallization temperature for polyoxymethylene crystallized from a variety of solvents is shown in Fig. 4.19(a). For a given solvent the lamellar thickness is found to increase with increasing crystallization temperature. This behaviour is typical of many crystalline polymers such as polyethylene or polystyrene for which the length of the fold period increases as the crystallization temperature is raised. The different behaviour in the various solvents displayed in Fig. 4.19(a) is again typical of solution crystallization and Fig. 4.19(b) is a plot of the lamellar thickness against the reciprocal of the supercooling $\Delta T (= T_s - T_c)$ which gives a master curve of all the data in Fig. 4.19(a). This

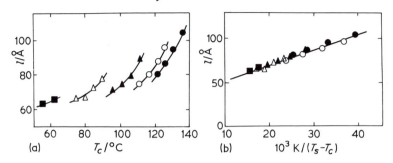

Fig. 4.19 (a) *Dependence of lamellar thickness, l, upon crystallization temperature, T_c for polyoxymethylene crystallized from different solvents.* (b) *Master curve for all the data in* (a) *plotted against the reciprocal of the supercooling. Solvents:* ■ *... phenol,* △ *... m-cresol,* ▲ *... furfuryl alcohol,* ○ *... benzyl alcohol,* ● *... acetophenone (after Magill).*

indicates that the crystallization process is controlled by the difference between the crystallization temperature T_c and solution temperature T_s, rather than the actual temperature of crystallization. This has important implications for the theories of polymer crystallization which will be discussed later (Section 4.3.3).

The lamellar thickness for melt-crystallized polymers is rather more difficult to determine to any great accuracy. SAXS curves can be obtained from such samples but they tend to be much weaker and less well-defined than their counterparts from single crystal mats. If the lamellar thickness is sufficiently large (>500 Å) it is possible to measure it by electron microscopic examination of sections of bulk samples. The variation of lamellar thickness with crystallization temperature for melt-crystallized polyethylene is shown in Fig. 4.20. The behaviour is broadly similar to that of solution-crystallized polymers in that the lamellar thickness increases as the crystallization temperature is raised. In fact it is found that at high supercoolings the thickness of solution and melt-crystallized lamellae are comparable. However, at low supercoolings there is a rapid rise in lamellar thickness with crystallization temperature which is not encountered with solution crystallization. It is thought that this is due to an isothermal thickening process whereby the crystals become thicker with time when held at a constant temperature close to the melting temperature. On the other hand there is very little effect of crystallization time upon the thickness of solution-grown lamellae.

As might be expected, the *molar mass* of the polymer only has a strong effect upon lamellar thickness when the length of the molecules is comparable to the crystal thickness. Normally the molecular length is many times greater than the fold length and so even large increases in molar mass produce only correspondingly small increases in lamellar thickness. Another variable which has a profound effect upon the crystallization of

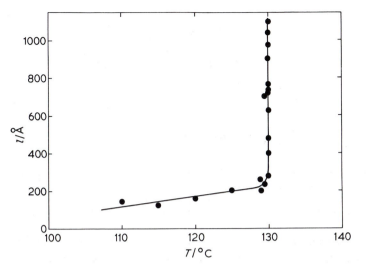

Fig. 4.20 *Lamellar thickness as a function of crystallization temperature for isothermally melt-crystallized polyethylene (after Wunderlich).*

polyethylene and a few other polymers from the melt is *pressure*. The melting temperature of polymers increases rapidly with pressure and it is found that crystallization of polyethylene at low supercoolings at pressures above about 3 kbar can produce crystal lamellae in which some of the molecules are in fully extended conformations. Crystals of up to 10 μm thick have been reported for high molar mass polymer. This chain-extended type of morphology can also be found in polytetrafluoroethylene crystallized slowly at ambient pressure and a micrograph of a replica of a fracture surface of this polymer is given in Fig. 4.21. The micrograph shows long tapering lamellae containing striations approximately perpendicular to the lamellar surfaces. The striations are caused by steps on the fracture surface and define the chain direction within each lamella. In the chain-extended lamellae seen in Fig. 4.21 it is thought that the molecules fold backwards and forwards several times as it is known that the molecular length is somewhat greater than the average crystal thickness. Chain-extended morphologies can have extremely high degrees of crystallinity, sometimes in excess of 95%. However, their mechanical properties are extremely disappointing. Chain-extended polyethylene is relatively stiff but absence of inter-crystalline link molecules causes it to be very brittle and it crumbles like cheese when deformed mechanically.

The effect of the different variables upon the fold length and lamellar thickness of polyethylene is summarized in Table 4.2. The number of positive or negative signs indicates the magnitude of the effect in each case.

Fig. 4.21 *Electron micrograph of a replica of a fracture surface of melt-crystallized polytetrafluoroethylene showing chain-extended crystals. The striations define the chain-direction in each crystal (Young, Chapter 7 in 'Developments in Polymer Fracture' edited by E. H. Andrews, Applied Science Publishers Ltd, 1979, reproduced with permission).*

4.2.4 *Crystallization with orientation*

If polymer melts or solutions are crystallized under stress, morphologies which are strikingly different from those obtained in the absence of stress are found. This observation has important technological consequences as many polymers are processed in the form of fibres, injection moulded or extruded and all of these processes involve the application of stress to the material during crystallization.

Some of the earliest observations on the effect of stress upon crystallization were made by Andrews upon thin stretched films of natural rubber. It was found that crystals formed transverse to the direction of the applied stress as shown in Fig. 4.22(a) for isotactic polystyrene. The

TABLE 4.2 *Effect of changing supercooling ΔT, pressure p, molar mass M and time t upon the lamellar thickness of polyethylene (after Wunderlich, reproduced with permission)*

	ΔT	p	M	t
Solution crystallized	$--$	0	0	0
Melt crystallized (High T)	$--$	$+++$	0	$+$
Melt crystallized (Low T)	$---$	$++++$	$+$	$++$
Annealed	$--$	$++$	$(+)$	$++$

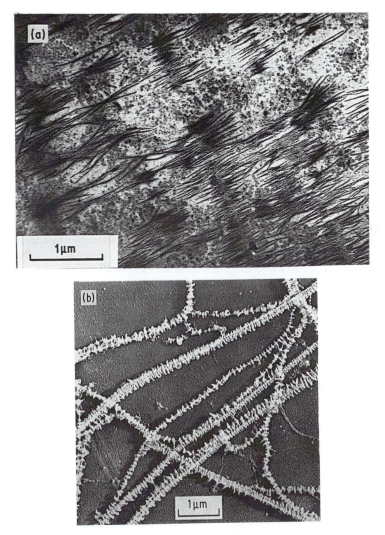

Fig. 4.22 *Electron micrographs showing the effect of applied stress upon the morphologies of crystalline polymers. (a) Row-nucleated structure in melt-crystallized isotactic polystyrene (courtesy of Dr J. Petermann). (b) Shish-kebabs obtained from a stirred solution of polyethylene (courtesy of Dr A. Pennings).*

crystals are found to nucleate on a central backbone and grow in perpendicular directions. Similar row-nucleated structures have since been found on crystallizing oriented melts of many other polymers.

A related structure is found on cooling rapidly stirred polymer solutions. Fig. 4.22(b) shows the type of morphology which results from the rapid stirring of a polyethylene solution and such structures have been termed *shish-kebab morphologies*. They consist of a central backbone (shish) with lamellar overgrowth (kebabs). It has been found by electron diffraction that in both the backbone and the lamellar overgrowths the polymer molecules are parallel to the shish-kebab axis. It is thought that the molecules are folded in the overgrowth, but that in the backbone there is a considerable degree of chain extension, although some folds may also be present.

Oriented polymer structures are extremely important as commercial materials since they allow the high unidirectional strength of the polymer molecules to be exploited.

4.2.5 *Defects in crystalline polymers*

It is clear that crystalline polymers are by no means perfect from a structural viewpoint. They contain crystalline and amorphous regions and probably also areas which are partially disordered. It has been recognized for many years that crystals of any material contain imperfections such as dislocations or point-defects and there is no fundamental reason why even the relatively well-ordered crystalline regions of crystalline polymers should not also contain such defects. In fact, it is now known that polymer crystals contain defects which are similar to those found in other crystalline solids, but in considering such imperfections it is essential to take into account the macromolecular structure of the crystals.

The presence of defects in a crystal will give rise to broadening of any diffraction maxima such as the rings seen in the X-ray diffraction pattern in Fig. 4.1. In principle, the degree of broadening can be used to determine the amount of disorder in the crystal caused by the presence of defects. However, broadening of the diffraction maxima can also be caused by the crystals having a finite size and since polymer crystals are normally relatively small most of the observed broadening is usually caused by the size effect. It is possible to estimate crystal size from the degree of broadening. If the crystals have a mean dimension normal to the (hkl) planes of L_{hkl} then the broadening of the X-ray diffraction peaks δs due to the size effect is given by*

$$\delta s = \frac{1}{L_{hkl}} \qquad (4.13)$$

The parameter δs is the breadth of the diffraction peak at half-height and is expressed in units of s, where s is defined through the equation

$$s = (2/\lambda) \sin \theta \qquad (4.14)$$

*e.g. J.M. Schultz, *Polymer Materials Science*.

where θ is the diffraction angle and λ is the wavelength of the X-ray used. It is possible to derive similar equations relating δs to the broadening due to defects but since it is quite difficult to measure δs with sufficient accuracy to usefully exploit the analysis, the reader is referred to more advanced texts for further details*.

In considering the types of defects that may be present in polymer crystals it is best to look at the different types separately. The defects which have been found to exist or postulated are as follows.

(1) *Point defects*

The presence of point defects such as vacancies or interstitial atoms or ions is well-established in atomic and ionic crystals. The situation is somewhat different in macromolecular crystals where the types of point defects are restricted by the long-chain nature of the polymer molecules. It is relatively easy to envisage the types of defects that may occur. They could include chain ends, short branches, folds or copolymer units. There is accumulated evidence that the majority of this chain disorder is excluded from the crystals and incorporated in non-crystalline regions. However, it is also clear that at least some of it must be present in the crystalline areas.

Another type of point defect which could occur and is unique for macromolecules is a molecular kink. For example, a kink could be generated in a planar zig-zag chain consisting of all-*trans* bonds by making two of the bonds *gauche*. The sequence of five consecutive bonds would be . . . tg^+tg^-t . . . rather than . . . $ttttt$ . . . Such a kinked chain could be incorporated into a crystal with only a relatively slight local distortion. One important aspect of such a kink defect is that in polyethylene it enables an extra —CH_2— group to be incorporated in the crystal and motion of the defect along the chain allows the transport of material across the crystal. It has been suggested that the motion of such defects may be the mechanism whereby crystals thicken during annealing (Section 4.3.4).

(2) *Dislocations*

It is found that in metal crystals slip can take place at stresses well below the theoretically-calculated shear stress. The movement of line defects known as dislocations was invoked to account for this observation and with the advent of electron microscopy the presence of such defects was finally proved. The two basic types of dislocations found in crystals, screw and edge, are shown in Fig. 4.23. The dislocation is characterized by its line and Burgers vector. If the Burgers vector is parallel to the line it is termed a *screw dislocation* and if it is perpendicular it is called an *edge dislocation*. In general the Burgers vector and dislocation line may be at any angle as

*e.g. J.M. Schultz—*Polymer Materials Science*.

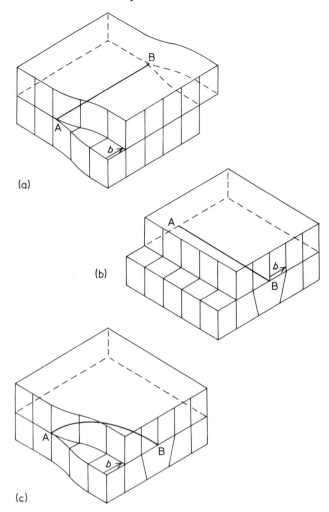

Fig. 4.23 *Schematic representation of dislocations in crystals.* (a) *Screw dislocation.* (b) *Edge dislocation.* (c) *Mixed screw and edge dislocation. The line of dislocation (AB) and the Burgers vector (b) are indicated in each case (after Kelly and Groves).*

shown in Fig. 4.23(c). When this is the case the dislocation has both edge and screw components and it is said to be *mixed*. The dislocations are drawn in Fig. 4.23 for an atomic crystal but there is no *a priori* reason why they cannot exist in polymer crystals. However, there are several important differences between polymeric and atomic crystals which lead to restrictions upon the type of dislocations that may occur in polymer crystals. The differences are as follows.

(a) There is a high degree of anisotropy in the bonding in polymer crystals. The covalent bonding in the chain direction is very much stronger than the relatively weak transverse secondary bonding.

(b) Polymer molecules strongly resist molecular fracture and so dislocations which tend to do this are not favoured.

(c) The crystal morphology can also affect the stability and motion of the dislocations. The occurrence of chain folding and the fact that polymer crystals are often relatively thin may affect matters.

The possibility of dislocations being involved in the deformation of polymer crystals has received considerable attention over recent years. This will be considered later (Section 5.5.5) and this section will be concerned with dislocations that may be present in undeformed crystals. The most obvious example of the occurrence of dislocations in polymer crystals is seen in Fig. 4.13 where the crystals contain growth spirals. It is thought that there is a screw dislocation with a Burgers vector of the size of the fold length (~ 100 Å) at the centre of the spiral. Such a dislocation is illustrated schematically in Fig. 4.24 and the Burgers vector and dislocation line are parallel to each other and the chain direction.

Dislocations with Burgers vectors of a similar size to the unit cell dimensions have been deduced from Moiré patterns obtained from overlapping lamellar single crystals observed in the electron microscope. The Moiré pattern is caused by double diffraction when the two crystals are slightly misaligned and they allow the presence of defects in the crystals to be established. This technique has allowed edge dislocations with Burgers vectors perpendicular to the chain direction to be identified in several different types of polymer crystals. Such dislocations are thought to be due to the presence of a terminated fold plane in crystals containing molecules folding by adjacent re-entry. If this is so it seems likely that the dislocation will not be able to move without breaking bonds at the fold surface.

With recent advances and improvements in the instrumentation of electron microscopy there have been numerous reports of the observation of dislocations in atomic, ionic and molecular crystals through direct imaging of the crystal lattice. Although this technique is difficult to apply to polymer crystals because of radiation damage, recently dislocations have been imaged directly in polymer crystals.

(3) *Other defects*

There are many other possible types of imperfections in crystalline polymers as well as point-defects and dislocations. The fold surface and chain folds can be considered as defects. It is also possible to consider the 'grain' boundaries between crystals as areas containing defects.

The effect of all the many types of defects discussed above is to introduce disorder into the polymer crystals. The result of their presence is to deform and distort the crystal lattice and produce broadening of the X-ray

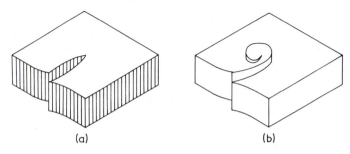

(a)   (b)

Fig. 4.24 *Schematic representation of screw dislocations with Burgers vectors parallel to the chain direction in lamellar polymer crystals.* (a) *Illustration of relation between the dislocation line and the chain direction (indicated by striations).* (b) *Screw dislocation leading to growth spiral.*

diffraction maxima (Fig. 4.1). This type of disorder in a crystalline structure has been termed *paracrystallinity*. The polymer paracrystal is modelled by allowing the unit cell dimensions to vary from one cell to another and this allows the broadening of the X-ray patterns, over and above that expected from crystal size effects, to be explained. The concept of paracrystallinity has proved extremely useful in characterizing the structures of semicrystalline polymers, but there is clearly more yet to be learned about the detailed structure of these materials.

4.3 Crystallization and melting

4.3.1 *General considerations*

Crystallization is the process whereby an ordered structure is produced from a disordered phase, usually a melt or dilute solution, and melting can be thought of as being essentially the opposite of this process. When the temperature of a polymer melt is reduced to the melting temperature there is a tendency for the random tangled molecules in the melt to become aligned and form small ordered regions. This process is known as *nucleation* and the ordered regions are called nuclei. These nuclei are only stable below the melting temperature of the polymer since they are disrupted by thermal motion above this temperature. The second step in the crystallization process is *growth* whereby the crystal nuclei grow by the addition of further chains. Crystallization is therefore a process which takes place by two distinct steps, nucleation and growth which may be considered separately.

Nucleation is classified as being either homogeneous or heterogeneous. During homogeneous nucleation in a polymer melt or solution it is envisaged that small nuclei form randomly throughout the melt. Although this process has been analysed in detail from a theoretical viewpoint it is thought that, in the majority of cases of crystallization from polymer melts and solutions, nucleation takes place heterogeneously on foreign bodies such as dust particles or the walls of the containing vessel. The number of nuclei formed depends, when all other factors are kept constant, upon the temperature of crystallization. At low undercoolings nucleation tends to be sporadic and during melt crystallization a relatively small number of large spherulites form. On the other hand when the undercooling is increased many more nuclei form and a large number of small spherulites are obtained.

The growth of a crystal nucleus can, in general, take place either one, two, or three dimensionally with the crystals in the form of rods, discs or spheres respectively. The growth of polymer crystals takes place by the incorporation of the macromolecular chains within crystals which are normally lamellar. Crystal growth will then be manifest as the change in the lateral dimensions of lamellae during crystallization from solution or the change of spherulite radius during melt crystallization. An important experimental observation which simplifies the theoretical analysis is that the change in linear dimensions of the growing entities at a given temperature of crystallization is usually linear with time. This means that the spherulite radius, r, will be related to time, t, through an equation of the form

$$r = vt \qquad (4.15)$$

where v is known as the growth rate. This equation is usually valid until the spherulites becomes so large that they touch each other. The change in the linear dimensions of lamellae tends to obey an equation of a similar form for solution crystallization. The growth rate v is strongly dependent upon the crystallization temperature as shown in Fig. 4.25. It is found that the growth rate is relatively low at crystallization temperatures just below the melting temperature of the polymer, but as the supercooling is increased there is a rapid increase in v. However, it is found that eventually there is a peak in v and further lowering of the crystallization temperature produces a reduction in v. Also, for a given polymer it is found that the growth rate at a particular temperature depends upon the molar mass with v increasing as M is reduced (Fig. 4.25). The reason for the peak in v is thought to be due to two competing effects. The thermodynamic driving force for crystallization will increase as the crystallization temperature is lowered. But as the temperature is reduced there will be an increase in viscosity and transport of material to the growth point will be more difficult and so v peaks and eventually decreases as the temperature is reduced even though the driving force continues to increase.

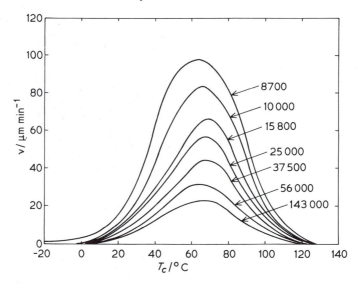

Fig. 4.25 *Dependence of crystal growth rate v upon crystallization temperature T_c, for different fractions of poly(tetramethyl-p-phenylene) siloxane (molar mass given in g mol^{-1}) (after Magill).*

4.3.2 *Overall crystallization kinetics*

The degree of crystallinity has an important effect upon the physical properties of a polymer. If a polymer melt is allowed to crystallize the crystallinity clearly increases as the spherulites form and grow and so the analysis of the crystallization of a polymer is of profound importance in understanding structure/property relationships in polymers. The nucleation and growth of spherulites in a polymer liquid can be readily analysed if a series of assumptions is made.

If a polymer melt of mass W_o is cooled below the crystallization temperature then spherulites will nucleate and grow over a period of time. It may be assumed that nucleation is homogeneous and that at a given temperature the number of nuclei formed per unit time per unit volume (i.e. the rate of nucleation) is a constant, N. The total number of nuclei formed in time interval, dt will then be $NW_o dt/\rho_L$ where ρ_L is the density of the liquid polymer. After a length of time, t, these nuclei will have grown into spherulites of radius, r. The volume of each spherulite will be $4\pi r^3/3$ or using Equation (4.15), $4\pi v^3 t^3/3$. If the density of spherulitic material is ρ_s then the mass of each spherulite will be $4\pi v^3 t^3 \rho_s/3$. The total mass of spherulitic material, dW_s, present at time t, and grown from the nuclei formed in the time interval, dt, will be given by

$$dW_s = \tfrac{4}{3}\pi v^3 t^3 \rho_s N W_o \frac{dt}{\rho_L} \tag{4.16}$$

The total mass of spherulitic material formed after time t, from all nuclei is then given by

$$W_s = \int_0^t \frac{4\pi v^3 \rho_s NW_o t^3}{3\rho_L} \mathrm{d}t \qquad (4.17)$$

which can be integrated to give

$$\frac{W_s}{W_o} = \frac{\pi N \dot{v}^3 \rho_s t^4}{3\rho_L} \qquad (4.18)$$

Alternatively this can be expressed in terms of the mass of liquid, W_L, remaining after time t, since $W_s + W_L = W_o$

i.e. $\quad \dfrac{W_L}{W_o} = 1 - \dfrac{\pi N v^3 \rho_s t^4}{3\rho_L} \qquad (4.19)$

This analysis is highly simplified and the equations are only valid for the early stages of crystallization. However, the essential features of spherulitic crystallization are predicted. It is expected that the mass fraction of the crystals should depend initially upon t^4. It also follows that if the nuclei are formed instantaneously then a t^3 dependence would be expected as only the change in spherulite volume is time dependent.

Equation (4.19) is only valid in the initial stages of crystallization and must be modified to account for impingement of the spherulites. Also the analysis is slightly incorrect because, during crystallization, there is a reduction in the overall volume of the system and the centre of the spherulites move closer to each other. However, it can be shown that when impingement is taken into account W_L/W_o is related to t through an equation of the form

$$W_L/W_o = \exp(-zt^4) \qquad (4.20)$$

This type of equation is generally known as an Avrami equation and when t is small Equations (4.19) and (4.20) have the same form. If types of nucleation and growth other than those considered here are found the Avrami equation can be expressed as

$$W_L/W_o = \exp(-zt^n) \qquad (4.21)$$

where n is called the Avrami exponent.

From an experimental viewpoint it is much easier to follow the crystallization process by measuring the change in specimen volume rather than the mass of spherulitic material. If the initial and final specimen volumes are defined as V_o and V_∞ respectively and the specimen volume at time t is given by V_t then it follows that

$$V_t = \frac{W_L}{\rho_L} + \frac{W_s}{\rho_s} = \frac{W_o}{\rho_s} + W_L \left(\frac{1}{\rho_L} - \frac{1}{\rho_s} \right) \qquad (4.22)$$

and since

$$V_o = \frac{W_o}{\rho_L} \quad \text{and} \quad V_\infty = \frac{W_o}{\rho_s}$$

then Equation (4.22) becomes

$$V_t = V_\infty + W_L \left(\frac{V_o}{W_o} - \frac{V_\infty}{W_o} \right) \tag{4.23}$$

Rearranging and combining Equations (4.21) and (4.23) gives

$$\frac{W_L}{W_o} = \frac{V_t - V_\infty}{V_o - V_\infty} = \exp(-zt^n) \tag{4.24}$$

This equation then allows the crystallization process to be monitored by measuring how the specimen volume changes with time. In practice this is normally done by dilatometry. The crystallizing polymer sample is enclosed in a dilatometer and the change in volume is monitored from the change in height of a liquid which is proportional to the specimen volume. In terms of heights measured in the dilatometer Equation (4.24) can be expressed as

$$\left(\frac{V_t - V_\infty}{V_o - V_\infty} \right) = \left(\frac{h_t - h_\infty}{h_o - h_\infty} \right) = \exp(-zt^n) \tag{4.25}$$

The Avrami exponent, n, can be determined from the slope of a plot of log $\{\ln[(h_t - h_\infty)/(h_o - h_\infty)]\}$ against log t. Fig. 4.26 shows an Avrami plot for polypropylene crystallizing at different temperatures. It is often difficult to estimate n from such plots because its value can vary with time. Also, non-integral values can be obtained and care must be exercised in using the Avrami analysis, as interpretation of the value of n in terms of specific nucleation and growth mechanisms can sometimes be ambiguous.

Serious deviations from the Avrami expression can be found particularly towards the later stages of crystallization because secondary crystallization often occurs and there is usually an increase in crystal perfection with time. They can take place relatively slowly and make estimation of V_∞ rather difficult.

4.3.3 *Molecular mechanisms of crystallization*

Although the Avrami analysis is fairly successful in explaining the phenomenology of crystallization it does not give any insight into the molecular process involved in the nucleation and growth of polymer crystals. There have been many attempts to develop theories to explain the important aspects of crystallization. Some of the theories are highly

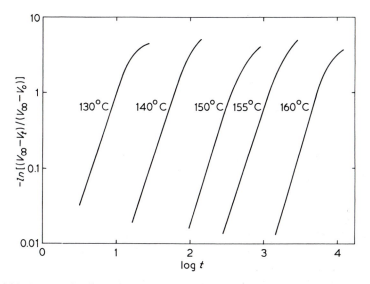

Fig. 4.26 *Avrami plot for polypropylene crystallizing from the melt at different indicated temperatures. The data points have been left off for clarity. (After Parrini and Corrieri, Makromol. Chem.* **62** *(1963), 83.)*

sophisticated and involve lengthy mathematical treatments. To date, none have been completely successful in explaining all the aspects of polymer crystallization. The important features that any theory must explain are the following characteristic experimental observations.

(a) Polymer crystals are usually thin and lamellar when crystallized from both dilute solution and the melt.

(b) A unique dependence is found between the lamellar thickness and crystallization temperature and, in particular, the lamellar thickness is found to be proportional to $1/\Delta T$ (cf. Fig. 4.19).

(c) Chain folding is known to occur during crystallization from dilute solution and probably occurs to a certain extent during melt crystallization.

(d) The growth rates of polymer crystals are found to be highly dependent upon the crystallization temperature and molar mass of the polymer.

There are major difficulties in testing the various theories such as the lack of reliable experimental data on well-characterized materials and an absence of some important thermodynamic data on the crystallization process. Nevertheless, it is worth considering the predictions of the various theories.

The most widely accepted approach is the kinetic description due to Hoffman, Lauritzen and others which has been used to explain effects observed during polymer crystallization. It is essentially an extension of the approach used to explain the kinetics of crystallization of small molecules. The process is divided up into two stages, nucleation and growth and the main parameter that is used to characterize the process is the Gibbs free energy, G, which has been defined in Equation (4.1). It follows from this equation that the change in free energy, ΔG on crystallization at a constant temperature, T, is given by

$$\Delta G = \Delta H - T\Delta S \tag{4.26}$$

where ΔH is the enthalpy change and ΔS is the change in entropy. It is envisaged that in the primary nucleation step a few molecules pack side-by-side to form a small cylindrical crystalline embryo. This process involves a change in free energy since the creation of a crystal surface, which has a surface energy will tend to cause G to increase whereas incorporation of molecules in a crystal causes a reduction in G (Section 4.1.1) which will depend upon the crystal volume. The result of these two competing effects upon G is illustrated schematically as a function of crystal size in Fig. 4.27. When the embryo is small the surface-to-volume ratio is high and so the overall value of G increases because of the rapid increase in surface energy. However, as the embryo becomes larger the surface-to-volume ratio decreases and there will be a critical size above which G starts to decrease and eventually the free energy will be less than that of the original melt. This concept of primary nucleation has been widely applied to many crystallizing systems. The main difference between polymers and small molecules or atoms is in the geometry of the embryos. It is expected that the form of the free-energy change in all cases will be as

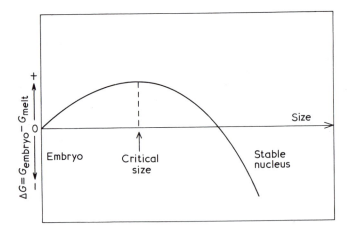

Fig. 4.27 *Schematic representation of change in free energy for the nucleation process during polymer crystallization.*

shown in Fig. 4.27. The peak in the curve may be regarded as an energy barrier and it is envisaged that at the crystallization temperature there will be sufficient thermal fluctuations to allow it to be overcome. Once the nucleus is greater than the critical size it will grow spontaneously as this will cause G to decrease.

The theories of crystallization envisage the growth of the polymer crystals as taking place by a process of secondary nucleation on a pre-existing crystal surface. This process is illustrated schematically in Fig. 4.28, whereby molecules are added to a molecularly smooth crystal surface. This process is similar to primary nucleation but differs somewhat because less new surface per unit volume of crystal is created than in the primary case and so the activation energy barrier is lower. The first step in the secondary nucleation process is the laying down of a molecular strand on an otherwise smooth crystal surface. This is followed by the subsequent addition of further segments through a chain-folding process. It is found that chain folding only occurs for flexible polymer molecules. Extended-chain crystals are obtained from more rigid molecules. It is thought that the folding only takes place when there is relatively free rotation about the polymer backbone. The basic energetics of the crystallization process can be developed if it is assumed that the polymer lamellae have a fold surface energy of γ_e, a lateral surface energy of γ_s and that the free-energy change on crystallization is ΔG_v per unit volume. If the secondary nucleation process illustrated in Fig. 4.28 is then considered the increase in surface

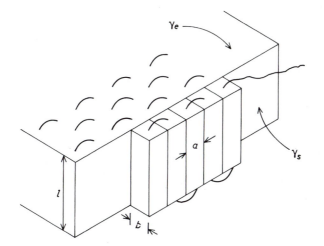

Fig. 4.28 *Model of the growth of a lamellar polymer crystal through the successive laying down of adjacent molecular strands. The different parameters are defined in the text (adapted from Magill).*

free energy involved in laying down n adjacent molecular strands of length l will be

$$\Delta G_n(\text{surface}) = 2bl\gamma_s + 2nab\gamma_e$$

when each strand has a cross-sectional area of ab. There will be a reduction in free energy because of the incorporation of molecular strands in a crystal which will be given by

$$\Delta G_n(\text{crystal}) = -n\,abl\,\Delta G_v$$

The overall change in free energy when n strands are laid down will then be

$$\Delta G_n = 2bl\gamma_s + 2nab\gamma_e - nabl\,\Delta G_v \tag{4.27}$$

The value of ΔG_v can be estimated from a simple calculation. It is assumed that polymer crystals have an 'equilibrium melting temperature', T_m^0, which is the temperature at which a crystal without any surface would melt. If unit volume of crystal is considered then at this temperature it follows from Equation (4.26) that

$$\Delta G_v = \Delta H_v - T_m^0 \Delta S_v \tag{4.28}$$

where ΔH_v and ΔS_v are the enthalpies and entropies of fusion per unit volume respectively. However, at T_m^0 there is no change in free energy for the idealized boundaryless crystal since melting and crystallization are equally probable and so $\Delta G_v = 0$ at this temperature which means that

$$\Delta S_v = \Delta H_v / T_m^0 \tag{4.29}$$

For crystallization below this temperature ΔG_v will be finite and it is envisaged that ΔS_v will not be very temperature dependent and so at a temperature, T $(<T_m^0)$ ΔG_v can be approximated to

$$\Delta G_v = \Delta H_v - T\Delta H_v / T_m^0 \tag{4.30}$$

Since the degree of undercooling ΔT is given by $(T_m^0 - T)$ then rearranging Equation (4.30) gives

$$\Delta G_v = \Delta H_v \Delta T / T_m^0 \tag{4.31}$$

Inspection of Equation (4.27) shows that for a given value of n, ΔG_n has a maximum value when l is small and decreases as l increases. Eventually a critical length of strand, l^0, will be achieved and the secondary nucleus will be stable ($\Delta G_n = 0$). Also, normally, n is large and so the term $2bl\gamma_s$ is negligible. A final approximate equation relating l^0 to ΔT can be obtained by combining Equations (4.27) and (4.31) which gives

$$l^0 \sim \frac{2\gamma_e T_m^0}{\Delta H_v \Delta T} \tag{4.32}$$

The inverse proportionality between l and ΔT which is observed experimentally (Fig. 4.19) is therefore predicted theoretically. The analysis outlined above is highly simplified, but it serves to illustrate the important aspects of the kinetic approach to polymer crystallization.

The kinetic approach can also be used to explain why chain folding is found for solution-grown crystals. The separation between individual molecules is relatively high in dilute solutions and so growth can take place most rapidly by the successive deposition of chain-folded strands of the same molecule once it starts to be incorporated in a particular crystal.

4.3.4 *Melting*

The melting of polymer crystals is essentially the reverse of crystallization, but it is more complicated than the melting of low molar mass crystals. There are several characteristics of the melting behaviour of polymers which distinguishes them from other materials. They can be summarized as follows.

(a) It is not possible to define a single melting temperature for a polymer sample as the melting generally takes place over a range of temperature.

(b) The melting behaviour depends upon the specimen history and in particular upon the temperature of crystallization.

(c) The melting behaviour also depends upon the rate at which the specimen is heated.

These observations are a reflection of the peculiar morphologies that polymer crystals can possess and, in particular, the fact that polymer crystals are normally thin. The concept of an equilibrium melting temperature T_m^0 (Section 4.3.3) is introduced because of the variability in the melting behaviour. This corresponds to the melting temperature of an infinitely large crystal. However, there is still a good deal of disagreement over the values of T_m^0 even for widely studied polymers such as polyethylene. This is particularly disturbing because accurate values of T_m^0 are required in order to test the theories of crystallization and melting quantitatively. The value of T_m^0 can be estimated by a simple extrapolation procedure. It is found that the observed melting temperature, T_m, for a polymer sample is always greater than the crystallization temperature, T_c and a plot of T_m versus T_c is usually linear as shown in Fig. 4.29. Since T_m can never be lower than T_c the line $T_m = T_c$ will represent the lower limit of the melting behaviour. The point at which the extrapolation of the upper line meets the $T_m = T_c$ line then represents the melting temperature of a polymer crystallized infinitely slowly and for which crystallization and melting would take place at the same temperature. This intercept therefore gives T_m^0.

There is found to be a strong dependence of the observed melting

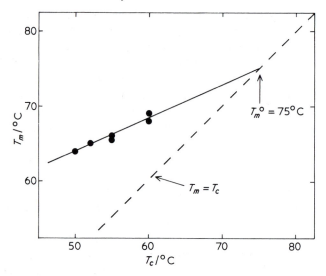

Fig. 4.29 *Plot of melting temperature, T_m, against crystallization temperature, T_c, for poly(dl-propylene oxide) (after Magill).*

temperature of a polymer crystal, T_m, upon the crystal thickness, l. This can be explained by considering the thermodynamics of melting a rectangular lamellar crystal with lateral dimensions x and y. If it is assumed that the crystal has side surface energy of γ_s and top and bottom surface energies of γ_e then melting causes a decrease in surface free energy of $(2xl\gamma_s + 2yl\gamma_s + 2xy\gamma_e)$. This is compensated by an increase in free energy of ΔG_v per unit volume due to molecules being incorporated in the melt rather than in a crystal. The overall change in free energy on melting the lamellar crystal is given by

$$\Delta G = xyl\Delta G_v - 2l(x + y)\gamma_s - 2xy\gamma_e \qquad (4.33)$$

The value of ΔG_v is given by Equation (4.31). For lamellar crystals the area of the top and bottom surfaces will be very much larger than the sides and so the term $2l(x + y)\gamma_s$ can be neglected. At the melting point of the crystal $\Delta G = 0$ and so it follows from Equations (4.31) and (4.33) that

$$T_m = T_m^0 - \frac{2\gamma_e T_m^0}{l\Delta H_v} \qquad (4.34)$$

where ΔH_v is the enthalpy of fusion per unit volume of the crystals. Inspection of this equation shows that for finite size crystals T_m will always be less than T_m^0. Also it allows T_m^0 and γ_e to be calculated if T_m is determined as a function of l. A plot of T_m against $1/l$ is predicted to have a slope of $-2\gamma_e T_m^0/\Delta H_v$ with T_m^0 as the intercept. The value of ΔH_v can be measured by calorimetry in a separate experiment.

A process which affects the melting behaviour of crystalline polymers and is of interest in its own right is *annealing*. This is a term which is normally used to describe the heat-treatment of metals and the annealing of polymers bears a similarity to that of metals. It is found that when crystalline polymers are heated to temperatures just below the melting temperature there is an increase in lamellar crystal thickness. The driving force is the reduction in free energy gained by lowering the surface area of a lamellar crystal when it becomes thicker and less wide. The lamellar thickening only happens at relatively high temperatures when there is sufficient thermal energy available to allow the necessary molecular motion to take place.

A certain amount of annealing usually takes place when a crystalline polymer sample is heated and melted. The increase in lamellar thickness, l, causes an increase in T_m (Equation 4.34). This means that the measured melting temperature will depend upon the heating rate because annealing effects will be lower for more rapid rates of heating. However, high heating rates can give rise to other problems such as with thermal conductivity and so great care is necessary when measuring the melting temperatures of polymer samples especially when the melting behaviour is being related to the as-crystallized structure which could change during heating.

4.3.5 *Factors affecting* T_m

The use of polymers in many practical applications is often limited by their relatively low melting temperatures. Because of this there has been considerable interest in determining the factors which control the value of T_m and in synthesizing polymers which have high melting temperatures. For a particular type of polymer the value of T_m depends upon the molar mass and degree of chain branching. This is because there is a higher proportion of chain ends in low-molar-mass polymers and the branches in non-linear polymers both have the effect of introducing defects into the crystals and so lower their T_m. If the molar mass of the polymer is sufficiently high that the polymer has useful mechanical properties the effect of varying M upon T_m is not strong. In contrast, the presence of branches in a high-molar-mass sample of polyethylene can reduce T_m by 30°C.

Chain ends and branches can be thought of as impurities which depress the melting points of polymer crystals. The behaviour can be analysed in terms of the chemical potentials (Section 3.2.3) per mole of the polymer repeat units in the crystalline state μ_u^c and in the pure liquid μ_u^0 (the standard state). For the *pure* polymer, which is a single-component system

$$\mu_u^c - \mu_u^0 = -\Delta G_u = -(\Delta H_u - T\Delta S_u) \tag{4.35}$$

where ΔG_u is the free energy of fusion per mole of repeat units and ΔH_u and ΔS_u are the enthalpy and entropy of fusion per mole of repeat units. It was shown in Section 4.3.3 that ΔH and ΔS would not be expected to be very temperature dependent between T and T_m^0 and so using Equation (4.29), Equation (4.35) becomes

$$\mu_u^c - \mu_u^0 = -\Delta H_u \left(1 - T/T_m^0\right) \tag{4.36}$$

For a multicomponent system in equilibrium the chemical potential of a component is given by Equation (3.25). Thus the difference in chemical potential in Equation (4.36) is given by

$$\mu_u^c - \mu_u^0 = \mathbf{R}T\ln a$$

where a is the activity of the crystalline phase which is less than unity due to the presence of the 'impurity' which depresses the melting point to T_m. Combining this equation with Equation (4.36) gives

$$\frac{1}{T_m} - \frac{1}{T_m^0} = -\frac{\mathbf{R}}{\Delta H_u}\ln a \tag{4.37}$$

If the mole fraction of crystallizable polymer is X_A and that of the impurity X_B then to a first approximation Equation (4.37) becomes, assuming ideal behaviour

$$\frac{1}{T_m} - \frac{1}{T_m^0} = -\frac{\mathbf{R}}{\Delta H_u}\ln X_A \tag{4.38}$$

and for small values of X_B it is easy to show that $-\ln X_A \sim X_B$. Hence it follows that

$$\frac{1}{T_m} - \frac{1}{T_m^0} = \frac{\mathbf{R}}{\Delta H_u}X_B \tag{4.39}$$

Linear polymers have two chain ends and so the mole fraction of chain ends is given approximately by $2/\bar{x}_n$. If these chain ends are considered to be the 'impurity' then Equation (4.39) becomes

$$\frac{1}{T_m} - \frac{1}{T_m^0} = \frac{\mathbf{R}}{\Delta H_u}\frac{2}{\bar{x}_n} \tag{4.40}$$

This equation shows clearly that $T_m \to T_m^0$ as $\bar{x}_n \to \infty$ and is found to give a reasonable prediction of the dependence of melting temperature upon molar mass. It is easy to modify Equation (4.40) to account for branches where the term $2/\bar{x}_n$ becomes $y/\bar{x}_n$ if there are y ends per chain.

The over-riding factor which determines the melting points of different polymers is their chemical structure. The melting points of several polymers are listed in Table 4.3. It is most convenient to consider the melting points of different polymers using polyethylene ($-CH_2-CH_2-$)$_n$

TABLE 4.3 *Approximate values of melting temperature, T_m, for various polymers.*

Repeat unit		T_m/K
$-CH_2-CH_2-$		410–419
$-CH_2-CH_2-O-$		340
$-CH_2-CH_2-CO-O-$		395
$-CH_2-\langle\bigcirc\rangle-CH_2-$		670
$-CH_2-CH_2-CO-NH-$		603
$-CH_2-CH_2-CH_2-CO-NH-$		533
$-CH_2-CH_2-CH_2-CH_2-CO-NH-$		531
	Side group (X)	
$-CH_2-CHX-$	$-CH_3$	460
	$-CH_2-CH_3$	398
	$-CH_2-CH_2-CH_3$	351
	$-CH_2-CH(CH_3)_2$	508
	$-\langle\bigcirc\rangle$	513

as a reference. The first factor which must be considered is the stiffness of the main polymer chain. This is controlled by the ease at which rotation can take place about the chemical bonds along the chain. In general, incorporation of groups such as $-O-$ or $-CO-O-$ in the main chain increases flexibility and so lowers T_m (Table 4.3). On the other hand the presence of a phenylene group in the main chain increases the stiffness and causes a large increase in T_m.

Another important factor which causes an increase in T_m is the presence of polar groups such as the amide linkage $-CONH-$ which allows intermolecular hydrogen-bonding to take place within the crystals. The presence of the hydrogen-bonding tends to stabilize the crystals and so raise their T_m. The melting points of different polyamides are very sensitive to the degree of intermolecular bonding and the value of the T_m is reduced as the number of $-CH_2-$ groups between the amide linkages is increased (Table 4.3).

A third factor which governs the value of T_m in different polymers is the type and size of any side groups present on the polymer backbone. This is most easily shown when the effect of having different aliphatic side groups in vinyl polymers of the type $(-CH_2-CHX-)_n$ is considered. The presence of a $-CH_3$ side group regularly placed along the polyethylene chain leads to a reduction in chain flexibility and means that polypropylene has a higher melting point than polyethylene. However, if the side group is long and flexible the T_m is lowered as the length is increased. On the other hand, an increase in the bulkiness of the side group restricts rotation about bonds in the main chain and so has the effect of raising T_m (Table 4.3).

It can be seen from the considerations outlined above that it is possible to exert a good deal of control upon the melting temperatures of different polymers. It must be borne in mind that, generally, factors which affect the T_m of a polymer also change the glass transition temperature (Section 4.4.3) and in general these two parameters cannot be varied independently of each other. Also, changing the structure of the polymer may affect the ease of crystallization and although the potential melting point of a crystalline phase may be high the amount of this phase that may form could be low.

The melting point of a polymer will also be affected by *copolymerization*. In the case of random or statistical copolymers (Section 1.2.3) the structure is very irregular and so crystallization is normally suppressed and the copolymers are usually amorphous. In contrast, in block and graft copolymers crystallization of one or more of the blocks may take place. It is possible to analyse the melting behaviour for a copolymer system in which there are a small number of non-crystallizable comonomer units incorporated in the chain, using Equation (4.39). These units will act as 'impurities' (cf. chain ends) and so the melting point of the copolymer will be given by

$$\frac{1}{T_m} - \frac{1}{T_m^0} = \frac{R}{\Delta H_u} X_B \tag{4.39}$$

where X_B is now the mole fraction of non-crystallizable comonomer units. However, it must be emphasized that this equation only holds for low values of X_B.

4.4 Amorphous polymers

We have so far been concerned principally with the structure of crystalline polymers which can readily be studied by using standard X-ray and electron diffraction methods. However, there is an important category of polymers which have not yet been considered which can be completely non-crystalline. They are generally termed *amorphous* and include the well-known polymer glasses and rubbers. Although the properties of these materials have been studied at length, very little is known about their structure. This is because there is no well-defined order in the structure of amorphous polymers and so they cannot be analysed very easily using standard diffraction techniques.

4.4.1 *Structure in amorphous polymers*

Amorphous polymers can be thought of simply as frozen polymer liquids. Over the years there have been many attempts to analyse the structure of liquids of small molecules and they have met with a varied success. As may

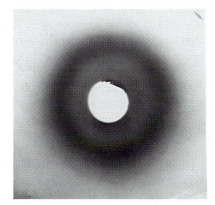

Fig. 4.30 *Flat-plate wide-angle X-ray diffraction pattern from atactic polystyrene. The polymer does not crystallize and shows only diffuse rings (cf. Fig. 4.1).*

be expected, the necessity of having long chains makes the problem more difficult to solve in the case of polymer melts and amorphous polymers. Investigation of the structure of amorphous polymers has yielded a limited amount of information. Figure 4.30 shows an X-ray diffraction pattern for an amorphous polymer. It has only one diffuse ring and can be compared with the corresponding pattern for a semi-crystalline polymer shown in Fig. 4.1 which consists of several relatively well-defined rings. In crystalline materials the discrete rings or spots on X-ray diffraction patterns correspond to Bragg reflections from particular crystallographic planes within the crystals. The lack of such discrete reflections in non-crystalline materials is taken as an indication of the lack of crystalline order. The very diffuse scattering observed in Fig. 4.30 corresponds simply to the regular spacing of the atoms along the polymer chain or in side groups and the variable separation of the atoms in adjacent molecules.

It was pointed out in Section 3.13 that small-angle neutron scattering has been used to show that in a pure amorphous polymer, the polymer molecules adopt their unperturbed dimensions (Section 3.3.3) as predicted originally by Flory.

There have been many suggestions over the years that both polymer melts and polymer glasses have domains of order. These are microscopic regions in which there is a certain amount of molecular order. This idea is not unique for polymers and similar ideas have been put forward to explain the structure of liquids and inorganic glasses. The main evidence for ordered regions in glassy polymers has come from examination of the structure of these materials by electron microscopy either by taking replicas of fracture surfaces or looking directly at the structure of thin films where a nodular structure on a scale of $\sim 1000\,\text{Å}$ can sometimes be seen. Similar structures have been seen in many glassy polymers and the nodules have been interpreted as domains of order. However, the presence of order has not been confirmed by X-ray diffraction which shows the structures to be random and some of the electron micrographic evidence has been criticized.

There are some polymers, such as poly(ethylene terephthalate) and isotactic polypropylene which normally crystallize rather slowly, that can be obtained in an apparently amorphous state by rapidly quenching the melt which does not allow sufficient time for crystals to develop by the normal processes of nucleation and growth. Crystallization can be induced by annealing the quenched polymer at an elevated temperature. The X-ray diffraction patterns which are obtained from the quenched polymer tend to be diffuse and ill-defined. This could be either because the samples are truly non-crystalline or due to the presence of very small and imperfect crystals. Crystallization is manifest during annealing by the diffraction patterns becoming more well defined. There must clearly be a certain amount of molecular rearrangement during annealing, but this cannot take place through large-scale molecular diffusion as the process occurs in the solid state. The mechanism of crystallization is therefore unclear and it is very difficult to decide whether the quenched polymers are truly amorphous or just contain very small crystals.

4.4.2 *The glass transition*

If the melt of a non-crystallizable polymer is cooled it becomes more viscous and flows less readily. If the temperature is reduced low enough it becomes rubbery and then as the temperature is reduced further it becomes a relatively hard and elastic polymer glass. The temperature at which the polymer undergoes the transformation from a rubber to a glass is known as the *glass transition temperature*, T_g. The ability to form glasses is not confined to non-crystallizable polymers. Any material which can be cooled sufficiently below its melting temperature without crystallizing will undergo a glass transition.

There is a dramatic change in the properties of a polymer at the glass transition temperature. For example, there is a sharp increase in the stiffness of an amorphous polymer when its temperature is reduced below T_g. This will be dealt with further in the section on mechanical properties (Section 5.2.6). There are also abrupt changes in other physical properties such as heat capacity and thermal expansion coefficient. One of the most widely used methods of demonstrating the glass transition and determining T_g is by measuring the specific volume of a polymer sample as a function of the temperature as shown in Fig. 4.31. In the regimens above and below the glass transition temperature there is a linear variation in specific volume

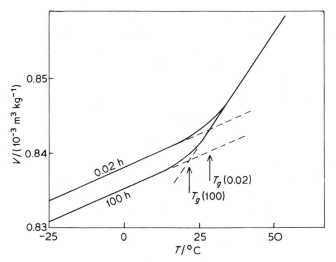

Fig. 4.31 *Variation of the specific volume of poly(vinyl acetate) with temperature taken after two different times (after Kovacs, (1958) J. Polym. Sci. 30 131).*

with temperature, but in the vicinity of the T_g there is a change in slope of the curve which occurs over several degrees. The T_g is normally taken as the point at which the extrapolations of the two lines meet. Another characteristic of the T_g is that the exact temperature depends upon the rate at which the temperature is changed. It is found that the lower the cooling rate the lower the value of T_g that is obtained. It is still a matter of some debate as to whether a limiting value of T_g would eventually be reached if the cooling rate were low enough. It is also possible to detect a glass transition in a semi-crystalline polymer, but the change in properties at T_g is usually less marked than for a fully amorphous polymer.

There have been attempts to analyse the glass transition from a thermodynamic viewpoint. Thermodynamic transitions are classified as being either first- or second-order. In the first-order transition there is an abrupt change in a fundamental thermodynamic property such as enthalpy, H or volume, V, whereas in a second-order transition only the first derivative of such properties changes. This means that during a first-order transition, such as melting, H and V will change abruptly whereas for a second-order transition changes will only be detected in properties such as heat capacity, C_p or volume thermal expansion coefficient, α which are defined as

$$C_p = \left(\frac{\partial H}{\partial T}\right)_p \quad \text{and} \quad \alpha = \frac{1}{V}\left(\frac{\partial V}{\partial T}\right)_p$$

respectively. As both of these parameters are found to change abruptly at the glass transition temperature it would appear that it may be possible to consider the glass transition as a second-order thermodynamic transition. However, more detailed thermodynamic measurements have shown that although the glass transition has the appearance of a thermodynamic second-order transition it cannot strictly be considered as one.

One of the most useful approaches to analysing the glass transition is to use the concept of *free volume*. This concept has been used in the analysis of liquids and it can be readily extended to the consideration of the glass transition in polymers. The free volume is the space in a solid or liquid sample which is not occupied by polymer molecules, i.e. the 'empty-space' between molecules. In the liquid state it is supposed that the free volume is high and so molecular motion is able to take place relatively easily because the unoccupied volume allows the molecules space to move and so change their conformations freely. A reduction in temperature will reduce the amount of thermal energy available for molecular motion. It is also envisaged that the free volume will be sensitive to the change in temperature and that most of the thermal expansion of the polymer rubber or melt can be accounted for by a change in the free volume. As the temperature of the melt is lowered the free volume will be reduced until eventually there will not be enough free volume to allow molecular rotation or translation to take place. The temperature at which this happens corresponds to T_g as below this temperature the polymer glass is effectively 'frozen'. The situation is represented schematically in Fig. 4.32. The free volume is represented as a shaded area. It is assumed to be constant at V_f^* below T_g and to

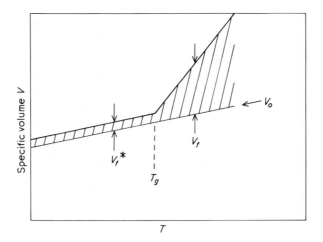

Fig. 4.32 *Schematic illustration of the variation of the specific volume, V, of a polymer with temperature, T. The free volume is represented by the shaded area.*

increase as the temperature is raised above T_g. The total sample volume V therefore consists of volume occupied by molecules V_o and free volume V_f such that

$$V = V_o + V_f$$

It is more convenient to talk in terms of fractional free volume, f, which is defined as $f = V_f/V$. At and below the T_g, f is given by $f_g = V_f^*/V$ and can be considered as being effectively constant. Above the T_g there will be an important contribution to V_f from the expansion of the melt. The free volume above T_g is then given by

$$V_f = V_f^* + (T - T_g)\left(\frac{\partial V}{\partial T}\right) \tag{4.41}$$

Dividing through by V gives

$$f = f_g + (T - T_g)\alpha_f \tag{4.42}$$

where α_f is the thermal expansion coefficient of the free volume which will be given, close to T_g, by the difference between the thermal expansion coefficients of the rubbery and glassy polymer. The equation for the fractional free volume will only be strictly valid over small increments of temperature above T_g. It is found for a whole range of different glassy polymers that f_g is remarkably constant and this concept of free volume has found important use in the analysis of the rate and temperature dependence of the viscoelastic behaviour of polymers between T_g and $T_g +$ 100 K (Section 5.2.7).

4.4.3 *Dependence of T_g upon chemical structure*

The onset of the glass transition marks a significant change in the physical properties of the polymer. A glassy polymer will lose its stiffness and have a tendency to flow above T_g. As many polymers are used in practical applications in the glassy state it is vital to know what factors control the T_g. At the glass transition the molecules which are effectively frozen in position in the polymer glass become free to rotate and translate and so it is not surprising that the value of the T_g will depend upon the physical and chemical nature of the polymer molecules. The effect of factors such as molar mass and branching will be considered in the next section and in this section the way in which the chemical nature of the polymer chain affects T_g will be examined.

The effect of the chemical nature of the polymer chain upon T_g is similar to the effect it has upon T_m (Section 4.3.5). The most important factor is chain flexibility which is governed by the nature of the chemical groups which constitute the main chain. Polymers such as polyethylene ($-CH_2-$

CH_2—$)_n$ and polyoxyethylene (—CH_2—CH_2—O—$)_n$ which have relatively flexible chains because of easy rotation about main chain bonds, tend to have low values of T_g. The value of the T_g of polyethylene is a matter of some dispute because samples are normally highly crystalline which makes measurement of T_g rather difficult. Values have been quoted between 140 K and 270 K. However, the incorporation of units which impede rotation and stiffen the chain clearly cause a large increase in T_g. For example, the presence of a *p*-phenylene ring in the polyethylene chain causes poly(*p*-xylylene) to have a T_g of the order of 80°C.

In vinyl polymers of the type (—CH_2—CHX—$)_n$ the nature of the side group has a profound effect upon T_g as can be seen in Table 4.4. The presence of side groups on the main chain has the effect of increasing T_g through a restriction of bond rotation. Large and bulky side groups tend to cause the greatest stiffening and if the side group itself is flexible the effect is not so marked. The presence of polar groups such as —Cl, —OH or —CN tends to raise T_g more than non-polar groups of equivalent size. This is because the polar interactions will restrict rotation further and so poly(vinyl chloride) (—CH_2—CHCl—$)_n$ has a higher T_g than polypropylene (—CH_2—$CHCH_3$—$)_n$ (Table 4.4).

It is clear that the chemical factors outlined above as controlling T_g are also ones which have an effect upon T_m (Section 4.3.5). In fact, the same factors tend to raise or lower both T_g and T_m as both are controlled principally by main-chain stiffness. It is not surprising, therefore, that a correlation is found between T_g and T_m for polymers which exhibit both

TABLE 4.4 *Approximate values of glass transition temperature, T_g, for various polymers*

Repeat unit		T_g/K
—CH_2—CH_2—		140–270
—CH_2—CH_2—O—		206
—⬡—O—		357
—CH_2—⬡—CH_2—		353
	Side group (X)	
—CH_2—CHX—	—CH_3	250
	—CH_2—CH_3	249
	—CH_2—CH_2—CH_3	233
	—CH_2—$CH(CH_3)_2$	323
	—⬡	373
	—Cl	354
	—OH	358
	—CN	370

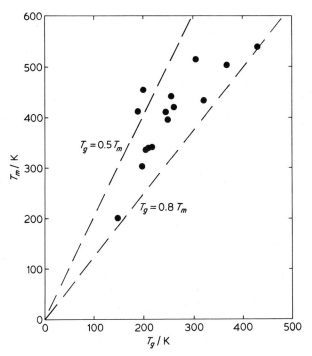

Fig. 4.33 *Plot of T_m against T_g for a variety of common polymers such as polyethylene, polypropylene, polystyrene, poly(ethylene oxide) etc. (data taken from Boyer, (1963) Rubber Chem. Tech.* **36,** *1303).*

types of transition. It is found that when they are expressed in Kelvins the value of T_g is generally between 0.5 and 0.8 T_m. This behaviour is shown schematically in Fig. 4.33. It demonstrates that it is not possible to control T_g and T_m independently for homopolymers. There are possibilities of having more control over T_g and T_m separately by making copolymers. For example, random copolymers of nylon 6.6 and nylon 6.10 can be made for which the T_g is very little different from that of the homopolymers, because the stiffness of the main chain will be virtually unchanged. On the other hand, the irregularity introduced into the main chain reduces the ability of the molecules to crystallize and so lowers T_m.

It is possible to analyse the glass transition behaviour of copolymers using the free volume concept outlined in the previous section. Block and graft copolymers normally have multiple glass transition temperatures which are near to the values of T_g for each constituent homopolymers. In contrast random or statistical copolymers usually have a single T_g between that of the corresponding homopolymers. In fact if a series of such copolymers are produced from two monomers by varying their relative proportions and T_g is plotted against composition, then it is found that the T_g of the copolymer lies on or more usually below a straight line joining the T_g's of the two homopolymers.

The dependence of the T_g of a random or statistical copolymer upon composition can be predicted if it is assumed that there is a characteristic amount of free volume associated with each type of repeat unit and that it is the same in a copolymer or homopolymer. If the two monomers are A and B then Equation (4.42) can be written as

$$f_A = f_{gA} + (T - T_{gA})\alpha_{fA} \tag{4.43}$$

$$f_B = f_{gB} + (T - T_{gB})\alpha_{fB} \tag{4.44}$$

and for the copolymer

$$f_{co} = f_{gco} + (T - T_{gco})\alpha_{fco} \tag{4.45}$$

where f is the fractional free volume at $T(>T_g)$, f_g is the fractional free volume at the glass transition temperature and α_f is the thermal expansion coefficient of the free volume. If it is assumed that in the copolymer the free volumes add in proportion to the weight fractions w_A and w_B of comonomers A and B then

$$f_{gco} = w_A f_{gA} + w_B f_{gB} \tag{4.46}$$

and

$$f_{co} = w_A f_A + w_B f_B \tag{4.47}$$

If the thermal expansion coefficients of the free volume add in a similar way then

$$\alpha_{fco} = w_A \alpha_{fA} + w_B \alpha_{fB} \tag{4.48}$$

Combining Equations (4.43)–(4.48) and rearranging leads to

$$T_{gco}(w_A + Cw_B) = T_{gA}w_A + CT_{gB}w_B \tag{4.49}$$

where $C = \alpha_{fB}/\alpha_{fA}$. If $C \sim 1$ then

$$T_{gco} = T_{gA}w_A + T_{gB}w_B \tag{4.50}$$

which gives a straight line relationship between T_{gco} and the copolymer composition. It is found that Equation (4.50) generally overestimates T_{gco} and other theoretical calculations have led to a relationship of the form

$$\frac{1}{T_{gco}} = \frac{w_A}{T_{gA}} + \frac{w_B}{T_{gB}} \tag{4.51}$$

This gives a curve below the straight line relationship and actual T_{gco} values tend to lie between values given by Equations (4.50) and (4.51).

4.4.4 *Dependence of T_g upon molecular architecture*

We have, so far, only been concerned with the way in which the chemical nature of the polymer molecules affects T_g. It is also known that the physical characteristics of the molecules such as molar mass, branching and crosslinking affect the temperature of the glass transition. The value of the T_g is found to increase as the molar mass of the polymer is increased and the behaviour can be approximated to an equation of the form

$$T_g^x = T_g + K/M \tag{4.52}$$

where T_g^x is the value of T_g for a polymer sample of infinite molar mass and K is a constant. The form of this equation can be justified using the concept of free volume if it is assumed that there is more free volume involved with each chain end than a corresponding segment in the middle part of the chain. In a polymer sample of density ρ and molar mass $\bar{M}_n$, the number of chains per unit volume is given by $\rho N_A/\bar{M}_n$ and so the number of chain ends per unit volume is $2\rho N_A/\bar{M}_n$ where N_A is the Avogadro number. If θ is the contribution of one chain end to the free volume then the total fractional free volume due to chain ends f_c is given by

$$f_c = 2\rho N_A \theta/\bar{M}_n \tag{4.53}$$

It can be reasoned that if a polymer with this value of f_c has a glass transition temperature of T_g then f_c will be equivalent to the increase in the free volume on expanding the polymer thermally between T_g and T_g^x. This means that

$$f_c = \alpha_f(T_g^x - T_g) \tag{4.54}$$

where α_f is the thermal expansion coefficient of the free volume (Section 4.4.2). Combining Equations (4.53) and (4.54) and rearranging leads to the final result

$$T_g = T_g^\infty - \frac{2\rho N_A \theta}{\alpha_f \bar{M}_n} \tag{4.55}$$

which is of exactly the same form as Equation (4.52) when $K = 2\rho N_A \theta/\alpha_f$. The observed dependence of the T_g upon molar mass is therefore predicted from considerations of free volume.

A small number of branches on a polymer chain are found to reduce the value of T_g. This can again be analysed using the free volume concept. It follows from the above analysis that the glass transition temperature of a chain possessing a total number of y ends per chain is given by an equation of the form

$$T_g = T_g^\infty - \frac{\rho y N_A \theta}{\alpha_f \bar{M}_n} \tag{4.56}$$

where T_g^∞ is again the glass transition temperature of a linear chain of infinite molar mass. Since a linear chain has two ends the number of branches per chain is $(y-2)$. Equation (4.56) is only valid when the number of branches is low. A high density of branching will have the same effect as side groups in restricting chain mobility and hence raising T_g.

The presence of chemical crosslinks in a polymer sample has the effect of increasing T_g although when the density of crosslinks is high the range of the transition region is broadened and the glass transition may not occur at all in highly crosslinked materials. Crosslinking tends to reduce the specific volume of the polymer which means that the free volume is reduced and so the T_g is raised because molecular motion is made more difficult.

4.5 Elastomers

The characteristics of *elastomers* were introduced in Section 1.2 in terms of the molecular structure of such polymers. It is worthwhile now to consider in detail the full structural requirements which are needed of a polymer before it can have elastomeric properties. The main three requirements are listed below.

(a) The polymer must be above its glass transition temperature, T_g.
(b) The polymer must have a very low degree of crystallinity ($\phi_c \to 0$).
(c) The polymer should be lightly crosslinked.

Although many polymers have some of these characteristics there are only relatively few which display true elastomeric behaviour. For example, polyethylene is above its T_g at room temperature but it is also highly crystalline and so it is not an elastomer. The copolymerization of ethylene with propylene destroys the crystallinity of the polymer and can lead to materials commonly known as ethylene/propylene rubbers. On crosslinking such polymers display elastomeric properties. The elastomeric properties of natural rubber, which is a naturally-occurring form of *cis*-1,4-polyisoprene, have been known for many years. Although it is rather tacky and also tends to flow in its natural state, on crosslinking it exhibits useful elastomeric behaviour.

The importance of crystallization can again be demonstrated by considering the physical properties of the two naturally-occurring isomers of polyisoprene the structure of which was described in Section 2.8.2. When the chains have a *cis* configuration crystallization is relatively difficult. The degree of crystallinity in natural rubber is fairly low and the melting point is just above ambient temperature ($\sim 35°C$). The chains in *cis*-1,4-polyisoprene can also coil relatively easily by rotation about single bonds and when the polymer is crosslinked it has good elastomeric properties. On the other hand in gutta percha or balata which are two forms of *trans*-1,4-polyisoprene the degree of crystallinity is higher because crystallization takes place more readily and the crystals are more stable ($T_m \sim 75°C$). The chains in the *trans* form are also less flexible. These factors make the material relatively hard and rigid with no useful elastomeric properties. Although natural rubber is used widely in a large number of applications synthetic elastomers are also often employed. Natural rubber is rather prone to chemical degradation by ozone (Section 5.6.7) and tends to swell in the presence of solvents. Synthetic elastomers are able to overcome some of these problems. Also elastomers are often used with a variety of additives such as carbon black. This reinforces the materials and improves stiffness, strength and abrasion resistance.

The process of crosslinking rubbers to produce elastomers is known, for historical reasons, as *vulcanization* (Section 1.1). *Sulphur vulcanization* was discovered first, specifically for crosslinking natural rubber, but can be used to crosslink most rubbers which possess sites of unsaturation in their structure. Sulphur is used at a level of 0.5–5 parts per hundred of rubber and the crosslinking reaction takes place upon heating during the shaping process, typically at temperatures in the range 120–180°C. Some examples of the types of crosslinks (i.e. junction points) formed in this highly complex reaction are shown below for a polymer with repeat units derived from butadiene

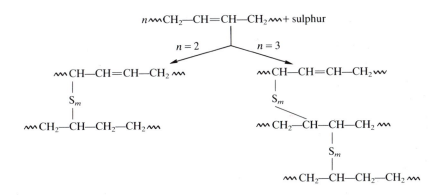

In the absence of other components the reaction is relatively slow and inefficient (*m* is large). Thus it is usual to include an *accelerator* (e.g. a mercaptobenzothiazole) and an *activator* (e.g. zinc stearate) to increase the rate of reaction and improve its efficiency ($m = 1$ or 2).

Methods of *non-sulphur vulcanization* were first developed for crosslinking rubbers which do not possess sites of unsaturation (e.g. poly(ethylene-

co-propylene)s and siloxanes), but now also are used to crosslink unsaturated rubbers. Crosslinking is effected using free-radical initiators, in particular peroxides (e.g. cumyl peroxide and benzoyl peroxide), at temperatures similar to those for sulphur vulcanization. In saturated rubbers, the free-radicals (R˙) generated from the initiator, abstract hydrogen atoms from C—H bonds in the rubber to give polymeric radicals which couple to produce the crosslinks

$$2R˙ + 2\text{---}CH_2\text{---}CH_2\text{---} \rightarrow 2\text{---}\overset{\displaystyle .}{C}H\text{---}CH_2\text{---} + 2R\text{---}H$$

coupling

$$\text{---}CH\text{---}CH_2\text{---}$$
$$|$$
$$\text{---}CH\text{---}CH_2\text{---}$$

A similar reaction involving the formation and coupling of allylic radicals occurs with unsaturated rubbers. Additionally, the allylic radicals can add to C=C bonds in other molecules to produce crosslinks

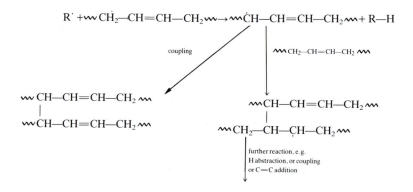

Other methods of non-sulphur vulcanization have been developed (e.g. for polychloroprene, and for room temperature vulcanization of siloxanes) but these are beyond the scope of this brief survey.

The properties of a particular elastomer are controlled by the nature of the crosslinked network. Ideally this can be envisaged as an amorphous network of polymer chains with junction-points from which at least three chains emanate. Every length of polymer chain can be thought of as being anchored at two separate sites. In reality this is a great oversimplification and the likely situation is sketched in Fig. 4.34. In a polymer of finite molar mass there will be some loose chain ends. Also it is possible that

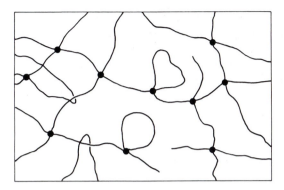

Fig. 4.34 *Schematic diagram of the molecules in an elastomer showing normal crosslinking, chain ends, intramolecular loops and entanglements.*

intramolecular crosslinking can take place leading to the presence of loops. Neither of these two structural irregularities will contribute to the strength of the network. On the other hand, entanglements can act as effective crosslinks, even in an unvulcanized rubber, at least during short-term loading.

Crystallization takes place only very slowly in natural rubber at ambient temperature since ΔT is small, but when a sample is stretched crystallization takes place relatively rapidly. The rate and degree of crystallization depends upon the extension of the sample and the length of time the extension is maintained. It is very rapid at strains of above about 300–400% and degrees of crystallinity of over 30% have been reported at the highest extensions. Crystallization is manifest by the appearance of arcs in the X-ray diffraction patterns. The unstretched material is non-crystalline and the diffraction pattern shows only diffuse rings. On deforming natural rubber the molecules tend to become aligned parallel to the stretching direction and since polymer molecules always pack side-by-side into crystals this alignment reduces the entropy change on crystallization (Section 4.1.1). The stretching process, therefore, aids the crystallization process and the observed arcing in the diffraction patterns follows from the molecular alignment which causes the crystallites to form with their c-axes approximately parallel to the stretching direction. On the other hand, when crystallization takes place in an unstretched sample the diffraction pattern contains only rings showing that the crystallites have random orientations.

4.5.1 *Thermoplastic elastomers*

Although it was emphasized in the previous section that (chemical) crosslinks are normally required for a polymer to display elastomeric behaviour, a novel category of polymers known as *thermoplastic elastomers* are capable of behaving as elastomers without the necessity of having chemical crosslinks. The materials can be processed as thermoplastics at elevated temperatures but when they are cooled down to ambient temperature they become elastomers. The 'crosslinks' in thermoplastic elastomers are physical rather than chemical in nature and these physical crosslinks anchor a network of flexible molecules in the material. This transition in behaviour from a thermoplastic melt to an elastomer is completely reversible and so, unlike conventional elastomers, the thermoplastic elastomers can be reprocessed without difficulty.

Several types of polymers have been developed for use as thermoplastic elastomers and the materials are best described by considering specific examples. Block copolymers are often employed and thermoplastic elastomers can be produced by both step and chain copolymerization.

Thermoplastic polyurethane elastomers are segmented copolymers prepared by the reaction of a diisocyanate with a prepolymer polyol and a short-chain diol as described in Section 2.15.2. The prepolymer polyol blocks form the *soft segments* and diisocyanate reacts with the diol to form polyurethane *hard segments*. The hard segments tend to aggregate to give ordered crystalline domains aided by the possibility of forming hydrogen bonds between adjacent hard segment units. In contrast the soft segments are above their T_g and remain amorphous at ambient temperature. The structure is shown schematically in Fig. 4.35 where it can be seen that the aggregates of hard segments act as junction points in a network of flexible molecules. Although the hard segment domains are relatively stable they can be disrupted by heating the material to its processing temperature or

Virtually crosslinked/extended network
of polymer primary chains

↑ ↓ Heat or solvent

Soft   Hard   Soft   Hard   Soft   Hard   Soft   Hard   Soft

Polymer primary chains

Fig. 4.35 *Schematic illustration of the formation of hard segments in a thermoplastic polyurethane elastomer. (After Schollenberger and Dinbergs (1975) J. Elast. Plast.* **1**, *65).*

by the action of a solvent. This means that the materials can be readily moulded into solid elastomeric artefacts or, by using a solvent, they can be applied as thin elastomeric coatings. Analogous behaviour is found for segmented polyester copolymers (Section 2.15.2).

It was pointed out in Section 2.16.9 that anionic living polymerisation can be used to prepare ABA tri-block copolymers suitable for use as thermoplastic elastomers. In such copolymers the A blocks are normally of a homopolymer which is glassy and the B block is of a rubbery homopolymer (e.g. a polydiene such as polybutadiene or polyisoprene). The characteristic properties of these materials stems from the fact that two polymers which contain repeat units of a different chemical type tend to be incompatible on the molecular level. Thus the block copolymers phase separate into domains which are rich in one or the other type of repeat unit. In the case of the polystyrene–polydiene–polystyrene types of tri-block copolymers used for thermoplastic elastomers (with about 25% by weight polystyrene blocks), the structure is phase-separated at ambient temperature into approximately spherical polystyrene-rich domains which are dispersed in a matrix of the polydiene chains. This type of structure is shown schematically in Fig. 4.36 where it can be seen that the polystyrene blocks are anchored in the spherical domains. At ambient temperature the polystyrene is below its T_g whereas the polydiene is above its T_g. Hence the material consists of a rubbery matrix containing a rigid dispersed phase.

However, since the flexible polydiene blocks are linked by covalent bonds to the polystyrene blocks at the boundaries of the rigid domains, the structure is effectively 'physically crosslinked' and the materials display elastomeric properties. On heating the material to elevated temperatures the polystyrene-rich domains become disrupted and the material can be

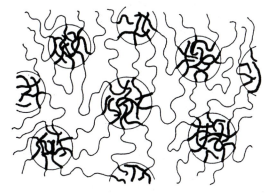

Fig. 4.36 *Schematic representation of the structure of an ABA tri-block copolymer of the polystyrene–polydiene–polystyrene type. The thicker lines represent the polystyrene blocks and the thinner lines the polydiene blocks.*

processed as a normal thermoplastic. Recent developments have led to more complex types of block copolymers with three or more arms with polydiene units in the central hub and polystyrene units in the outer arms.

Further reading

Alexander, L.E. (1969), *X-ray Diffraction Methods in Polymer Science*, Wiley-Interscience, New York.

Bassett, D.C. (1981), *Principles of Polymer Morphology*, Cambridge University Press, Cambridge.

Bassett, D.C. (1982), *Developments in Crystalline Polymers-1*, Applied Science Publishers, London.

Bassett, D.C. (1988), *Developments in Crystalline Polymers-2*, Elsevier Applied Science, London.

Blackley, D.C. (1983), *Synthetic Rubbers*, Applied Science Publishers, London.

Booth, C. and Price, C. (eds) (1989), *Comprehensive Polymer Science*, Vol. 2, Pergamon Press, Oxford.

Hall, I.H. (1984), *Structure of Crystalline Polymers*, Elsevier Applied Science, London.

Haward, R.N. (ed.) (1973), *The Physics of Glassy Polymers*, Applied Science Publishers, London.

Kelly, A. and Groves, G.W. (1970), *Crystallography and Crystal Defects*, Longman, London.

Magill, J.H. (1977), 'Morphogenesis of Solid Polymers' in *Treatise on Materials Science and Technology*, Vol. 10A (ed. J.M. Schultz), Academic Press, New York.

Schultz, J.M. (1974), *Polymer Materials Science*, Prentice-Hall Inc., New Jersey.

Treloar, L.R.G. (1975), *The Physics of Rubber Elasticity*, 3rd edn, Clarendon Press, Oxford.

Wunderlich, B. (1973), *Macromolecular Physics* Vol. 1: Academic Press, London.

Wunderlich, B. (1976), *Macromolecular Physics* Vol. 2: Academic Press, London.

Wunderlich, B. (1980), *Macromolecular Physics* Vol. 3: Academic Press, London.

Problems

Unit cell dimensions for polyethylene may be assumed to be as follows:

$a = 7.41$ Å, $b = 4.96$ Å, $c = 2.54$ Å, $\alpha = \beta = \gamma = 90°$
The wavelength of CuKα radiation is 1.542 Å.

4.1 A flat-plate X-ray diffraction pattern was obtained from an unoriented sample of polyethylene using CuKα radiation. It consisted of three rings of radius 22.2 mm, 36.6 mm and 19.7 mm. The specimen-to-film distance was 50 mm. Calculate the d-spacing of the planes giving rise to these reflections. It is thought that these reflections are from (hk0) type planes. Draw a scale diagram of the unit cell of polyethylene projected along the *c*-axis. Measure the spacing of the low-index (hk0) type planes and hence identify the planes giving rise to the observed reflections.

4.2 A flat-plate X-ray diffraction pattern was obtained from an oriented polymer fibre using CuKα radiation. It consisted of a series of spots arranged along a horizontal line containing the central spot and first-order layer lines above and below the central line. It was found that with a specimen-to-film distance of 30 mm the first order lines were 22.9 mm above and below the central spot.

(a) Calculate the spacing of the chain repeat c of the polymer crystals.

(b) Explain why only first-order layer lines are observed for this polymer and suggest how higher order lines may be obtained.

(c) Determine the angle through which the specimen would have to be tilted in order to obtain the (001) reflection on the diffraction pattern.

4.3 The crystalline density of polyethylene is 1000 kg m^{-3} and the density of amorphous polyethylene is 865 kg m^{-3}. Calculate the mass fraction crystallinity in a sample of linear polyethylene of density 970 kg m^{-3} and in a sample of branched polyethylene of density 917 kg m^{-3}. Why do the two samples have considerably different crystallinities?

4.4 A linear homopolymer was crystallized from the melt at crystallization temperatures (T_c) within the range 270 K to 330 K. Following complete crystallization the melting temperatures (T_m) were measured by differential scanning calorimetry.

T_c/K	T_m/K
270	300.0
280	306.5
290	312.5
300	319.0
310	325.0
320	331.0
330	337.5

Determine graphically the equilibrium melting temperature, $T_m°$. Small-angle X-ray scattering experiments using CuKα radiation gave the positions of the first maxima as:

T_c/K	Θ/degrees
270	0.44
280	0.39
290	0.33
300	0.27
310	0.22
320	0.17
330	0.11

Using the Bragg equation calculate the values of the long period for each crystallization temperature.

The degree of crystallinity was measured to be 45% for all samples. Calculate the lamellar thickness in each case and hence determine graphically the fold surface energy, γ_e, if the enthalpy of fusion of the polymer per unit volume ΔH_V is $1.5 \times 10^8 \mathrm{J\,m^{-3}}$.

4.5 The melting temperature (T_m) of a sample of poly(decamethylene adipate) with a number-average degree of polymerization ($\bar{x}_n$) of 3 was found to be 65°C. T_m was found to be 75°C for a sample of the same polymer with $\bar{x}_n = 10$. Estimate the equilibrium melting temperature T_m^0 for poly(decamethylene adipate)

4.6 The refractive index, $\bar{n}$, of a polymer was determined as a function of temperature. The results are given below:

$T/°C$	$\bar{n}$
20	1.5913
30	1.5898
40	1.5883
50	1.5868
60	1.5853
70	1.5838
80	1.5822
90	1.5801
100	1.5766
110	1.5725
120	1.5684
130	1.5643

Determine the glass transition temperature of the polymer.

4.7 It is intended to toughen poly(methyl methacrylate) (PMMA) by the addition of particles of rubbery poly[(n-butyl acrylate)-*co*-styrene] made by emulsion polymerization. In order to retain the good optical clarity of the PMMA it is necessary to match the refractive indices, $\bar{n}$, of the PMMA and copolymer particles. Estimate the relative proportions of the two comonomers required in the particles given that at room temperature the refractive index of PMMA is 1.489 and $\bar{n}$ is 1.466 for poly(n-butyl acrylate) and 1.591 for polystyrene. Would you expect the material to retain its clarity if the temperature were changed from ambient?

4.8 Discuss the reasons for the differences in glass transition temperature for the following pairs of polymers with similar chemical structures.

(a) Polyethylene (~150 K) and polypropylene (~250 K).
(b) Poly(methyl acrylate) (283 K) and poly(vinyl acetate) (305 K).
(c) Poly(but-1-ene) (249 K) and poly(but-2-ene) (200 K).
(d) Poly(ethylene oxide) (232 K) and poly(vinyl alcohol) (358 K).
(e) Poly(ethyl acrylate) (249 K) and poly(methyl methacrylate) (378 K).

4.9 The following data were obtained for the values of glass transition temperature, T_g of poly[(vinylidene fluoride)-*co*-chlorotrifluoroethylene] as a function of the weight fraction w_A of vinylidene fluoride

w_A	0	0.14	0.35	0.4	0.54	1.0
T_g/K	319	292	270	265	258	235

By fitting these data to a suitable curve estimate the value of T_g for a copolymer with $w_A = 0.75$.

4.10 The glass transition temperature, T_g, of a linear polymer with number-average molar mass $\bar{M}_n = 2300$ was found to be 121°C. The value of T_g increased to 153°C for a sample of the same linear polymer with $\bar{M}_n = 9000$. A branched version of the same polymer with $\bar{M}_n = 5200$ was found to have a T_g of 115°C. Determine the average number of branches on the molecules of the branched polymer.

5 Mechanical properties

5.1 **General considerations**

Polymers are often thought of as being mechanically weak and their mechanical properties have consequently been somewhat ignored in the past. These days many polymers are used in structural engineering applications and are subjected to appreciable stresses. This increase in use has been due to several factors. One of the most important being that, although on an absolute basis their mechanical strength and stiffness may be relatively low compared with metals and ceramics, when the low density of polymers is taken into account their specific strength and stiffness become more comparable with those of conventional materials. Also the fabrication costs of polymeric components are usually considerably lower than for other materials. Polymers melt at relatively low temperature and can be readily moulded into quite intricate components using a single moulding.

Perhaps the most over-riding reason for using polymers is that they can display unique mechanical properties. One polymer that has been used for many years is natural rubber which is elastomeric. It can be stretched to very high extensions and will snap back immediately the stress is removed. Polyolefins are widely used in domestic containers and in packaging because of their remarkable toughness and flexibility and ease of fabrication. Other polymers are employed as adhesives. Epoxy resins can produce bonds in metals which can be as strong as those made by conventional joining methods such as welding or rivetting.

With the now increasingly more widespread use of polymers an understanding of their mechanical properties is becoming essential. In this chapter the mechanical properties have been considered from a pheno-menological viewpoint and in terms of the molecular deformation processes which occur. This latter approach has been facilitated by a better understanding of polymer structure which was outlined in the previous chapter.

5.1.1 *Stress*

Force can be exerted externally on a body in two particular ways. Gravity and inertia can be thought of as *body forces* since they act directly on all the individual particles in the body. The other type are *surface* or *contact forces* which act only on the surface of the body, but their effect is transmitted to the particles inside the body through the atomic and molecular bonds. In consideration of the mechanical properties of polymers we are mainly interested in the effect of applying surface forces such as stress or pressure to the material but it must be remembered that certain polymers such as non-crosslinked rubbers will flow under their own weight, a simple response to the body force of gravity.

Although the response of these forces reflects the displacement of the individual particles within the body the system is normally considered from a macroscopic viewpoint and it is regarded as a continuum. In order to define the state of stress at a point within the body we consider the surface forces acting on a small cube of material around that point as shown in Fig. 5.1. Each of these surface forces is divided by the area upon which it acts and then it is resolved into components which are parallel to the three co-ordinate axes. In total there will be nine stress components given as

$$\begin{matrix} \sigma_{11} & \sigma_{12} & \sigma_{13} \\ \sigma_{21} & \sigma_{22} & \sigma_{23} \\ \sigma_{31} & \sigma_{32} & \sigma_{33} \end{matrix}$$

where the first subscript gives the normal to the plane on which the stress acts and the second subscript defines the direction of the stress. The components σ_{11}, σ_{22} and σ_{33} are known as normal stresses as they are perpendicular to the plane on which they act and the others are shear stresses. The nine stress components are not independent of each other. If the cube is in equilibrium and not rotating it is necessary that for the shear stresses

$$\sigma_{12} = \sigma_{21}, \sigma_{23} = \sigma_{32}, \sigma_{31} = \sigma_{13}$$

and so the state of stress at a point in a body in equilibrium can be defined by a stress tensor σ_{ij} which has six independent components

$$\sigma_{ij} = \begin{pmatrix} \sigma_{11} & \sigma_{12} & \sigma_{13} \\ \sigma_{21} & \sigma_{22} & \sigma_{23} \\ \sigma_{31} & \sigma_{32} & \sigma_{33} \end{pmatrix} \tag{5.1}$$

The knowledge of the state of stress at a point provides enough information to calculate the stress acting on any plane within the body.

It is possible to express the stress acting at a point in terms of three *principal stresses* acting along principal axes. In this case the shear stresses ($i \neq j$) are all zero and the only terms remaining in the stress tensor are σ_{11}, σ_{22} and σ_{33}. It is often possible to determine the principal axes from simple inspection of the body and the state of stress. For example, two principal axes always lie in the plane of a free surface. In the consideration of the

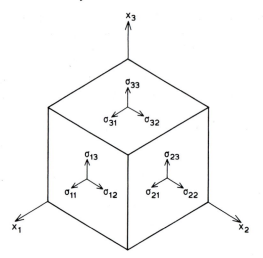

Fig. 5.1 *Schematic illustration of the nine stress components for a rectangular coordinate system.*

mechanical properties of polymers it is often useful to divide the stress tensor into its hydrostatic and deviatoric components. The hydrostatic pressure p is given by

$$p = \tfrac{1}{3}(\sigma_{11} + \sigma_{22} + \sigma_{33}) \tag{5.2}$$

and the deviatoric stress tensor σ'_{ij} is found by subtracting the hydrostatic stress components from the overall tensor such that

$$\sigma'_{ij} = \begin{pmatrix} (\sigma_{11} - p) & \sigma_{12} & \sigma_{13} \\ \sigma_{21} & (\sigma_{22} - p) & \sigma_{23} \\ \sigma_{31} & \sigma_{32} & (\sigma_{33} - p) \end{pmatrix} \tag{5.3}$$

5.1.2 *Strain*

When forces are applied to a material the atoms change position in response to the force and this change is known as *strain*. A simple example is that of a thin rod of material of length, l, which is extended a small amount δl by an externally applied stress. In this case the strain e could be represented by

$$e = \delta l / l \tag{5.4}$$

For a general type of deformation consisting of extension and compression in different directions and shears the situation is very much more complicated. However, the analysis can be greatly simplified by assuming that any strains are small and terms in e^2 can be neglected. In practice the analysis can only normally be strictly applied to the elastic deformation of crystals and a rather different approach has to be applied in the case of, for example, elastomer deformation where very high strains are involved.

In the analysis of *infinitesimal strains* the displacements of particular points in a body are considered. It is most convenient to use a rectangular co-ordinate system with axes x_1, x_2 and x_3 meeting at the origin 0. The analysis is most easily explained by looking at the two-dimensional case as shown in Fig. 5.2. If P is a point in the body with co-ordinates (x_1, x_2) then when the body is deformed it will move to a point P′ given by co-ordinates $(x_1 + u_1, x_2 + u_2)$. However, another point Q which is infinitesimally close to P will not be displaced by the same amount. If Q originally has co-ordinates of $(x_1 + dx_1, x_2 + dx_2)$ its displacement to Q′ will have components $(u_1 + du_1, u_2 + du_2)$. Since dx_1 and dx_2 are infinitesimal we can write

$$du_1 = \frac{\partial u_1}{\partial x_1}dx_1 + \frac{\partial u_1}{\partial x_2}dx_2 \tag{5.5}$$

and

$$du_2 = \frac{\partial u_2}{\partial x_1}dx_1 + \frac{\partial u_2}{\partial x_2}dx_2 \tag{5.6}$$

It is necessary to define the partial differentials and this can be done by writing them as

$$e_{11} = \frac{\partial u_1}{\partial x_1} \qquad e_{12} = \frac{\partial u_1}{\partial x_2}$$

$$e_{21} = \frac{\partial u_2}{\partial x_1} \qquad e_{22} = \frac{\partial u_2}{\partial x_2}$$

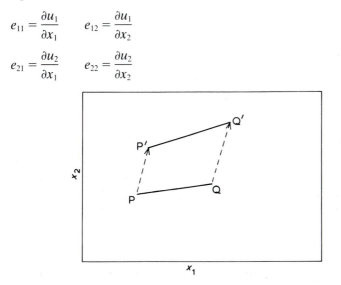

Fig. 5.2 *General displacement in two dimensions of points P and Q in a body (after Kelly and Groves).*

Equations (5.5) and (5.6) can be written more neatly as

$$du_i = e_{ij} \, dx_j \qquad (j = 1,2)$$

Since du_i and dx_j are vectors then e_{ij} must be a tensor and the significance of the various e_{ij} can be seen by taking particular examples of states of strain.

The simplest case is where the strains are small and the vector PQ can be taken as first parallel to the x_1 axis and then parallel to the x_2 axis as shown in Fig. 5.3(a). This enables the distortion of a rectangular element to be determined by considering the displacement of each vector separately. Since PQ_1 is parallel to x_1 then $dx_2 = 0$ and so

$$du_1 = \frac{\partial u_1}{\partial x_1} dx_1 = e_{11} dx_1 \tag{5.7}$$

$$du_2 = \frac{\partial u_2}{\partial x_1} dx_1 = e_{21} dx_1 \tag{5.8}$$

The significance of these equations can be seen from Fig. 5.3(a). When e_{11} and e_{21} are small the term e_{11} corresponds to the extensional strain of PQ_1 resolved along x_1 and e_{21} gives the anticlockwise rotation of PQ_1. It also follows that e_{22} will correspond to the extensional strain of PQ_2 resolved along x_2 and e_{12} the clockwise rotation of PQ_2.

The simple tensor notation of strain outlined above is not an entirely satisfactory representation of the strain in a body as it is possible to have certain components of $e_{ij} \neq 0$ without the body being distorted. This situation is illustrated in Fig. 5.3(b). In this case $e_{11} = e_{22} = 0$ but a small rotation, ω of the body can be given by $e_{12} = -\omega$ and $e_{21} = \omega$. This problem can be overcome by expressing the general strain tensor e_{ij} as a sum of an anti-symmetrical tensor and a symmetrical tensor such that

$$\left. \begin{array}{l} e_{ij} = \tfrac{1}{2}(e_{ij} - e_{ji}) + \tfrac{1}{2}(e_{ij} + e_{ji}) \\ \text{or} \quad e_{ij} = \qquad \omega_{ij} \qquad + \qquad \varepsilon_{ij} \end{array} \right\} \tag{5.9}$$

The term ω_{ij} then gives the rotation and ε_{ij} represents the pure strain.

This analysis can be readily extended to three dimensions and the strain tensor e_{ij} has nine rather than four components. The tensor ε_{ij} which represents the pure strain is again given by

$$\varepsilon_{ij} = \tfrac{1}{2}(e_{ij} + e_{ji})$$

and it can be written more fully as

$$\begin{pmatrix} \varepsilon_{11} & \varepsilon_{12} & \varepsilon_{13} \\ \varepsilon_{21} & \varepsilon_{22} & \varepsilon_{23} \\ \varepsilon_{31} & \varepsilon_{32} & \varepsilon_{33} \end{pmatrix} = \begin{pmatrix} e_{11} & \tfrac{1}{2}(e_{12} + e_{21}) & \tfrac{1}{2}(e_{13} + e_{31}) \\ \tfrac{1}{2}(e_{21} + e_{12}) & e_{22} & \tfrac{1}{2}(e_{23} + e_{32}) \\ \tfrac{1}{2}(e_{31} + e_{13}) & \tfrac{1}{2}(e_{32} + e_{23}) & e_{33} \end{pmatrix} \tag{5.10}$$

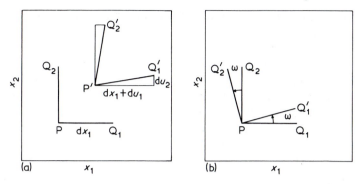

Fig. 5.3 *Two-dimensional displacements used in the definition of strain.* (a) *General displacements.* (b) *Rotation when* $e_{11} = e_{22} = 0$ *but* $e_{12} = -\omega$ *and* $e_{21} = \omega$ *(after Kelly and Groves).*

Inspection of this equation shows that $\varepsilon_{12} = \varepsilon_{21}$, $\varepsilon_{23} = \varepsilon_{32}$, $\varepsilon_{31} = \varepsilon_{13}$ which means that the strain tensor ε_{ij} has only six independent components (cf. stress tensor σ_{ij} in the last section). The diagonal components of ε_{ij} correspond to the tensile strains parallel to the axes and the off-diagonal ones are then shear strains.

As with stresses it is possible to choose a set of axes such that the shear strains are all zero and the only terms remaining are the *principal strains*, ε_{11}, ε_{22} and ε_{33}. In fact, for an isotropic body the axes of principal stress and principal strain coincide. This type of deformation can be readily demonstrated by considering what happens to a unit cube with the principal axes parallel to its edges. On deformation it transforms into a rectangular brick of dimensions $(1 + \varepsilon_{11})$, $(1 + \varepsilon_{22})$ and $(1 + \varepsilon_{33})$.

5.1.3 Relationship between stress and strain

For many materials the relationship between stress and strain can be expressed, at least at low strains, by *Hooke's law* which states that stress is proportional to strain. It must be borne in mind that this is not a fundamental law of nature. It was discovered by empirical observation and certain materials, particularly polymers, tend not to obey it. Hooke's law enables us to define the Young's modulus, E of a material which for simple uniaxial extension or compression is given by

$E = \text{stress/strain}$

It will be readily appreciated that this simple relationship cannot be applied without modification to complex systems of stress. In order to do this Hooke's law must be generalized and the relationship between σ_{ij} and ε_{ij} is obtained in its most general form by assuming that every stress component is a linear function of every strain component.

i.e. $\sigma_{11} = A\varepsilon_{11} + B\varepsilon_{22} + C\varepsilon_{33} + D\varepsilon_{12} \ldots$

and $\sigma_{22} = A'\varepsilon_{11} + B'\varepsilon_{22} + C'\varepsilon_{33} + D'\varepsilon_{12} \ldots$

This can be expressed most neatly using tensor notation as

$$\sigma_{ij} = c_{ijkl}\varepsilon_{ij} \tag{5.11}$$

where c_{ijkl} is a fourth-rank tensor containing all the stiffness constants. In principle, there are $9 \times 9 = 81$ stiffness constants but the symmetry of σ_{ij} and ε_{ij} reduces them to $6 \times 6 = 36$.

The notation used is often simplified to a contracted form

$$\sigma_{11} \rightarrow \sigma_1, \quad \sigma_{22} \rightarrow \sigma_2, \quad \sigma_{33} \rightarrow \sigma_3$$
$$\sigma_{23} \rightarrow \sigma_4, \quad \sigma_{13} \rightarrow \sigma_5, \quad \sigma_{12} \rightarrow \sigma_6$$
$$\varepsilon_{11} \rightarrow \varepsilon_1, \quad \varepsilon_{22} \rightarrow \varepsilon_2, \quad \varepsilon_{33} \rightarrow \varepsilon_3$$
$$2\varepsilon_{23} \rightarrow \gamma_4, \quad 2\varepsilon_{13} \rightarrow \gamma_5, \quad 2\varepsilon_{12} \rightarrow \gamma_6$$

It is conventional to replace the tensor shear strains by the angles of shear γ, which are twice the corresponding shear component in the strain tensor. There is also a corresponding contraction in c_{ijkl} which can then be written as

$$c_{ij} = \begin{pmatrix} c_{11} & c_{12} & c_{13} & c_{14} & c_{15} & c_{16} \\ c_{21} & c_{22} & c_{23} & c_{24} & c_{25} & c_{26} \\ c_{31} & c_{32} & c_{33} & c_{34} & c_{35} & c_{36} \\ c_{41} & c_{42} & c_{43} & c_{44} & c_{45} & c_{46} \\ c_{51} & c_{52} & c_{53} & c_{54} & c_{55} & c_{56} \\ c_{61} & c_{62} & c_{63} & c_{64} & c_{65} & c_{66} \end{pmatrix} \tag{5.12}$$

Only 21 of the 36 components of the matrix are independent since strain energy considerations show that $c_{ij} = c_{ji}$. Also if the strain axes are chosen to coincide with the symmetry axes in the material there may be a further reduction in the number of independent components. For example, in a cubic crystal the x_1, x_2 and x_3 axes are equivalent and so

$$c_{11} = c_{22} = c_{33}$$
$$c_{12} = c_{23} = c_{31}$$
$$c_{44} = c_{55} = c_{66}$$

and it can be proved further that all the other stiffness constants are zero. This means that there are only three independent elastic constants for cubic crystals, c_{11} c_{12} and c_{44}. Polymer crystals are always of lower symmetry and consequently there are more independent constants. Orthorhombic polyethylene (Table 4.1) requires nine stiffness constants to fully characterize its elastic behaviour.

In the case of elastically-isotropic solids there are only two independent elastic constants, c_{11} and c_{12} because $2c_{44} = c_{11} - c_{12}$. Since glassy polymers and randomly oriented semi-crystalline polymers fall into this category it is worth considering how the stiffness constants can be related to quantities such as Young's modulus, E, Poisson's ratio v, shear modulus G and bulk modulus K which are measured directly. The *shear modulus* G relates the shear stress σ_4 to the angle of shear γ_4 through the equation

$$\sigma_4 = G\gamma_4 \tag{5.13}$$

and since $c_{44} = \sigma_4/\gamma_4$ then we have the simple result that the shear modulus is given by

$$G = c_{44} \tag{5.14}$$

The situation is not so simple in the case of *uniaxial tension* where all stresses other than σ_1 are equal to zero and $\varepsilon_2 = \varepsilon_3$. The generalized Hooke's law can be expressed as

$$\begin{aligned}
\sigma_1 &= c_{11}\varepsilon_1 + c_{12}\varepsilon_2 + c_{13}\varepsilon_3 \\
&= c_{12}(\varepsilon_1 + \varepsilon_2 + \varepsilon_3) + (c_{11} - c_{12})\varepsilon_1 \\
&= c_{12}\Delta + 2c_{44}\varepsilon_1
\end{aligned} \tag{5.15}$$

where $\Delta = \varepsilon_1 + \varepsilon_2 + \varepsilon_3$ represents, at low strains, the fractional volume change or dilatation. For triaxial deformation $(\sigma_1, \sigma_2, \sigma_3 \neq 0)$ the *bulk modulus* K relates the hydrostatic pressure p to Δ through

$$p = K\Delta \tag{5.16}$$

Since $p = \frac{1}{3}(\sigma_1 + \sigma_2 + \sigma_3)$ using variants of Equation (5.15)

$$p = (c_{12} + \tfrac{2}{3}c_{44})\Delta \tag{5.17}$$

Comparing Equations (5.16) and (5.17) enables K to be expressed as

$$K = c_{12} + \tfrac{2}{3}c_{44} \tag{5.18}$$

This equation then allows an expression for the Young's modulus E to be obtained. The Young's modulus relates the stress and strain in uniaxial tension through

$$\sigma_1 = E\,\varepsilon_1 \tag{5.19}$$

In uniaxial tension $\sigma_2 = \sigma_3 = 0$ and so

$$\sigma_1 = 3p = (3c_{12} + 2c_{44})\Delta \tag{5.20}$$

Substituting for σ_1 from Equation (5.15) and rearranging gives

$$\varepsilon_1 = \frac{(c_{12} + c_{44})\Delta}{c_{44}} \tag{5.21}$$

and so the modulus is given by

$$E = \frac{\sigma_1}{\varepsilon_1} = \frac{c_{44}(3c_{12} + 2c_{44})}{(c_{12} + c_{44})} \tag{5.22}$$

The final constant that can be derived is *Poisson's ratio* v which is the ratio of the lateral strain to the longitudinal extension measured in a uniaxial tensile test. It is then given by

$$v = -\frac{\varepsilon_2}{\varepsilon_1} = -\frac{\varepsilon_3}{\varepsilon_1} \tag{5.23}$$

This can be related to Δ through

$$v = \tfrac{1}{2}(1 - \Delta/\varepsilon_1) \tag{5.24}$$

which using Equation (5.21) gives

$$v = \frac{c_{12}}{2(c_{12} + c_{44})} \tag{5.25}$$

It is clear that since the four constants, G, K, E and v can be expressed in terms of three stiffness constants, only two of which are independent, then G, K, E and v must all be related to each other. It is a matter of simple algebraic manipulation to show that, for example

$$G = \frac{E}{2(1 + v)} \quad \text{and} \quad K = \frac{E}{3(1 - 2v)} \quad \text{etc.}$$

The final point of this exercise is to show how an isotropic body will respond to a general system of stresses. A stress of σ_1 will cause a strain of σ_1/E along the x_1 direction, but will also lead to strains of $-v\sigma_1/E$ along the x_2 and x_3 directions. Similarly a stress of σ_2 will cause strains of σ_2/E along x_2 and of $-v\sigma_2/E$ in the other directions. If it is assumed that the strains due to the different stresses are additive then using the original notation it follows that

$$\varepsilon_{11} = \frac{\sigma_{11}}{E} - \frac{v}{E}(\sigma_{22} + \sigma_{33}) = e_{11} \tag{5.26}$$

$$\varepsilon_{22} = \frac{\sigma_{22}}{E} - \frac{v}{E}(\sigma_{33} + \sigma_{11}) = e_{22} \tag{5.27}$$

$$\varepsilon_{33} = \frac{\sigma_{33}}{E} - \frac{v}{E}(\sigma_{11} + \sigma_{22}) = e_{33} \tag{5.28}$$

The shear strains can be obtained directly from Equation (5.13) and are $\varepsilon_{12} = \sigma_{12}/2G$, $\varepsilon_{23} = \sigma_{23}/2G$, $\varepsilon_{31} = \sigma_{31}/2G$.

5.1.4 *Theoretical shear stress*

In the consideration of the resistance of materials to plastic flow it is useful to estimate theoretically the stress that may be required to shear the structure. It is necessary to do this before a sensible comparison can be made of the relative strengths of different types of materials such as metals and polymers. The general situation to be considered is shown in Fig. 5.4 where one sheet of atoms or molecules is allowed to slip over another under the action of a shear stress σ. The separation of the sheets is h and the repeat distance in the direction of shear is b. It is envisaged that the shear stress will vary periodically with the shear displacement x in an approximately sinusoidal manner and that the maximum resistance to shear will be when $x = b/4$. The variation of σ with the displacement x can then be written as

$$\sigma = \sigma_u \sin(2\pi x/b) \tag{5.29}$$

The initial slope of the curve of σ as a function of x will be equal to the shear modulus G of the material and so at small strains

$$\sigma = \sigma_u 2\pi x/b = Gx/h \tag{5.30}$$

and so the ideal shear strength is given by

$$\sigma_u = \frac{Gb}{2\pi h} \tag{5.31}$$

In a face-centred cubic metal $h = a/\sqrt{3}$ and $b = a/\sqrt{6}$ where a is the lattice parameter and so the theoretical shear strength is predicted to be $\sigma_u \approx G/9$. For chain direction slip on the (020) planes in the polyethylene crystal $b = 2.54$ Å and h is equal to the separation of the (020) planes which is 2.47 Å. The theoretical shear stress would be expected to be of the order of $G/6$. However, more sophisticated calculations of Equation (5.31) lead to lower estimates of σ_u which come out to be of the order of $G/30$ for most materials. Even so, this estimate of the stress required to shear the structure is very much higher than the values that are normally measured and the discrepancy is due to the presence of defects such as *dislocations* within the crystals. The high values are only realized for certain crystal whiskers and other perfect crystals.

5.1.5 *Theoretical tensile strength*

Another parameter of interest in the consideration of the mechanical behaviour of materials is the stress that is expected to be required to cause

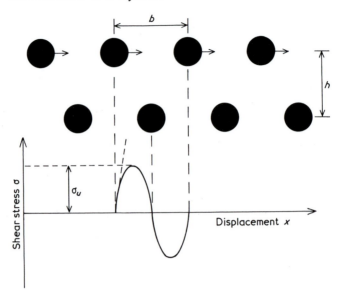

Fig. 5.4 *Schematic representation of the shear of layers of atoms in crystal and the expected variation of shear stress with displacement.*

cleavage through the breaking of bonds. The situation is visualized in Fig. 5.5 in which a bar of unit cross-sectional area, made of a perfectly elastic material is pulled in tension. Work is done in stressing the material and since fracture involves the creation of two new surfaces thermodynamic considerations demand that the work cannot be less than the surface energy of the new surfaces. The expected dependence of the stress σ upon the displacement x is shown in Fig. 5.5 and it is envisaged that this can be approximated to a sine function of wavelength λ and of the form

$$\sigma = \sigma_t \sin(2\pi x/\lambda) \tag{5.32}$$

where σ_t is the maximum tensile stress. This can be written for low strains as

$$\sigma = \sigma_t 2\pi x/\lambda \tag{5.33}$$

but another equation can be derived relating σ to x since the material is perfectly elastic. If the equilibrium separation of the atomic planes perpendicular to the tensile axis is h then for small displacements Hooke's law gives

$$\sigma = Ex/h \tag{5.34}$$

and comparing Equations (5.33) and (5.34) leads to

$$\sigma_t = \frac{\lambda E}{2\pi h} \tag{5.35}$$

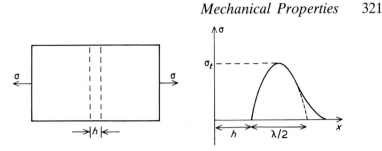

Fig. 5.5 *Model used to calculate the theoretical cleavage stress of a crystal.*

The work done when the bar is fractured is given approximately by the area under the curve and equating this to the energy required to create the two new surfaces leads to

$$\int_{0}^{\lambda/2} \sigma_t\sin(2\pi x/\lambda)\mathrm{d}x = 2\gamma$$

or
$$\gamma = \lambda\sigma_t/2\pi$$

Substituting for $\lambda/2\pi$ from Equation (5.35) then gives

$$\sigma_t = \sqrt{\frac{E\gamma}{h}} \tag{5.36}$$

Putting typical values of E, γ and h into this equation for a variety of materials shows that it is generally expected that

$$\sigma_t \approx E/10 \tag{5.37}$$

This result is particularly interesting in the context of polymers because although their moduli are generally low there are certain circumstances in which E can be made large. Oriented amorphous and semi-crystalline polymers have relatively high moduli and the moduli of polymer crystals in the chain direction are comparable to those of metals. High tensile strengths might therefore be expected to be displayed when polymers are deformed in ways which exploit these high moduli.

In general, it is found that the tensile strengths of materials are very much lower than $E/10$ because brittle materials fracture prematurely due to the presence of flaws (Section 5.6.2) and ductile materials undergo plastic deformation through the motion of dislocations. However, fine whiskers of glass, silica and certain polymer crystals which do not contain any flaws and are not capable of plastic deformation have values of fracture strength which are close to the theoretically predicted ones.

5.2 **Viscoelasticity**

A distinctive feature of the mechanical behaviour of polymers is the way in which their response to an applied stress or strain depends upon the rate or time period of loading. This dependence upon rate and time is in marked contrast to the behaviour of elastic solids such as metals and ceramics which, at least at low strains, obey *Hooke's law* and the stress is proportional to the strain and independent of loading rate. On the other hand the mechanical behaviour of viscous liquids is time dependent. It is possible to represent their behaviour at low rates of strain by *Newton's law* whereby the stress is proportional to the strain-rate and independent of the strain. The behaviour of most polymers can be thought of as being somewhere between that of elastic solids and liquids. At low temperatures and high rates of strain they display elastic behaviour whereas at high temperatures and low rates of strain they behave in a viscous manner, flowing like a liquid. Polymers are therefore termed *viscoelastic* as they display aspects of both viscous and elastic types of behaviour.

Polymers used in engineering applications are often subjected to stress for prolonged periods of time, but it is not possible to know how a polymer will respond to a particular load without a detailed knowledge of its viscoelastic properties. There have been many studies of the viscoelastic properties of polymers and several textbooks written specifically about the subject of viscoelasticity. It is intended to give a brief introduction to what is really a vast subject and to leave the reader to consult more advanced texts for further details.

5.2.1 *General time-dependent behaviour*

The exact nature of the time dependence of the mechanical properties of a polymer sample depends upon the type of stress or straining cycle employed. The variation of the stress σ and strain e with time t is illustrated schematically in Fig. 5.6 for a simple polymer tensile specimen subjected to four different deformation histories. In each case the behaviour of an elastic material is also given as a dashed line for comparison. During *creep* loading a constant stress is applied to the specimen at $t = 0$ and the strain increases rapidly at first, slowing down over longer time periods. In an elastic solid the strain stays constant with time. The behaviour of a viscoelastic material during *stress relaxation* is shown in Fig. 5.6(b). In this case the strain is held constant and the stress decays slowly with time whereas in an elastic solid it would remain constant. The effect of deforming a viscoelastic material at a *constant stress-rate* is shown in Fig. 5.6(c). The increase in strain is not linear and the curve becomes steeper with time and also as the stress-rate is increased. If different *constant strain-*

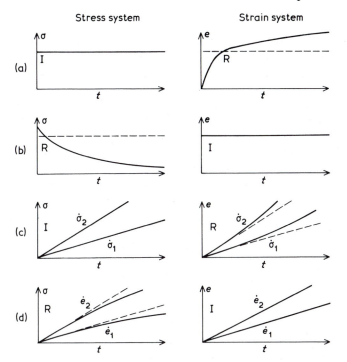

Fig. 5.6 *Schematic representation of the variation of stress and strain with time indicating the input (I) and responses (R) for different types of loading. (a) Creep, (b) Relaxation, (c) Constant stressing-rate, (d) Constant straining-rate (After Williams).*

rates are used the variation of stress with time is not linear. The slope of the curve tends to decrease with time, but it is steeper for higher strain-rates. The variation of both strain and stress with time is linear for constant stress- and strain-rate tests upon elastic materials.

5.2.2 Viscoelastic mechanical models

It is clear that any theoretical explanations of the above phenomena should be able to account for the dependence of the stress and strain upon time. Ideally it should be possible to predict, for example, the stress relaxation behaviour from knowledge of the creep curve. In practice, with real polymers this is somewhat difficult to do but the situation is often simplified by assuming that the polymer behaves as a *linear viscoelastic material*. It can be assumed that the deformation of the polymer is divided into an elastic component and a viscous component and that the deformation of the polymer can be described by a combination of Hooke's law and Newton's law. The linear elastic behaviour is given by Hooke's law as

$$\sigma = Ee \tag{5.38}$$

$$\text{or} \quad d\sigma/dt = E\, de/dt \tag{5.39}$$

where E is the elastic modulus and Newton's law describes the linear viscous behaviour through the equation

$$\sigma = \eta \frac{de}{dt} \tag{5.40}$$

where η is the viscosity and de/dt the strain rate. It should be noted that these equations only apply at small strains. A particularly useful method of formulating the combination of elastic and viscous behaviour is through the use of mechanical models. The two basic components used in the models are an elastic spring of modulus E which obeys Hooke's law (Equation 5.38) and a viscous dashpot of viscosity, η, which obeys Newton's law (Equation 5.40). The various models that have been proposed involve different combinations of these two basic elements.

(i) Maxwell model

This model was proposed in the 19th century by Maxwell to explain the time-dependent mechanical behaviour of viscous materials, such as tar or pitch. It consists of a spring and dashpot in series as shown in Fig. 5.7(a). Under the action of an overall stress σ there will be an overall strain e in the system which is given by

$$e = e_1 + e_2 \tag{5.41}$$

where e_1 is the strain in the spring and e_2 the strain in the dashpot. Since the elements are in series the stress will be identical in each one and so

$$\sigma_1 = \sigma_2 = \sigma \tag{5.42}$$

Equations (5.39) and (5.40) can be written therefore as

$$\frac{d\sigma}{dt} = E\frac{de_1}{dt} \quad \text{and} \quad \sigma = \eta\frac{de_2}{dt}$$

for the spring and dashpot respectively. Differentiation of Equation (5.41) gives

$$\frac{de}{dt} = \frac{de_1}{dt} + \frac{de_2}{dt} \tag{5.43}$$

and so for the Maxwell model

$$\frac{de}{dt} = \frac{1}{E}\frac{d\sigma}{dt} + \frac{\sigma}{\eta} \tag{5.44}$$

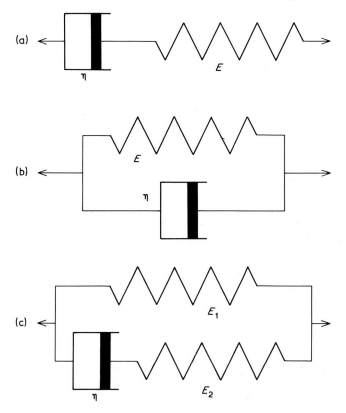

Fig. 5.7 *Mechanical models used to represent the viscoelastic behaviour of polymers.* (a) *Maxwell model,* (b) *Voigt model,* (c) *Standard linear solid.*

At this stage it is worth considering how closely the Maxwell model predicts the mechanical behaviour of a polymer. In the case of creep the stress is held constant at $\sigma = \sigma_0$ and so $d\sigma/dt = 0$. Equation (5.44) can therefore be written as

$$\frac{de}{dt} = \frac{\sigma_0}{\eta} \tag{5.45}$$

Thus, the Maxwell model predicts Newtonian flow as shown in Fig. 5.8(a). The strain is expected to increase linearly with time which is clearly not the case for a viscoelastic polymer (Fig. 5.6) where de/dt decreases with time.

The model is perhaps of more use in predicting the response of a polymer during stress relaxation. In this case a constant strain $e = e_0$ is imposed on the system and so $de/dt = 0$. Equation (5.44) then becomes

$$0 = \frac{1}{E}\frac{d\sigma}{dt} + \frac{\sigma}{\eta} \tag{5.46}$$

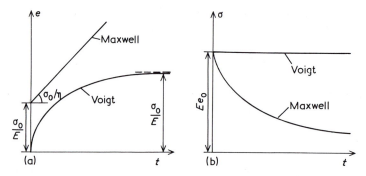

Fig. 5.8 *The behaviour of the Maxwell and Voigt models during different types of loading.* (a) *Creep (constant stress σ_0), (b) Relaxation (constant strain e_0) (after Williams).*

and hence

$$\frac{d\sigma}{\sigma} = -\frac{E}{\eta}dt \tag{5.47}$$

This can be readily integrated if at $t = 0$, $\sigma = \sigma_0$ and so

$$\sigma = \sigma_0 \exp\left(-\frac{Et}{\eta}\right) \tag{5.48}$$

where σ_0 is the initial stress. The term η/E will be constant for a given Maxwell model and it is sometimes referred to as a 'relaxation time', τ_0. The equation can then be written as

$$\sigma = \sigma_0 \exp(-t/\tau_0) \tag{5.49}$$

and it predicts an exponential decay of stress as shown in Fig. 5.8(b). This is a rather better representation of polymer behaviour than the Maxwell model gave for creep. However, Equation (5.49) predicts that the stress relaxes completely over long a period of time which is not normally the case for a real polymer.

(ii) *Voigt model*

This is also known as the Kelvin model. It consists of the same elements as are in the Maxwell model but in this case they are in parallel (Fig. 5.7(b)) rather than in series. The parallel arrangement of the spring and the dashpot means that the strains are uniform

i.e. $e = e_1 = e_2$ \hfill (5.50)

and the stresses in each component will add to make an overall stress of σ such that

$$\sigma = \sigma_1 + \sigma_2 \tag{5.51}$$

The individual stresses σ_1 and σ_2 can be obtained from Equations (5.38) and (5.40) and are

$$\sigma_1 = Ee \quad \text{and} \quad \sigma_2 = \eta\frac{de}{dt}$$

for the spring and the dashpot respectively. Putting these expressions into Equation (5.51) and rearranging gives

$$\frac{de}{dt} = \frac{\sigma}{\eta} - \frac{Ee}{\eta} \tag{5.52}$$

The usefulness or otherwise of the model can again be determined by examining its response to particular types of loading. The Voigt or Kelvin model is particularly useful in describing the behaviour during creep where the stress is held constant at $\sigma = \sigma_0$. Equation (5.52) then becomes

$$\frac{de}{dt} + \frac{Ee}{\eta} = \frac{\sigma_0}{\eta} \tag{5.53}$$

This simple differential equation has the solution

$$e = \frac{\sigma_0}{E}\left[1 - \exp\left(-\frac{Et}{\eta}\right)\right] \tag{5.54}$$

The constant ratio η/E can again be replaced by τ_0, the relaxation time and so the variation of strain with time for a Voigt model undergoing creep loading is given by

$$e = \frac{\sigma_0}{E}[1 - \exp(-t/\tau_0)] \tag{5.55}$$

This behaviour is shown in Fig. 5.8(a) and it clearly represents the correct form of behaviour for a polymer undergoing creep. The strain rate decreases with time and $e \to \sigma_0/E$ as $t \to \infty$. On the other hand, the Voigt model is unsuccessful in predicting the stress relaxation behaviour of a polymer. In this case the strain is held constant at e_0 and since $de/dt = 0$ then Equation (5.52) becomes

$$\frac{\sigma}{\eta} = \frac{Ee_0}{\eta} \tag{5.56}$$

$$\text{or} \quad \sigma = Ee_0 \tag{5.57}$$

which is the linear elastic response indicated in Fig. 5.8(b).

(iii) *Standard linear solid*
It has been demonstrated that the Maxwell model describes the stress relaxation of a polymer to a first approximation whereas the Voigt model

similarly describes creep. A logical step forward is therefore to find some combination of these two basic models which can account for both phenomena. A simple model which does this is known as the standard linear solid, one example of which is shown in Fig. 5.7(c). In this case a Maxwell element and spring are in parallel. The presence of the second spring will stop the tendency of the Maxwell element undergoing simple viscous flow during creep loading, but will still allow the stress relaxation to occur.

There have been many attempts at devising more complex models which can give a better representation of the viscoelastic behaviour of polymers. As the number of elements increases the mathematics becomes more complex. It must be stressed that the mechanical models only give a mathematical representation of the mechanical behaviour and as such do not give much help in interpreting the viscoelastic properties on a molecular level.

5.2.3 *Boltzmann superposition principle*

A corner-stone of the theory of linear viscoelasticity is the *Boltzmann superposition principle*. It allows the state of stress or strain in a viscoelastic body to be determined from knowledge of its entire deformation history. The basic assumption is that during viscoelastic deformation in which the applied stress is varied, the overall deformation can be determined from the algebraic sum of strains due to each loading step. Before the use of the principle can be demonstrated it is necessary, first of all, to define a parameter known as the *creep compliance J(t)* which is a function only of time. It allows the strain after a given time $e(t)$ to be related to the applied stress σ for a linear viscoelastic material since

$$e(t) = J(t)\sigma \tag{5.58}$$

The use of the superposition principle can be demonstrated by considering the creep deformation caused by a series of step loads as shown in Fig. 5.9. If there is an increment of stress $\Delta\sigma_1$ applied at a time τ_1 then the strain due this increment at time t is given by

$$e_1(t) = \Delta\sigma_1 J(t - \tau_1) \tag{5.59}$$

If further increments of stress, which may be either positive or negative, are applied then the principle assumes that the contributions to the overall strain $e(t)$, from each increment are additive so that

$$e(t) = e_1(t) + e_2(t) + \ldots$$
$$= \Delta\sigma_1 J(t - \tau_1) + \Delta\sigma_2 J(t - \tau_2) + \ldots$$
$$= \sum_{n=0}^{n} J(t - \tau_n)\Delta\sigma_n \tag{5.60}$$

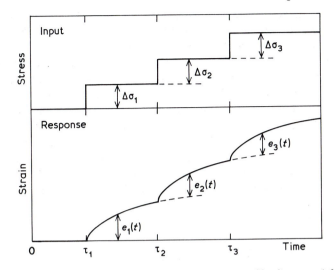

Fig. 5.9 *The response of a linear viscoelastic material to a series of loading steps (after Ward).*

It is possible to represent this summation in an integral form such that

$$e(t) = \int_{-\infty}^{t} J(t - \tau)d\sigma(t) \tag{5.61}$$

The integral is taken from $-\infty$ to t because it is necessary to take into account the entire deformation history of the viscoelastic sample as this will affect the subsequent behaviour. This equation can be used to determine the strain after any general loading history. It is normally expressed as a function of τ which gives

$$e(t) = \int_{-\infty}^{t} J(t - \tau)\frac{d\sigma(\tau)}{d\tau}d\tau \tag{5.62}$$

The use of the Boltzmann superposition principle can be seen by considering specific simple examples. A trivial case is for a single loading step of σ_0 at a time $\tau = 0$. In this case $J(t - \tau) = J(t)$ and so $e(t) = \sigma_0 J(t)$. A more useful application is demonstrated in Fig. 5.10 whereby a stress of σ_0 is applied at time $\tau = 0$ and taken off at time $\tau = t_1$. The strain due to loading, e_1, is given by

$$e_1 = \sigma_0 J(t) \tag{5.63}$$

and that due to unloading is similarly

$$e_2 = -\sigma_0 J(t - t_1) \tag{5.64}$$

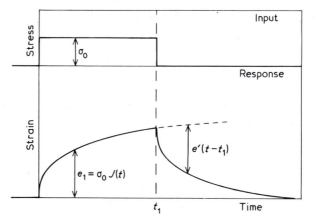

Fig. 5.10 *The response of a linear viscoelastic material to loading followed by unloading, illustrating the Boltzmann superposition principle.*

The total strain after time $t(>t_1)$, $e(t)$ is given by

$$e(t) = \sigma_0 J(t) - \sigma_0 J(t - t_1) \tag{5.65}$$

In this region $e(t)$ is decreasing and this process is known as recovery. If the recovered strain $e'(t - t_1)$ is defined as the difference between the strain that would have occurred if the initial stress had been maintained and the actual strain then

$$e'(t - t_1) = \sigma_0 J(t) - e(t) \tag{5.66}$$

Combining Equations (5.65) and (5.66) gives the recovered strain as

$$e'(t - t_1) = \sigma_0 J(t - t_1) \tag{5.67}$$

which is identical to the creep strain expected for a stress of $+\sigma_0$ applied after a time t_1. It is clear, therefore, that the extra extension or recovery that results from each loading or unloading event is independent of the previous loading history and is considered by the superposition principle as a series of separate events which add up to give the total specimen strain.

It is possible to analyse stress relaxation in a similar way using the Boltzmann superposition principle. In this case it is necessary to define a stress relaxation modulus $G(t)$ which relates the time-dependent stress $\sigma(t)$ to the strain e through the relationship

$$\sigma(t) = G(t)e \tag{5.68}$$

If a series of incremental strains Δe_1, Δe_2 etc. are applied to the specimen at times τ_1, τ_2 etc., the total stress will be given by

$$\sigma(t) = \Delta e_1 G(t - \tau_1) + \Delta e_2 G(t - \tau_2) + \ldots \tag{5.69}$$

after time t. This equation is analogous to Equation (5.60) and can be given in a similar way as an integral (cf. Equation 5.62)

$$\sigma(t) = \int_{-\infty}^{t} G(t - \tau)\frac{de(\tau)}{d\tau}d\tau \tag{5.70}$$

This equation can then be used to predict the overall stress after any general straining programme.

5.2.4 *Dynamic mechanical testing*

So far only the response of polymers during creep and stress relaxation has been considered. Often, they are subjected to variable loading at a moderately high frequency and so it is pertinent to consider their behaviour during this type of deformation. The situation is most easily analysed when an oscillating sinusoidal load is applied to a specimen at a particular frequency. If the applied stress varies as a function of time according to

$$\sigma = \sigma_0 \sin \omega t \tag{5.71}$$

where ω is the angular frequency (2π times the frequency in Hz), the strain for an elastic material obeying Hooke's law would vary in a similar manner as

$$e = e_0 \sin \omega t \tag{5.72}$$

However, for a viscoelastic material the strain lags somewhat behind the stress (e.g. during creep). This can be considered as a damping process and the result is that when a stress defined by Equation (5.71) is applied to the sample the strain varies in a similar sinusoidal manner, but out of phase with the applied stress. Thus, the variation of stress and strain with time can be given by expressions of the type

$$\left.\begin{array}{l} e = e_0 \sin \omega t \\ \sigma = \sigma_0 \sin(\omega t + \delta) \end{array}\right\} \tag{5.73}$$

where δ is the 'phase angle' or 'phase lag' i.e. the relative angular displacement of the stress and strain (Fig. 5.11). The equation for stress can be expanded to give

$$\sigma = \sigma_0 \sin \omega t \cos \delta + \sigma_0 \cos \omega t \sin \delta \tag{5.74}$$

The stress can therefore be considered as being resolved into two components; one of $\sigma_0 \cos \delta$ which is in phase with the strain and another $\sigma_0 \sin \delta$ which is $\pi/2$ out of phase with the strain. Hence it is possible to define two dynamic moduli; E_1 which is in phase with the strain and E_2,

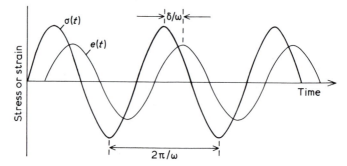

Fig. 5.11 *The variation of stress and strain with time for a viscoelastic material.*

$\pi/2$ out of phase with the strain. Since $E_1 = (\sigma_0/e_0)\cos \delta$ and $E_2 = (\sigma_0/e_0)\sin \delta$, Equation (5.74) becomes

$$\sigma = e_0 E_1 \sin \omega t + e_0 E_2 \cos \omega t \qquad (5.75)$$

The phase angle δ is then given by

$$\tan \delta = E_2/E_1 \qquad (5.76)$$

A complex notation is often favoured for the representation of the dynamic mechanical properties of viscoelastic materials. The stress and strain are given as

$$e = e_0 \exp i\omega t$$

$$\text{and} \quad \sigma = \sigma_0 \exp i(\omega t + \delta) \qquad (5.77)$$

where $i = (-1)^{1/2}$. The overall complex modulus $E^* = \sigma/e$ is then given by

$$E^* = \frac{\sigma_0}{e_0} \exp i\delta = \frac{\sigma_0}{e_0}(\cos \delta + i \sin \delta) \qquad (5.78)$$

From the definitions of E_1 and E_2 it follows that

$$E^* = E_1 + i E_2 \qquad (5.79)$$

and because of this, E_1 and E_2 are sometimes called the *real* and *imaginary* parts of the modulus respectively.

At this stage it is worth considering how the dynamic mechanical properties of a polymer can be measured experimentally. This can be done quite simply by using a *torsion pendulum* similar to the one illustrated schematically in Fig. 5.12. It consists of a cylindrical rod of polymer connected to a large inertial disc. The apparatus is normally arranged such that the weight of the disc is counter-balanced so that the specimen is not under any uniaxial tension or compression. The system can be set into oscillation by twisting and releasing the inertial disc and it undergoes sinusoidal oscillations at a frequency ω which depends upon the dimensions

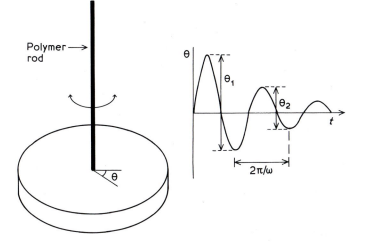

Fig. 5.12 *Schematic diagram of a torsion pendulum and the damping of the oscillations for a viscoelastic material.*

of the system and the properties of the rod. If the system is completely frictionless it will oscillate indefinitely when the rod is perfectly elastic. For a viscoelastic material such as a polymer the oscillations are damped and their amplitude decreases with time as shown in Fig. 5.12. The apparatus is sufficiently compact to allow measurements to be made over a wide range of temperature (typically $-200°C - +200°C$) but since the dimensions of the specimen cannot be varied greatly ω is limited to a relatively narrow range (typically 0.01–10Hz). Where a large range of frequency is required measurements of viscoelastic properties are usually made using forced-vibration techniques.

The motion of a torsion pendulum can be analysed relatively simply. If a circular cross-sectioned rod of radius a and length l is used then the torque T in the rod when one end is twisted by an angle Θ relative to the other is given by

$$T = \frac{G \pi a^4}{2l} \Theta \qquad (5.80)$$

where G is the shear modulus of the rod. If the material is perfectly elastic then the equation of motion of the oscillating system is

$$I\ddot{\Theta} + T = 0 \qquad (5.81)$$

where I is the moment of inertia of the disc. Hence

$$I\ddot{\Theta} + \frac{G \pi a^4}{2l} \Theta = 0 \qquad (5.82)$$

describes the motion of the system. An equation of this form has the standard solution

$$\Theta = \Theta_0\cos(\omega t + \alpha)$$

where

$$\omega = \left(\frac{G\pi a^4}{2lI}\right)^{1/2} \tag{5.83}$$

The shear modulus for an elastic rod is therefore given by

$$G = \frac{2lI\omega^2}{\pi a^4} \tag{5.84}$$

It was shown earlier that the Young's modulus of a viscoelastic material can be considered as a complex quantity E^*. It is possible to define the complex shear modulus in a similar way as

$$G^* = G_1 + iG_2 \tag{5.85}$$

where G_1 and G_2 are the real and imaginary parts of the shear modulus. If a viscoelastic material is used in the rod the equation of motion (Equation 5.82) then becomes

$$I\ddot{\Theta} + \frac{\pi a^4}{2l}(G_1 + iG_2)\Theta = 0 \tag{5.86}$$

If this equation is assumed to have a solution

$$\Theta = \Theta_0 e^{-\lambda t}e^{i\omega t} \tag{5.87}$$

then equating real and imaginary parts gives

$$G_1 = \frac{2lI}{\pi a^4}(\omega^2 - \lambda^2) \tag{5.88}$$

and

$$G_2 = \frac{2lI}{\pi a^4}2\omega\lambda \tag{5.89}$$

Also the phase angle δ is given by

$$\tan \delta = E_2/E_1 = G_2/G_1 = 2\omega\lambda/(\omega^2 - \lambda^2) \tag{5.90}$$

A quantity that is normally measured for a viscoelastic material in the torsion pendulum is the logarithmic decrement Λ. This is defined as the natural logarithm of the ratio of the amplitude of two successive cycles.

i.e. $\Lambda = \ln(\Theta_n/\Theta_{n+1})$ $\tag{5.91}$

The determination of Λ is shown in Fig. 5.12. Since the time period between two successive cycles is $2\pi/\omega$ then combining Equations (5.87) and (5.91) gives

$$\Lambda = 2\pi\lambda/\omega \qquad (5.92)$$

Hence Equation (5.88) becomes

$$G_1 = \frac{2ll\omega^2}{\pi a^4}\left(1 - \frac{\Lambda^2}{4\pi^2}\right) \qquad (5.93)$$

and Equation (5.89)

$$G_2 = \frac{2ll\omega^2}{\pi a^4}\frac{\Lambda}{\pi} \qquad (5.94)$$

and δ is given by

$$\tan\delta = \frac{\Lambda/\pi}{1 - \Lambda^2/4\pi^2} \qquad (5.95)$$

Usually Λ is small ($\ll 1$) and so the term $\Lambda^2/4\pi^2$ is normally neglected making Equation (5.93) become

$$G_1 \simeq \frac{2ll\omega^2}{\pi a^4} = G \qquad (5.96)$$

and Equation (5.95) approximates to

$$\tan\delta \simeq \Lambda/\pi \qquad (5.97)$$

when the damping is small.

The real part of the complex modulus G_1 is often called the *storage modulus* because it can be identified with the in-phase elastic component of the deformation. Elastic materials 'store' energy during deformation and release it on unloading. The imaginary part G_2 is sometimes called the *loss modulus* since it gives a measure of the energy dissipated during each cycle through its relation with Λ (Equation 5.94) and tan δ. The real and imaginary parts of E^* can be treated in an analogous way.

5.2.5 *Frequency dependence of viscoelastic behaviour*

It is found experimentally that when the values of storage modulus E_1, loss modulus E_2 and tan δ are measured for a polymer at a fixed temperature their values depend upon the testing frequency or rate. Generally it is found that tan δ and E_2 are usually small at very low and very high frequencies and their values peak at some intermediate frequency. On the other hand, E_1 is high at high frequencies when the polymer is displaying glassy behaviour and low at low frequencies when the polymer is rubbery. The value of E_1 changes rapidly at intermediate frequencies in the viscoelastic region where the damping is high and tan δ and E_2 peak (normally at slightly different frequencies).

The frequency dependence of the viscoelastic properties of a polymer can be demonstrated very simply using the Maxwell model (Section 5.2.2). The mechanical behaviour of the model is represented by Equation (5.44) as

$$\frac{de}{dt} = \frac{1}{E}\frac{d\sigma}{dt} + \frac{\sigma}{\eta} \tag{5.44}$$

where E is the modulus of the spring and η is the viscosity of the dashpot (Fig. 5.7(a)). It was shown earlier that the model has a characteristic relaxation time $\tau_0 = \eta/E$ and so Equation (5.44) becomes

$$E\tau_0\frac{de}{dt} = \tau_0\frac{d\sigma}{dt} + \sigma \tag{5.98}$$

It was shown in Section 5.2.4 that if the stress on a viscoelastic material is varied sinusoidally at a frequency ω then the variation of stress and strain with time can be represented by complex equations of the form

$$\left.\begin{array}{l} e = e_0\exp i\omega t \\ \sigma = \sigma_0\exp i(\omega t + \delta) \end{array}\right\} \tag{5.77}$$

where δ is the 'phase lag'. Substituting the relationships for σ and e given in Equation (5.77) into Equation (5.98) gives

$$\frac{E\tau_0 i\omega}{(\tau_0 i\omega + 1)} = \frac{\sigma_0\exp i(\omega t + \delta)}{e_0\exp i\omega t} = \frac{\sigma}{e} \tag{5.99}$$

The ratio σ/e is the complex modulus E^* and so

$$E_1 + iE_2 = E\tau_0 i\omega/(\tau_0 i\omega + 1) \tag{5.100}$$

Equating the real and imaginary parts then gives

$$E_1 = \frac{E\tau_0^2\omega^2}{(\tau_0^2\omega^2 + 1)} \quad \text{and} \quad E_2 = \frac{E\tau_0\omega}{(\tau_0^2\omega^2 + 1)} \tag{5.101}$$

and since tan $\delta = E_2/E_1$ it follows that

$$\tan \delta = 1/\tau_0\omega \tag{5.102}$$

The variation of E_1, E_2 and tan δ with frequency for a Maxwell model with a fixed relaxation time is sketched in Fig. 5.13. The behaviour of E_1 and E_2 is predicted quite well by the analysis with a peak in E_2 at $\tau_0\omega = 1$, but tan δ

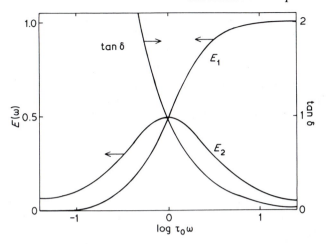

Fig. 5.13 *Variation of E_1, E_2 and tan δ with frequency ω for a Maxwell model with a spring modulus $E = 1$ and a fixed relaxation time τ_0.*

drops continuously as ω increases and the peak that is found experimentally does not occur. Better predictions can be obtained by using more complex models, such as the standard linear solid, and combinations of several models which have a range of relaxation times.

One of the main reasons for studying the frequency dependence of the dynamic mechanical properties is that it is often possible to relate peaks in E_2 and tan δ to particular types of molecular motion in the polymer. The peaks can be regarded as a 'damping' effect and occur at the frequency of some molecular motion in the polymer structure. There is particularly strong damping at the *glass transition*. We have, so far, considered this a purely static phenomenon (Section 4.4.2) being observed through changes in properties such as specific volume or heat capacity. The large increase in free volume which occurs above the T_g allows space for molecular motion to occur more easily. Significant damping occurs at T_g when the applied test frequency equals the natural frequency for main chain rotation. At higher frequencies there will be insufficient time for chain uncoiling to occur and the material will be relatively stiff. On the other hand at lower frequencies the chains will have more time to move and the polymer will appear to be soft and rubbery.

Peaks in E_2 can also be found at the *melting temperature, T_m,* in semi-crystalline polymers. This is due to the greater freedom of molecular motion possible when the molecules are no longer packed regularly into crystals.

It is possible that other types of molecular motion can occur such as the rotation of side groups as these are often reflected in other damping occurring at somewhat different frequencies. The magnitude of the peaks in E_2 and tan δ in this case are usually smaller than those obtained at T_g and such phenomena are known as 'secondary transitions'. The fact that dynamic mechanical testing can be used to follow main chain and side-group motion in polymers makes it a powerful technique for the characterization of polymer structures.

The exact frequency at which the T_g and secondary transitions occur depends upon the temperature of testing as well as the structure of the polymer. In general, the frequency at which the transition takes place increases as the testing temperature is increased and for a given testing frequency it is possible to induce a transition by changing the testing temperature. From a practical viewpoint, it is normally much easier to keep the frequency fixed and vary the testing temperature rather than do measurement at a variable frequency and so most of the data available on the dynamic mechanical behaviour of polymers has been obtained at a fixed frequency and over a range of temperature. For a complete picture of the dynamic mechanical behaviour of a polymer it is desirable that measurements should be made over as wide a range of frequency and temperature as possible and it is sometimes possible to relate data obtained at different frequencies and temperatures through a procedure known as time–temperature superposition (see Section 5.2.7).

5.2.6 *Transitions and polymer structure*

The transition behaviour of a very large number of polymers has been studied widely as a function of testing temperature using dynamic mechanical testing methods. The types of behaviour observed are found to depend principally upon whether the polymers are amorphous or crystalline. Fig. 5.14 shows the variation of the shear modulus G_1 and tan δ with temperature for atactic polystyrene which is typical for amorphous polymers. The shear modulus decreases as the testing temperature is increased and drops sharply at the glass transition where there is a corresponding large peak in tan δ. Fig. 5.14(b) shows the variation of tan δ with temperature and it is possible to see minor peaks at low temperatures. These correspond to secondary transitions. The peak which occurs at the highest temperature is normally labelled α and the subsequent ones are called β, γ etc. In an amorphous polymer the α-transition or α-relaxation corresponds to the glass transition which can be detected by other physical testing methods (Section 4.4.2) and so the temperature of the α-relaxation depends upon the chemical structure of the polymer. In general, it is found that the transition temperature is increased as the main chain is made stiffer. Side groups raise the T_g or α-relaxation temperature if they are polar or large and bulky and reduce it if they are long and flexible.

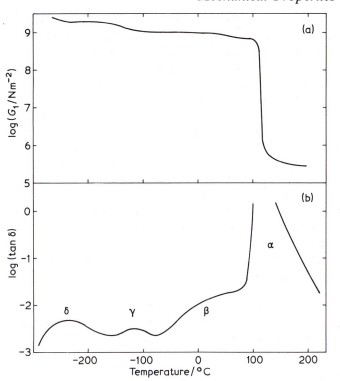

Fig. 5.14 *Variation of shear modulus G_1 and tan δ with temperature for polystyrene (after Arridge).*

The assignment of particular mechanisms to the secondary β, γ etc. transitions can sometimes be difficult since their position and occurrence depends upon which polymer is being studied. At least three secondary transitions have been reported for atactic polystyrene (Fig. 5.14b). The position of the β-peak at about 50°C is rather sensitive to testing frequency and it merges with the α-relaxation at high frequencies. It has been assigned to rotation of phenyl groups around the main chain or alternatively the co-operative motion of segments of the main chain containing several atoms. It is thought that the γ-peak may be due to the occurrence of head-to-head rather than head-to-tail polymerization (Section 2.4.3) and the δ-peak has been assigned to rotation of the phenyl group around its link with the main chain.

The interpretation of the relaxation behaviour of semi-crystalline polymers can be extremely difficult, because of their complex two-phase structure (Section 4.2). It is sometimes possible to consider the crystalline and amorphous regions as separate entities with the amorphous areas having a glass transition. It is clear, however, that as particular molecules can traverse both the amorphous and crystalline regions, this picture may be too simplistic. The relaxation behaviour of polyethylene has probably been more widely studied than that of any other semi-crystalline polymer and so this will be taken as an example, even though it is not completely understood. Fig. 5.15 shows the variation of tan δ with temperature for two samples of polyethylene, one high density (linear) and the other low density (branched). In polyethylene there are four transition regions designated α', α, β and γ. The γ transition is very similar in both the high- and low-density samples whereas the β-relaxation is virtually absent in the high density polymer. The α and α' peaks are also somewhat different in the pre-melting region.

The presence of branches upon the polyethylene molecules produces a difference in polymer morphology as well as in the molecules themselves. The degree of crystallinity and crystal size and perfection are all reduced by the branching and this can lead to complications in interpretation of the relaxation behaviour. The intensities of the α'- and α-relaxation decrease as the degree of crystallinity is reduced, implying that they are associated with motion within the crystalline regions. On the other hand, the intensity of the γ-relaxation increases with a reduction in crystallinity indicating that it is associated with the amorphous material and it has been tentatively assigned to a glass transition in the non-crystalline regions. The disappearance of the β-transition with the absence of branching has been taken as a strong indication that it is associated with relaxations at the branch-points. In general, the assignment of the peaks in crystalline polymers to particular types of molecular motion is sometimes difficult and often a matter for some debate.

5.2.7 *Time–temperature superposition*

It has been suggested earlier that it may be possible to interrelate the time and temperature dependence of the viscoelastic properties of polymers. It is thought that there is a general equivalence between time and temperature. For instance, a polymer which displays rubbery characteristics under a given set of testing conditions can be induced to show glassy behaviour by either reducing the temperature or increasing the testing rate or frequency. This type of behaviour is shown in Fig. 5.16 for the variation of the shear compliance J of a polymer with testing frequency measured at different temperatures in the region of the T_g. The material is rubber-like with a high compliance at high temperatures and low frequencies and becomes glassy with a low compliance as the temperature is reduced and the frequency increased.

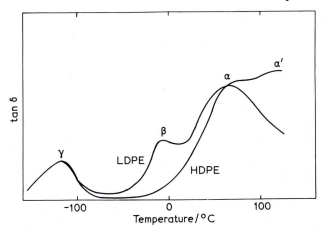

Fig. 5.15 *Variation of tan δ with temperature for high- and low-density polyethylene (adapted from Ward).*

It was found empirically that all the curves in Fig. 5.16 could be superposed by keeping one fixed and shifting all the others by different amounts horizontally parallel to the logarithmic frequency axis. If an arbitrary reference temperature T_s is taken to fix one curve then if ω_s is the frequency of a point on the curve at T_s with a particular compliance and ω is the frequency of a point with the same compliance on a curve at a different temperature then the amount of shift required to superpose the two curves is a displacement of $(\log \omega_s - \log \omega)$ along the log frequency axis. The 'shift factor' a_T is defined by

$$\log a_T = \log \omega_s - \log \omega = \log(\omega_s/\omega) \qquad (5.103)$$

and this parameter is a function only of temperature. The values of $\log a_T$ which must be used to superpose the curves in Fig. 5.16 are given in the insert. The density of the polymer changes with temperature and so this will also affect the compliance of the polymer. It is normally unnecessary to take this into account unless a high degree of accuracy is required. Much of the early work upon time–temperature superposition was done by Williams, Landel and Ferry and they proposed that a_T could be given by an equation of the form

$$\log a_T = \frac{-C_1(T - T_s)}{C_2 + (T - T_s)} \qquad (5.104)$$

where C_1 and C_2 are constants and T_s is the reference temperature. This is normally called the WLF equation and was originally developed empirically. It holds extremely well for a wide range of polymers in the vicinity of

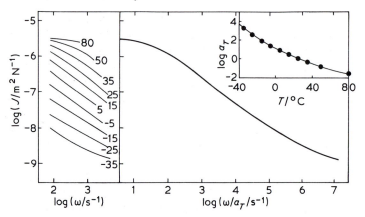

Fig. 5.16 *Example of the time–temperature superposition principle using shear compliance, J, data for polyisobutylene. The curves of J as a function of frequency obtained at different temperatures can all be shifted to lie on a master curve using the shift factor, a_T. (Using data of Fitzgerald, Grandine and Ferry, J. Appl. Phys. 24 (1953) 650.)*

the glass transition and if T_s is taken as T_g, measured by a static method such as dilatometry then

$$\log a_T = \frac{-C_1^g(T - T_g)}{C_2^g + (T - T_g)} \tag{5.105}$$

and the new constants C_1^g and C_2^g become 'universal' with values of 17.4 and 51.6 K respectively. In fact, the constants vary somewhat from polymer to polymer, but it is often quite safe to assume the universal values as they usually give shift factors which are close to measured values.

The WLF master curve produced by superposing data obtained at different temperatures is a very useful way of presenting the mechanical behaviour of a polymer. The WLF equation is also useful in predicting the mechanical behaviour of a polymer outside the range of temperature and frequency (or time) for which experimental data are available.

Although the WLF equation was developed originally by curve fitting it is possible to justify it theoretically from considerations of free volume. The concept of free volume was introduced in Section 4.4.2 in the context of the variation of the specific volume of a polymer with temperature and the abrupt change in slope of the $V - T$ curve at T_g. The fraction free volume f in a polymer is given as (Equation 4.42)

$$f = f_g + (T - T_g)\alpha_f \tag{5.106}$$

where f_g is the fractional free volume at T_g and α_f is the thermal expansion coefficient of the free volume. There are several ways of deriving a relationship of the form of the WLF equation, but perhaps the simplest is by assuming that the polymer behaves as a viscoelastic model having a relaxation time, τ_0 (Section 5.2.2). For the Maxwell model $\tau_o = \eta/E$ where η is the viscosity of the dashpot and E is modulus of the spring. It can be assumed that E is independent of temperature and only η varies with temperature. If the shifts in the time–temperature superposition are thought of as a process of matching the relaxation times then the shift factor is given by

$$a_T = \frac{\tau_0(T)}{\tau_0(T_g)} = \frac{\eta(T)}{\eta(T_g)} \tag{5.107}$$

when the T_g is used as the reference temperature. It is possible to relate the viscosity to the free volume through a semi-empirical equation developed by Doolittle from the study of the viscosities of liquids. He was able to show that for a liquid η is related to the free volume V_f through an equation of the form

$$\ln \eta = \ln A + B(V - V_f)/V_f \tag{5.108}$$

where V is the total volume and A and B are constants. This equation can be rearranged to give

$$\ln \eta(T) = \ln A + B(1/f - 1) \tag{5.109}$$

and at the T_g

$$\ln \eta(T_g) = \ln A + B(1/f_g - 1) \tag{5.110}$$

Substitution for f from Equation (5.106) in Equation (5.109) and subtracting Equation (5.110) from (5.109) gives

$$\ln \frac{\eta(T)}{\eta(T_g)} = B\left(\frac{1}{f_g + \alpha_f(T - T_g)} - \frac{1}{f_g}\right) \tag{5.111}$$

which can be rearranged to give

$$\log \frac{\eta(T)}{\eta(T_g)} = \log a_T = \frac{-(B/2.303f_g)(T - T_g)}{f_g/\alpha_f + (T - T_g)} \tag{5.112}$$

which has an identical form as the WLF equation (5.104). The universal nature of C_1^g and C_2^g in the WLF equation implies that f_g and α_f should also be the same for different polymers. It is found that f_g is of the order of 0.025 and α_f approximately $4.8 \times 10^{-4} \mathrm{K}^{-1}$ for most amorphous polymers.

5.2.8 Non-linear viscoelasticity

An unfortunate aspect of the analysis of the viscoelastic behaviour of polymers is that for most polymers deformed in practical situations, the theory of linear viscoelasticity does not apply. It was emphasized earlier (Section 5.2.2) that Hooke's law and Newton's law are only obeyed at low strains. However, many polymers and especially semi-crystalline ones do not obey the Boltzmann superposition principle even at low strains and it is found that the exact loading route affects the final state of stress and strain. In this case empirical approaches are normally used and the mechanical properties are evaluated for the range of conditions of service expected (e.g. stress, strain, time and temperature) and are tabulated or displayed graphically for use by design engineers and anyone who requires the information.

5.3 Deformation of elastomers

Crosslinked rubbers, *elastomers*, possess the remarkable ability of being able to stretch to five to ten times their original length and then retract rapidly to near their original dimensions when the stress is removed. This behaviour sets them apart from materials such as metals and ceramics where the maximum reversible strains that can be tolerated are usually less than one percent. Most non-crystallizable polymers are capable of being obtained in the rubber-like state. Fig. 5.17 shows the variation of the Young's modulus E of an amorphous polymer with temperature for a given rate of frequency testing. At low temperatures the polymer is glassy with a high modulus ($\sim 10^9 \text{N m}^{-2}$). The modulus then falls through the glass transition region where the polymer is viscoelastic and the modulus becomes very rate- and temperature-dependent and at a sufficiently high temperature the polymer becomes rubbery. However, all rubbery polymers do not show useful elastomeric properties as they will flow irreversibly on loading unless they are crosslinked. The modulus of a linear polymer drops effectively to zero at a sufficiently high temperature when it behaves like a liquid, but if a polymer is crosslinked it will remain approximately constant at $\sim 10^6 \text{N m}^{-2}$ as the temperature is increased.

The properties of elastomeric materials are controlled by their molecular structure which has been discussed earlier (Section 4.5). They are basically all amorphous polymers above their glass transition and normally crosslinked. Their unique deformation behaviour has fascinated scientists for many years and there are even reports of investigations into the deformation of natural rubber from the beginning of the nineteeth century. Elastomer deformation is particularly amenable to analysis using thermodynamics, as an elastomer behaves essentially as an 'entropy spring'. It is even possible to derive the form of the basic stress-strain relationship from first principles by considering the statistical thermodynamic behaviour of the molecular network.

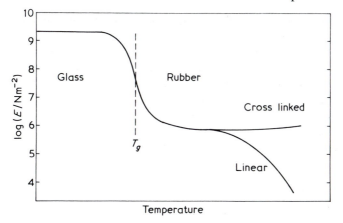

Fig. 5.17 *Typical variation of Young's modulus, E, with temperature for a polymer showing the effect of cross-linking upon E in the rubbery state.*

5.3.1 Thermodynamics of elastomer deformation

As well as having the ability to sustain large reversible amounts of extension there are several other more subtle aspects of the deformation of elastomers which are rather unusual and merit further consideration. A well-known property is that when an elastomer is stretched rapidly it becomes warmer and conversely when it is allowed to equilibrate for a while at constant extension and then allowed to contract rapidly it cools down. This can be readily demonstrated by putting an elastic band to ones lips, which are quite sensitive to changes in temperature, and stretching it quickly. Another property is that the length of an elastomer sample held at constant stress decreases as the temperature is increased whereas an unstressed specimen exhibits conventional thermal expansion behaviour. It is possible to rationalize these aspects of the thermomechanical behaviour by considering the deformation of elastomers in thermodynamic terms. This allows the basic relationship between force, length and temperature to be established for an elastomeric specimen in terms of parameters such as internal energy and entropy. One important experimental observation which allows the analysis to be simplified is that the deformation of elastomers takes place approximately at constant volume.

The first law of thermodynamics establishes the relationship between the change in internal energy of a system dU and the heat $đQ$ absorbed by the system and work $đW$ done by the system as

$$dU = đQ - đW \tag{5.113}$$

The bars indicate that $đQ$ and $đW$ are inexact differentials because Q and W, unlike U, are not macroscopic functions of the system. If the length of an elastic specimen is increased a small amount dl by a tensile force f then an amount of work $f\,dl$ will be done *on* the system. There is also a change in volume dV during elastic deformation and work $P\,dV$ is done against the pressure, P. Since elastomers deform at roughly constant volume the contribution of $P\,dV$ to $đW$ at ambient pressure will be small and so the work done *by* the system (i.e. the specimen) when it is extended is given by

$$đW = -f\,dl \qquad (5.114)$$

The deformation of elastomers can be considered as a reversible process and so $đQ$ can be evaluated from the second law of thermodynamics which states that for a reversible process

$$đQ = T\,dS \qquad (5.115)$$

where T is the thermodynamic temperature and dS is the change in entropy of the system. Combining Equations (5.113), (5.114) and (5.115) gives

$$f\,dl = dU - T\,dS \qquad (5.116)$$

Most of the experimental investigations on elastomers have been done under conditions of constant pressure. The thermodynamic function which can be used to describe equilibrium under these conditions is the Gibbs free energy (Equation 4.1), but since elastomers tend to deform at constant volume it is possible to use the Helmholtz free energy, A in the consideration of equilibrium. It is defined as

$$A = U - TS \qquad (5.117)$$

and for a change taking place under conditions of constant temperature

$$dA = dU - TdS \qquad (5.118)$$

Comparing Equations (5.116) and (5.118) gives

$$fdl = dA \quad (const.\ T)$$

and so

$$f = (\partial A/\partial l)_T \qquad (5.119)$$

Combining Equations (5.118) and (5.119) enables the force to be expressed as

$$f = \left(\frac{\partial U}{\partial l}\right)_T - T\left(\frac{\partial S}{\partial l}\right)_T \qquad (5.120)$$

The first term refers to the change in internal energy with extension and the second to the change in entropy with extension. An expression of this form could be used approximately to describe the response of any solid to an applied force. For most materials the internal energy term is dominant, but in the case of elastomers the change in entropy gives the largest contribution to the force. This can be demonstrated after modifying the entropy term in Equation (5.120). The Helmholtz free energy can be expressed in a differential form for any general change as

$$dA = dU - TdS - SdT \tag{5.121}$$

and combining this equation with Equation (5.116) gives

$$dA = fdl - SdT \tag{5.122}$$

and partial differentiation under conditions first of all of constant temperature and then of constant length gives

$$\left.\begin{array}{l} (\partial A/\partial l)_T = f \\ \text{and} \quad (\partial A/\partial T)_l = -S \end{array}\right\} \tag{5.123}$$

But we have the standard relation for partial differentiation

$$\frac{\partial}{\partial l}\left(\frac{\partial A}{\partial T}\right)_l = \frac{\partial}{\partial T}\left(\frac{\partial A}{\partial l}\right)_T \tag{5.124}$$

and applying this to Equations (5.123) gives

$$(\partial S/\partial l)_T = -(\partial f/\partial T)_l \tag{5.125}$$

Equation (5.120) then becomes

$$f = \left(\frac{\partial U}{\partial l}\right)_T + T\left(\frac{\partial f}{\partial T}\right)_l \tag{5.126}$$

These last two equations contain parameters which can be measured experimentally and so allow the entropy and internal energy terms to be calculated. This can be done by measuring how the force required to hold an elastomer at constant length varies with the absolute temperature. Fig. 5.18 shows the results of an early experiment of this type. The stress in the elastomer drops as the temperature is reduced until about 220 K at which point it starts to rise again. This change in slope corresponds to the transition to a glassy state when the material is no longer elastomeric. The curve in the rubbery region is virtually linear and can be extrapolated approximately to zero tension at absolute zero. The curve in Fig. 5.18 is of great significance and can be related directly to Equation (5.126). The slope of the curve at any point $(\partial f/\partial T)_l$ gives the variation of entropy with extension at that temperature. Since the curve is linear it suggests that $(\partial S/\partial l)_T$ is fairly temperature independent and as the intercept is approximately zero it also implies that $(\partial U/\partial l)_T$ is small. This means that there is very little change in internal energy during the extension of a rubber and that the deformation is dominated by the change in entropy.

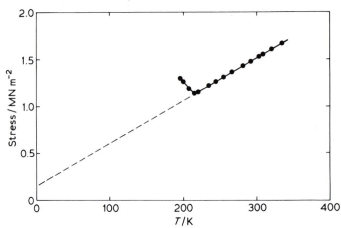

Fig. 5.18 *Variation of the stress in an elastomer (vulcanized natural rubber), as a function of temperature, held at a fixed extension of 350% (data of Meyer and Ferri reported by Treloar). Treloar).*

The data in Fig. 5.18 were obtained at a fairly high extension (350 percent). When the experiment is repeated at lower extensions the slope of the curve decreases and below about 10% extension it becomes negative. This is caused by a reduction in stress due to thermal expansion of the material as the temperature is increased and it is known as the *thermo-elastic inversion effect*. If the effective change in extension due to thermal expansion is allowed for then the thermo-elastic inversion effect disappears and the stress increases proportionately with temperature at low extensions as well as high extensions.

All the considerations of thermo-elastic effects have so far been concerned with changes in stress at constant extension, but they can be extended directly to tests done at constant tension. In this case the tendency of the stress to increase as the temperature is raised will cause a specimen subjected to a constant stress to reduce its length as the temperature is increased. This can be demonstrated readily by hanging a weight on a length of an elastomer and changing the temperature. The weight moves upwards when the elastomer is heated and it goes downwards when the elastomer is cooled.

Another thermo-elastic effect that was mentioned earlier is the tendency of an elastomer to become warm when stretched rapidly. This type of behaviour is illustrated in Fig. 5.19 for the adiabatic extension of an elastomer. Some of the heating can be explained by crystallization which can also occur when natural rubber is stretched (Section 4.5) but a temperature rise is found even in non-crystallizable elastomers. This can be explained by consideration of the thermodynamic equilibrium. The rise in

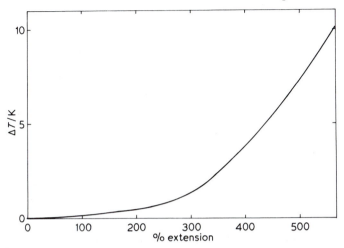

Fig. 5.19 *Increase in temperature, ΔT, upon the adiabatic extension of an elastomer (vulcanized natural rubber) (data of Dart, Anthony and Guth reported by Treloar).*

temperature as the length is increased adiabatically (isentropically) is characterized by $(\partial T/\partial l)_s$ which is given by the standard relation

$$\left(\frac{\partial T}{\partial l}\right)_s = -\left(\frac{\partial T}{\partial S}\right)_l\left(\frac{\partial S}{\partial l}\right)_T \qquad (5.127)$$

The first factor on the right-hand side can be expressed as

$$\left(\frac{\partial T}{\partial S}\right)_l = \left(\frac{\partial T}{\partial H}\right)_l\left(\frac{\partial H}{\partial S}\right)_l \qquad (5.128)$$

But

$$\left(\frac{\partial T}{\partial H}\right)_l = \frac{1}{C_l} \quad \text{and} \quad \left(\frac{\partial H}{\partial S}\right)_l = T$$

and so Equation (5.127) becomes

$$\left(\frac{\partial T}{\partial l}\right)_s = -\frac{T}{C_l}\left(\frac{\partial S}{\partial l}\right)_T \qquad (5.129)$$

where C_l is the heat capacity of the elastomer held at constant length. Combining Equations (5.125) and (5.129) gives

$$\left(\frac{\partial T}{\partial l}\right)_s = \frac{T}{C_l}\left(\frac{\partial f}{\partial T}\right)_l \qquad (5.130)$$

This equation shows clearly that the temperature of an elastomer will rise when it is stretched adiabatically as long as $(\partial f/\partial T)_l$ is positive as is normally the case (Fig. 5.18). It is a direct consequence of the fact that when the elastomer is stretched, work is transformed reversibly into heat.

5.3.2 *Statistical theory of elastomer deformation*

Although the thermodynamic approach shows clearly that when an elastomer is deformed it behaves like an entropy spring, no specific deformation mechanisms are implied. In fact, none of the thermodynamic equations depend upon the macromolecular nature of the material although the behaviour obviously stems from the presence of polymer molecules in the elastomer). The statistical approach looks directly at how the molecular structure changes during deformation and allows the change in entropy during deformation to be calculated and the stress–strain curve derived from first principles. It is similar to the calculation of the pressure of an ideal gas using statistical thermodynamics and it makes use of the calculations of the conformations of freely-jointed chains outlined in Section 3.3.1.

(i) *Entropy of an individual chain*
The material is assumed to be made up of a network of crosslinked polymer chains with the individual lengths of chain in their most random conformations. When the network is extended the molecules become uncoiled and their entropy is reduced. The force required to deform the elastomer can therefore be related directly to this change in entropy. The first step in the analysis involves the calculation of the entropy of an individual polymer molecule. This can be done with the help of Equation (3.51) which gives the probability per unit volume $W(x, y, z)$ of finding one end a freely-jointed chain at a point (x, y, z) a distance r from the other end which is fixed at the origin. This probability can be expressed as

$$W(x, y, z) = (\beta/\pi^{1/2})^3 \exp(-\beta^2 r^2) \tag{3.51}$$

and the function has been plotted graphically in Fig. 3.4(a). The parameter β is defined as $\beta = (3/(2nl^2))^{1/2}$ where n is the number of links of length l in the chain and β can be considered to be characteristic for a particular chain. Equation (3.51) allows the entropy of a single chain to be calculated using the Boltzmann relationship from statistical thermodynamics which is

$$S = \mathbf{k} \ln \Omega \tag{5.131}$$

where $\mathbf{k}$ is Boltzmann's constant and Ω is the number of possible conformations the chain can adopt. It can be assumed that Ω will be proportional to $W(x, y, z)$. This is clearly a reasonable assumption since when r is small $W(x, y, z)$ is large (Fig. 3.4a) and there will be many conformations available to the chain. $W(x, y, z)$ drops as r is increased and the number of possible conformations is reduced as the molecule becomes extended. The entropy, S, of a single chain can then be expressed by an equation of the form

$$S = c - k\beta^2 r^2 \qquad (5.132)$$

where c is a constant.

(ii) *Deformation of the polymer network*

Having obtained an expression for the entropy of a single chain it is a relatively simple matter to determine the change in entropy on deforming an elastomeric polymer network. It is assumed that when the elastomer is in either the strained or unstrained state the junction points can be regarded as being fixed at their mean positions and that the lengths of chain between the points behave as freely-jointed chains so that Equation (5.132) can be applied. It is also assumed that when the material is deformed the change in the components of the displacement vector, r, of each chain are proportional to the corresponding change in specimen dimensions. In the consideration of the deformation of elastomers large strains are encountered and it is more convenient to use *extension ratios* λ_1, λ_2, λ_3 rather than infinitesimal strains. The extension ratio, λ, in a particular direction is defined as the deformed length of the specimen in that direction divided by the original length. The axes of these ratios are chosen to coincide with the rectangular coordinate system used to characterize the conformations of the original lengths of chain. If an individual chain is considered, first of all, then if one junction point is fixed at the origin, O, a general deformation of the elastomer will displace the other junction point from (x, y, z) to (x', y', z') as shown in Fig. 5.20. The assumption that the change in the components of the displacement vector will be proportional to the overall change in specimen dimensions means that

$$x' = \lambda_1 x, \quad y' = \lambda_2 y, \quad z' = \lambda_3 z \qquad (5.133)$$

Since r^2 is equal to $(x^2 + y^2 + z^2)$ then it follows from Equation (5.132) that the entropies of the individual chain will be

$$S = c - k\beta^2(x^2 + y^2 + z^2) \qquad \text{(before deformation)}$$
$$S' = c - k\beta^2(\lambda_1^2 x^2 + \lambda_2^2 y^2 + \lambda_3^2 z^2) \qquad \text{(after deformation)}$$

and so the change in entropy of an individual chain which has its end displaced from (x, y, z) to (x', y', z') will be given by

$$\Delta S_i = S' - S = -k\beta^2[(\lambda_1^2 - 1)x^2 + (\lambda_2^2 - 1)y^2 + (\lambda_3^2 - 1)z^2] \qquad (5.134)$$

The polymer consists of a network of many such chains with a range of displacement vectors. If the number of these chains per unit volume is

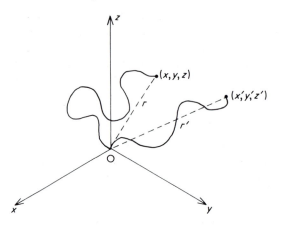

Fig. 5.20 *Schematic representation of the displacement of a junction point in an elastomeric network from (x, y, z) to (x', y', z') during a general deformation (after Treloar).*

defined as N, the number of dN which have ends initially in the small volume element dx dy dz at the point (x, y, z) can be determined from the Gaussian distribution function as

$$dN = NW(x,y,z)dx\, dy\, dz$$

or $$dN = N(\beta/\pi^{1/2})^3 \exp[-\beta^2(x^2 + y^2 + z^2)]dx\, dy\, dz \qquad (5.135)$$

On deformation the change in entropy of these chains is given by $\Delta S_i dN$ and the total entropy change ΔS per unit volume of the sample during deformation is given by the sum of the entropy changes for all chains. This can be expressed in an integral form as

$$\Delta S = \int \Delta S_i\, dN$$

or $$\Delta S = \int_{-\infty}^{\infty}\int_{-\infty}^{\infty}\int_{-\infty}^{\infty} -\frac{Nk\beta^5}{\pi^{3/2}}[(\lambda_1^2 - 1)x^2 + (\lambda_2^2 - 1)y^2$$
$$+ (\lambda_3^2 - 1)z^2]\exp[-\beta^2(x^2 + y^2 + z^2)]dx\, dy\, dz \qquad (5.136)$$

which when evaluated* gives the simple result

$$\Delta S = \tfrac{1}{2}Nk(\lambda_1^2 + \lambda_2^2 + \lambda_3^2 - 3)$$

*The integrals in Equation (5.136) are of the types

$$\int_{-\infty}^{\infty} \exp(-\beta^2 x^2)x^2 dx = \pi^{1/2}/2\beta^3$$

and $$\int_{-\infty}^{\infty} \exp(-\beta^2 x^2)dx = \pi^{1/2}/\beta$$

This equation relates the change in entropy to the extension ratios and the number of chains between crosslinks per unit volume. It was shown earlier (Section 5.3.1) that there is ideally no change in internal energy U when an elastomer is deformed and since the deformation takes place at constant volume then the change in the Helmholtz free energy per unit volume ΔA is

$$\Delta A = -T\Delta S = \tfrac{1}{2}NkT(\lambda_1^2 + \lambda_2^2 + \lambda_3^2 - 3) \tag{5.137}$$

for isothermal deformation. This will be identical with the isothermal reversible work of deformation w per unit volume and hence

$$w = \tfrac{1}{2}NkT(\lambda_1^2 + \lambda_2^2 + \lambda_3^2 - 3) \tag{5.138}$$

(iii) *Limitations and use of the theory*
The statistical theory is remarkable in that it enables the macroscopic deformation behaviour of an elastomer to be predicted from considerations of how the molecular structure responds to an applied strain. However, it is important to realize that it is only an approximation to the actual behaviour and has significant limitations. Perhaps the most obvious problem is with the assumption that end-to-end distances of the chains can be described by the Gaussian distribution. This problem has been highlighted earlier in connection with solution properties (Section 3.3) where it was shown that the distribution cannot be applied when the chains become extended. It can be overcome to a certain extent with the use of more sophisticated distribution functions, but the use of such functions is beyond the scope of this present discussion. Another problem concerns the value of N. This will be governed by the number of junction points in the polymer network which can be either chemical (crosslinks) or physical (entanglements) in nature. The structure of the chain network in an elastomer has been discussed earlier (Section 4.5). There will be chain ends and loops which do not contribute to the strength of the network, but if their presence is ignored it follows that if all network chains are anchored at two crosslinks then the density, ρ, of the polymer can be expressed as

$$\rho = NM_c/\mathbf{N}_A \tag{5.139}$$

where M_c is the number average molar mass of the chain lengths between cross-links and $\mathbf{N}_A$ is the Avogadro number. It follows that N is given by

$$N = \rho\mathbf{N}_A/M_c = \rho\mathbf{R}/M_c\mathbf{k} \tag{5.140}$$

and Equation (5.138) then becomes

$$w = \frac{\rho\mathbf{R}T}{2M_c}(\lambda_1^2 + \lambda_2^2 + \lambda_3^2 - 3) \tag{5.141}$$

This is normally written as

$$w = \tfrac{1}{2}G(\lambda_1^2 + \lambda_2^2 + \lambda_3^2 - 3) \tag{5.142}$$

where G is given by

$$G = \rho \mathbf{R} T / M_c \tag{5.143}$$

The parameter G relates w, the work of deformation per unit volume which has dimensions of stress to the extension ratios and so G also has dimensions of stress. It is therefore often referred to as the modulus of the elastomer. Inspection of Equation (5.143) shows that G has some interesting properties. As might be expected, G increases as the length of chain between cross-links is reduced. This means that the material becomes stiffer as the crosslink density increases and the network becomes tighter. A rather more surprising prediction of the equation, which has been substantiated experimentally, is that, unlike almost every other material, the modulus of an elastomer increases as the temperature is increased. This is yet another consequence of the deformation of elastomers being dominated by changes in entropy rather than of internal energy.

5.3.3 *Stress–strain behaviour of elastomers*

The statistical theory allows the stress-strain behaviour of an elastomer to be predicted. The calculation is greatly simplified when the observation that elastomers tend to deform at constant volume is taken into account. This means that the product of the extension ratios must be unity

i.e. $\lambda_1 \lambda_2 \lambda_3 = 1$ \hfill (5.144)

The form of the stress–strain behaviour depends upon the loading geometry employed. The present analysis will be restricted to simple uniaxial tension or compression when specimens are deformed to an extension ratio λ in the direction of the applied stress. It is possible to replace λ_1 by λ but Equation (5.144) requires that $\lambda_2 \lambda_3 = 1/\lambda_1$ and so $\lambda_2 = \lambda_3 = 1/\lambda^{1/2}$. Equation (5.142) can therefore be written as

$$w = \tfrac{1}{2}G(\lambda^2 + 2/\lambda - 3) \tag{5.145}$$

for uniaxial tension or compression. If the specimen has an initial cross-sectional area of A_0 and length l_0 and it is extended δl by a uniaxial tensile force, f, then the amount of work δW done *on* the specimen is given by

$$\delta W = f \delta l$$

Equation (5.145) gives an expression for w which is the work per unit volume. The volume of the specimen is $A_0 l_0$ and so

$$\delta w = \delta W / A_0 l_0 = (f/A_0)\,(\delta l/l_0) \tag{5.146}$$

But f/A_0 is the force per unit initial cross-sectional area, i.e. the nominal stress σ_n and $\delta l/l_0 = \delta e \approx \delta\lambda$. A relationship between σ_n and λ can then be derived from Equation (5.145) since rearranging Equation (5.146) and expressing it in a differential form gives

$$\sigma_n = dw/d\lambda = G(\lambda - 1/\lambda^2) \qquad (5.147)$$

It is possible to compare the relationship between σ_n and λ predicted by this equation with experimental results by fitting the theoretical curve to experimental data by choosing a suitable value of G. Fig. 5.21(a) shows the results of such an exercise for a sample of vulcanized natural rubber deformed in tension. The value of G chosen was $\sim 0.4\,\mathrm{MNm}^{-2}$ and it can be seen that the agreement between experiment and theory is good at low elongations ($\lambda < 1.5$). However, the theoretical curve falls below the experimental one as the strain is increased further, but eventually rises above it at elongations of $\lambda > 6$. It can be seen that the theory predicts the basic form of the relationship between σ_n and λ especially at low strains when the Gaussian approximation may be expected to hold well. The lack of agreement at high strains is thought to be due to at least two factors. The Gaussian formula will no longer apply and crystallization can also occur in the material (Section 4.4.5). The theory predicts the deformation behaviour rather better in compression. Fig. 5.21(b) shows the variation of σ_n with λ for the same elastomer as used in the tensile experiment, but in this case deformed into the compressive region ($\lambda < 1$). Good agreement between experiment and theory (Equation 5.147) is found between $\lambda = 0.4$ and 1.3. In this regime the Gaussian formula would be expected to be a better approximation of the behaviour of the chains and crystallization does not occur.

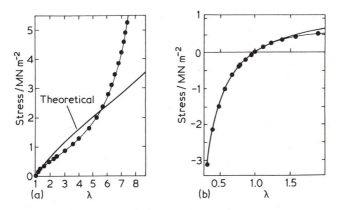

Fig. 5.21 *Relationship between stress and extension ratio λ for vulcanized natural rubber. The theoretical curves for a value of $G = 0.4\,\mathrm{MN\,m}^{-2}$ are given as heavy lines* (a) *Extension,* (b) *Compressive deformation (after Treloar).*

Another aspect of the stress–strain behaviour of elastomers which is not predicted by the statistical theory is the phenomenon of *mechanical hysteresis* which is encountered when a deformed rubber is allowed to relax. This type of behaviour is shown in Fig. 5.22 for a strain-crystallizing elastomer. In this case although all the strain is recovered upon removing the load the unloading curve does not follow the same path as the loading curve. The area between the two curves corresponds to the energy dissipated within the cycle. Hysteresis is found to be more prevalent in strain-crystallizing and filled elastomers and in certain applications it can lead to undesirable consequences. For example, if an elastomer exhibiting a high hysteresis is cycled rapidly between low and high strains the dissipation of energy can lead to a large heat build up which may cause a deterioration in the properties of the material.

5.4 **Yield in polymers**

Determination of the stress–strain behaviour of a material is particularly useful as it gives information concerning important mechanical properties such as Young's modulus, yield strength and brittleness. These are parameters which are vital in design considerations when the material is used in a practical situation. Stress–strain curves can be readily obtained for polymers by subjecting a specimen to a tensile force applied at a constant rate of testing. The stress–strain behaviour of polymers is not fundamentally different from that of conventional materials, the main difference being that polymers show a marked time or rate dependence (cf. viscoelasticity). The situation can be greatly simplified by making measurements at a fixed testing rate, but it is important to bear in mind the underlying time dependence that exists.

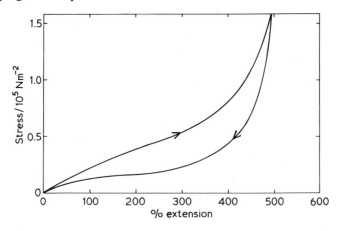

Fig. 5.22 *Illustration of mechanical hysteresis for a strain-crystallizing elastomer (after Andrews).*

5.4.1 *Phenomenology of yield*

Fig. 5.23 shows an idealized stress–strain curve for a ductile polymer sample. In this case the nominal stress σ_n is plotted against the strain e. The change in the cross-section of a parallel-sided specimen is also sketched schematically at different stages of the deformation. Initially the stress is proportional to the strain and Hooke's law is obeyed. The tensile modulus can be obtained from the slope. As the strain is increased the curve decreases in slope until it reaches a maximum. This is conventionally known as the *yield point* and the yield stress and yield strain, σ_y and e_y, are indicated on the curve. The yield point for a polymer is rather difficult to define. It should correspond to the point at which permanent plastic deformation takes place, but for polymers a 'permanent set' can be found in specimens loaded to a stress, below the maximum, where the curve becomes non-linear. The situation is further complicated by the observation that even for specimens loaded well beyond the yield strain the plastic deformation can sometimes be completely recovered by annealing the specimen at elevated temperature. In practice, the exact position of the yield point is not of any great importance and the maximum point on the curve suffices as a definition of yield. The value of the yield strain for polymers is typically of the order of 5–10% which is very much higher than that of metals and ceramics. Yield in metals normally occurs at strains below 0.1%.

During elastic deformation the cross-sectional area of the specimen decreases uniformly as length increases, but an important change occurs at the yield point. The cross-sectional area starts to decrease more rapidly

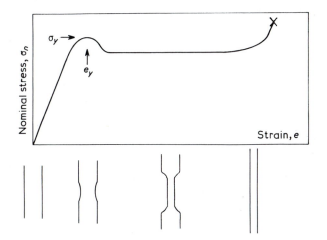

Fig. 5.23 *Schematic representation of the stress–strain behaviour of a ductile polymer and consequent change in specimen dimensions ($\times$, final fracture).*

at one particular point along the gauge length as a 'neck' starts to form. The nominal stress falls after yield and settles at a constant value as the neck extends along the specimen. Eventually, when the whole specimen is necked, strain hardening occurs and the stress rises until fracture eventually intervenes. The process whereby the neck extends is known as *cold drawing*. It was originally thought that this was due to local heating of the specimen by the energy expended during deformation. Detailed measurements have shown that a neck can form even at very low strain rates when the heat is easily dissipated.

Polymers differ in their ability to form a stable neck and some do not neck at all. The form of the neck varies from polymer to polymer and with the testing conditions for a given polymer. The degree of necking is characterized by the 'draw ratio' which is defined as the length of a fully necked specimen divided by the original length. Since cold drawing takes place at approximately constant volume it is also equal to the ratio of original specimen cross-sectional area to that of the drawn specimen.

The exact form of the stress–strain curve for a particular polymer varies with both the temperature and rate of testing. A series of stress–strain curves for poly(methyl methacrylate) tested at different temperatures is given in Fig. 5.24. The modulus, as given by the initial slope of the curves, increases as the temperature is reduced. On the other hand, the ductility of the polymer decreases and below about 320 K the polymer does not draw and behaves in a brittle manner. The exact temperature at which this *ductile-to-brittle transition* takes place depends upon the rate of testing. Its temperature decreases as the rate of testing increases. There is a similar interdependence of the properties upon temperature and rate as is found for viscoelasticity (Section 5.2). In general, it is found that the effect of increasing the rate of testing upon properties such as the Young's modulus or yield stress is the same as reducing the testing temperature.

5.4.2 *Necking and the Considère construction*

It is clear that some polymers have the ability under certain conditions to form a stable neck and undergo cold-drawing. This ability is shared by certain metals such as mild steel whereas other metals form an unstable neck which continues to thin-down until fracture ensues. The phenomenon of necking and cold drawing can best be considered using the *Considère construction* which involves plotting the true stress σ against nominal strain e. This must be contrasted with the curves in Fig. 5.23 and 5.24 where the nominal stress σ_n is plotted against the nominal strain e.

The *true stress* σ is defined as the load L divided by the instantaneous cross-sectional area A and so

$$\sigma = L/A \tag{5.148}$$

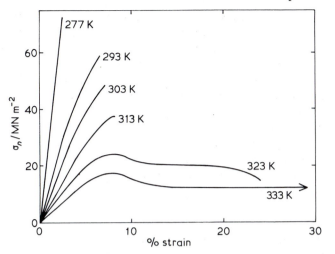

Fig. 5.24 *Variation of the stress–strain behaviour of poly(methyl methacrylate) with temperature (after Andrews).*

It may be assumed that deformation takes place at approximately constant volume and so A can be related to the original cross-sectional area A_0 through the equation

$$Al = A_0 l_0 \tag{5.149}$$

where l and l_0 are the instantaneous and original specimen lengths respectively. The nominal strain is defined as

$$e = (l - l_0)/l_0 \qquad l/l_0 = (1 + e) \tag{5.150}$$

and the nominal stress as

$$\sigma_n = L/A_0 \tag{5.151}$$

Combining Equations (5.148) to (5.151) gives the relationship between the nominal and true stress as

$$\sigma_n = \sigma/(1 + e) \tag{5.152}$$

This means that for finite tensile strains σ will always be greater than σ_n.

When a neck starts to form, the load on the specimen ceases to rise as the strain is increased. This corresponds to the condition that $dL/de = 0$ or $d\sigma_n/de = 0$ (Fig. 5.23). Applying this condition to Equation (5.152) gives

$$\frac{d\sigma_n}{de} = 0 = \frac{1}{(1 + e)} \frac{d\sigma}{de} - \frac{\sigma}{(1 + e)^2}$$

$$\text{or} \quad \frac{d\sigma}{de} = \frac{\sigma}{(1 + e)} \tag{5.153}$$

as the condition for a neck to form. The use of this equation is illustrated in Fig. 5.25. The point at which necking occurs can be determined by drawing a tangent to the curve of true stress against nominal strain from the point $e = -1$. Equation (5.153) will then be satisfied at the point at which the tangent meets the curve. This construction can also be used to determine whether or not the neck will be stable. In the curve drawn in Fig. 5.25(a) only one tangent can be drawn and after the neck forms the sample continues to thin down in the necked region until fracture occurs. A stable neck will only form if a second tangent can be drawn to the curve as shown in Fig. 5.25(b). The point where the second tangent meets the σ–e curve corresponds to a minimum in the nominal stress-nominal strain curve (Fig. 5.23) which is necessary for the neck to be stable. The condition for both necking *and* cold-drawing to occur is therefore that $d\sigma/de$ must be equal to $\sigma/(1 + e)$ at *two* points on the σ–e curve.

So far the mechanisms whereby a stable neck forms in a polymer have not been considered. In metals it is known to be due to the multiplication of dislocations which lead to a phenomenon known as strain-hardening. The metal becomes 'harder' as the strain is increased and if the strain-hardening takes place to a sufficient extent the material in the neck, which supports a smaller area, does not strain further. The material outside the neck is less deformed and hence 'softer' and continues to deform even though the cross-sectional area is larger and the true stress locally lower. Strain-hardening in polymers is thought to be due principally to the effects of orientation. It has been shown by X-ray diffraction that the molecules are aligned parallel to the stretching direction in the cold-drawn regions of both amorphous and crystalline polymers. Since the anisotropic nature of the chemical bonding in the molecules causes an oriented polymer to be very much stronger and stiffer than an isotropic one, the material in the necked region is capable of supporting a much higher true stress than that outside the neck. Hence, polymers tend to form stable necks and undergo cold-drawing if they do not first undergo brittle fracture.

5.4.3 *Yield criteria*

We have so far in this section on the yield behaviour of polymers only considered tensile deformation. In order to obtain a complete idea of the yield process it is necessary to know under what conditions yield occurs for any general combination of stresses. For example, glassy polymers are usually brittle in tension when the temperature of testing is sufficiently below T_g whereas when they are deformed in compression at similar temperatures they can undergo considerable plastic deformation. Also a knowledge of yield behaviour under general stress systems is important in engineering structures where components are subjected to a variety of

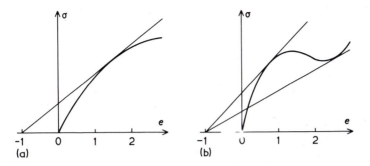

Fig. 5.25 *Schematic curves of true stress, σ against nominal strain, e for polymers showing the Considère construction.* (a) *Unstable neck,* (b) *stable neck.*

tensile, compressive and shear stresses. A description of the conditions under which yield can occur under a general stress system is called a *yield criterion*. Several criteria have been suggested for materials in general, but before they can be described it is necessary to consider how the state of stress on a body can be represented. If the body is isotropic then the stresses can in general be given in terms of a tensor (Section 5.1.1) as

$$\sigma_{ij} = \begin{pmatrix} \sigma_{11} & \sigma_{21} & \sigma_{31} \\ \sigma_{21} & \sigma_{22} & \sigma_{23} \\ \sigma_{31} & \sigma_{23} & \sigma_{33} \end{pmatrix}$$

It is possible to choose a set of three mutually orthogonal axes such that the shear stress are all zero and the stress system can be described in terms of three normal stress such that the tensor becomes

$$\begin{pmatrix} \sigma_1 & 0 & 0 \\ 0 & \sigma_2 & 0 \\ 0 & 0 & \sigma_3 \end{pmatrix}$$

These three *principal stresses* σ_1, σ_2 and σ_3 are used in the formulation of the different yield criteria which are outlined below.

(i) *Tresca yield criterion*
This was one of the earliest criteria developed to describe the yield behaviour of metals and Tresca proposed that yield occurs when a critical value of the maximum shear stress σ_s is reached. If $\sigma_1 > \sigma_2 > \sigma_3$ then the criterion can be given as

$$\tfrac{1}{2}(\sigma_1 - \sigma_3) = \sigma_s \tag{5.154}$$

In a simple tensile test σ_1 is equal to the applied stress and $\sigma_2 = \sigma_3 = 0$ and so at yield

$$\sigma_s = \sigma_1/2 = \sigma_y/2 \qquad (5.155)$$

where σ_y is the yield stress in tension.

(ii) *Von Mises yield criterion*

Although the relative simplicity of the Tresca criterion is rather attractive it is found that the criterion suggested by von Mises gives a somewhat better prediction of the yield behaviour of most materials. The criterion corresponds to the condition that yield occurs when the shear-strain energy in the material reaches a critical value and it can be expressed as a symmetrical relationship between the principal stresses of the form

$$(\sigma_1 - \sigma_2)^2 + (\sigma_2 - \sigma_3)^2 + (\sigma_3 - \sigma_1)^2 = \text{constant} \qquad (5.156)$$

If the case of simple tension is considered then $\sigma_2 = \sigma_3 = 0$ and if the tensile yield stress is again σ_y then the constant is equal to $2\sigma_y^2$. Equation (5.156) can be therefore expressed as

$$(\sigma_1 - \sigma_2)^2 + (\sigma_2 - \sigma_3)^2 + (\sigma_3 - \sigma_1)^2 = 2\sigma_y^2 \qquad (5.157)$$

If the experiment is done in pure shear then $\sigma_1 = -\sigma_2$ and $\sigma_3 = 0$ and Equation (5.157) becomes

$$\sigma_1 = \sigma_y/\sqrt{3} \qquad (5.158)$$

Hence the shear yield stress is predicted to be $1/\sqrt{3}$ times the tensile yield stress. This should be compared with the Tresca criterion which predicts that the shear yield stress is $\sigma_y/2$. On the other hand, it can be readily seen that both criteria predict that the yield stresses measured in uniaxial tension and compression will be equal.

It is useful to represent the yield criteria graphically. This is done for the case of plane stress deformation ($\sigma_3 = 0$) in Fig. 5.26 where the stress axes are normalized with respect to the tensile yield stress. In plane stress the von Mises criterion (Equation 5.157) reduces to

$$\left(\frac{\sigma_1}{\sigma_y}\right)^2 + \left(\frac{\sigma_2}{\sigma_y}\right)^2 - \left(\frac{\sigma_1}{\sigma_y}\right)\left(\frac{\sigma_2}{\sigma_y}\right) = 1 \qquad (5.159)$$

which describes the ellipse shown in Fig. 5.26. It also shows that under certain conditions the applied tensile stress can be larger than σ_y. The Tresca criterion can also be represented on the same plot since when $\sigma_3 = 0$ Equations (5.154) and (5.155) can be combined to give

$$\tfrac{1}{2}(\sigma_1 - \sigma_2) = \sigma_s = \tfrac{1}{2}\sigma_y$$

$$\text{or} \quad \frac{\sigma_1}{\sigma_y} - \frac{\sigma_2}{\sigma_y} = 1 \qquad (5.160)$$

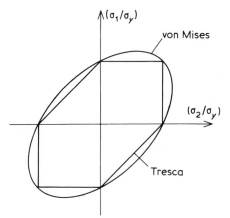

Fig. 5.26 *Schematic representation of the Tresca and von Mises yield criteria for plane stress deformation.*

This equation is applicable in the quadrants where σ_1 and σ_2 have different signs. Where they have the same signs the maximum shear stress criterion requires that

$$\sigma_1/\sigma_y = 1 \quad \text{and} \quad \sigma_2/\sigma_y = 1$$

It can be seen that the Tresca criterion gives a surface consisting of connected straight lines which inscribe the von Mises ellipse.

In practice materials are subjected to triaxial stress systems and the form of the criterion in three-dimensional principal stress space is of interest. They both plot out as infinite cylinders parallel to the [111] direction (cf. crystallography). The Tresca cylinder has a regular hexagonal cross-section and the von Mises one is circular in cross-section. It is found that metals tend to obey the von Mises criterion rather than Tresca although Tresca is often assumed in calculations because of its simpler form.

5.4.4 *Pressure-dependent yield behaviour*

In order to determine which is the most appropriate yield criterion for a particular material it is necessary to follow the yield behaviour by using a variety of different combinations of multiaxial stress. However, with polymers rather unusual features are revealed when the yield stress of the same polymer is measured just in simple uniaxial tension and compression. Both the Tresca and von Mises criteria predict that the yield stress should be the same in both cases and this is what is found for metals. But for polymers the compressive yield stress is usually higher than the tensile one. This difference between the compressive and tensile yield stress can be taken as an indication that the hydrostatic component of the applied stress is exerting an influence upon the yield process. This can be demonstrated more directly by measuring the tensile yield stress under the action of an overall hydrostatic pressure. The results of such measurements on different polymers are shown in Fig. 5.27 and it can be seen that there is a clear increase in σ_y with hydrostatic pressure. In general, it is found that the yield stresses of amorphous polymers show larger pressure dependences than those of crystalline polymers. There is no significant change in the yield stresses of metals at similar pressures, but it is quite reasonable to expect that the physical properties of polymers will change at these relatively low pressures. They tend to have much more lower bulk moduli than metals and so undergo significant volume changes on pressurization.

It is possible to modify the yield criteria described in the last section to take into account the pressure dependence. The hydrostatic pressure p is given by

$$p = \tfrac{1}{3}(\sigma_1 + \sigma_2 + \sigma_3) \tag{5.2}$$

and the effect of hydrostatic stress can be incorporated into Equation (5.156) by writing it as

$$A[\sigma_1 + \sigma_2 + \sigma_3] + B[(\sigma_1 - \sigma_2)^2 + (\sigma_2 - \sigma_3)^2 + (\sigma_3 - \sigma_1)^2] = 1 \tag{5.161}$$

where A and B are constants. It is possible to define A and B in terms of the simple uniaxial tensile and compressive yield stresses, σ_{yt} and σ_{yc}, since when $\sigma_2 = \sigma_3 = 0$ then

$$A\sigma_{yt} + 2B\sigma_{yt}^2 = 1 \quad \text{(tension)}$$
$$-A\sigma_{yc} + 2B\sigma_{yc}^2 = 1 \quad \text{(compression)}$$

and so

$$\left. \begin{array}{l} A = (\sigma_{yc} - \sigma_{yt})/\sigma_{yc}\sigma_{yt} \\ \text{and} \quad B = 1/(2\sigma_{yc}\sigma_{yt}) \end{array} \right\} \tag{5.162}$$

Equation (5.161) can therefore be written as

$$\begin{aligned} &2(\sigma_{yc} - \sigma_{yt})[\sigma_1 + \sigma_2 + \sigma_3] \\ &+ [(\sigma_1 - \sigma_2)^2 + (\sigma_2 - \sigma_3)^2 + (\sigma_3 - \sigma_1)^2] = 2\sigma_{yc}\sigma_{yt} \end{aligned} \tag{5.163}$$

If the magnitude of the tensile yield stress is the same as that for compression (i.e. $\sigma_{yt} = \sigma_{yc}$) the normal von Mises criterion is recovered.

It is possible to modify the Tresca maximum shear-stress criterion in several ways. The simplest way is to make the critical shear stress a function of the hydrostatic pressure p and so σ_s can be expressed by an equation of form

$$\sigma_s = \sigma_s^0 - \mu p \tag{5.164}$$

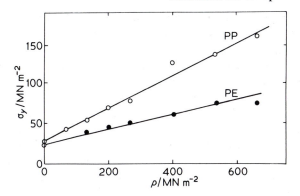

Fig. 5.27 *Variation of yield stress of polyethylene and polypropylene with hydrostatic pressure, p (data taken from Mears, Pae and Sauer, J. Appl. Phys.* **40** *(1969) 4229).*

where σ_s^0 is the shear yield stress in the absence of any overall hydrostatic pressure and μ is a material constant. The hydrostatic pressure p is taken to be positive for uniaxial tensile loading and negative in compression. The yield stress in uniaxial tension is therefore given by combining Equations (5.2), (5.155) and (5.164) as

$$\sigma_{yt} = 2\sigma_s^0 - 2\mu\sigma_{yt}/3$$

$$\text{or} \quad \sigma_{yt} = 2\sigma_s^0/(1 + 2\mu/3) \tag{5.165}$$

and in compression ($\sigma_1 = -\sigma_{yc}$) as

$$\sigma_{yc} = 2\sigma_s^0 + 2\mu\sigma_{yc}/3$$

$$\text{or} \quad \sigma_{yc} = 2\sigma_s^0/(1 - 2\mu/3) \tag{5.166}$$

Combining these two equations leads to

$$\mu = \frac{3}{2}\left(\frac{\sigma_{yc} - \sigma_{yt}}{\sigma_{yc} + \sigma_{yt}}\right) \tag{5.167}$$

which shows that $\mu > 0$ when the magnitude of the compressive yield stress is greater than the one measured in tension.

   In order to determine which is the most appropriate yield criterion for a particular polymer it is necessary to follow the yield behaviour under a variety of states of stress. This is most conveniently done by working in plane stress ($\sigma_3 = 0$) and making measurements in pure shear ($\sigma_1 = -\sigma_2$) and biaxial tension ($\sigma_1, \sigma_2 > 0$) as well as in the simple uniaxial cases. The results of such experiments on glassy polystyrene are shown in Fig. 5.28. The modified von Mises and Tresca envelopes are also plotted. In both cases they have been fitted to the measured uniaxial tensile and compressive yield stresses, σ_{yt} and σ_{yc}. It can be seen that the von Mises

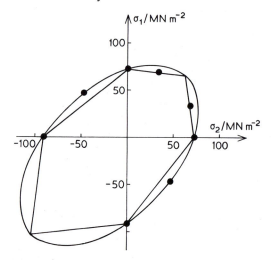

Fig. 5.28 *Modified von Mises and Tresca criteria fitted to experimental data upon the yield behaviour of polystyrene. (Data taken from Whitney and Andrews, J. Polym. Sci.,* **C-16** *(1967) 2981.)*

criterion plots out as a distorted ellipse which is inscribed by the modified Tresca hexagon. In general it is found that under most conditions polymers tend to follow a pressure-dependent von Mises criterion rather than the modified Tresca.

In three-dimensional principal stress space the modified Tresca cylinder becomes a hexagonal pyramid and the pressure-dependent von Mises cylinder becomes a cone. The significance of the apices of the pyramid and cone is that they define the conditions for which there can be yielding under the influence of hydrostatic stress alone. This is something which cannot happen for materials which obey the unmodified criteria (Section 5.4.3).

5.4.5 *Rate and temperature dependence*

So far we have considered only deformation which takes place at constant rate and temperature, but plastic deformation, like other aspects of the mechanical behaviour of polymers, has a strong dependence upon the testing rate and temperature. Typical behaviour is illustrated in Fig. 5.29 for a glassy thermoplastic deformed in tension. At a given strain-rate the yield stress drops as the temperature is increased and σ_y falls approximately linearly to zero at the glass-transition temperature when the polymer glass becomes a rubber. If the strain-rate is increased and the temperature held constant the yield stress increases (cf. time–temperature superposition (Section 5.2.7)).

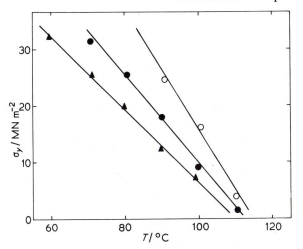

Fig. 5.29 *Variation of the yield stress of poly(methyl methacrylate) with temperature. Strain rates* ▲, 0.002 *min*$^{-1}$, ●, 0.02 *min*$^{-1}$, ○, 0.2 *min*$^{-1}$, *(Data of Langford, Whitney and Andrews reported by Ward.)*

The behaviour of semi-crystalline polymers is rather similar with the main difference being that the yield stress drops to zero at the melting temperature of the crystals rather than at the T_g. Between T_g and T_m the non-crystalline areas are rubbery and the material gains its strength from the crystalline regions which reinforce the rubbery matrix. If the temperature of a semi-crystalline polymer is reduced below T_g it behaves more like a glassy polymer and the crystals do not have such a significant strengthening effect.

5.4.6 *Craze yielding*

Consideration of the plastic deformation of polymers has, so far, been concerned only with 'shear yielding'. This occurs at constant volume and can take place uniformly throughout the sample. Certain polymers, particularly thermoplastics in the glassy state, are capable of undergoing a localized form of plastic deformation known as *crazing*. This is found to take place only when there is an overall hydrostatic tensile stress (i.e. $p >$ 0) and the formation of crazes causes the material to undergo a significant increase in volume. Fig. 5.30 shows a deformed specimen of polycarbonate which has undergone crazing. The crazes appear as small crack-like entities which are usually initiated on the specimen surface and are oriented perpendicular to the tensile axis. Closer examination shows that they are regions of cavitated polymer and so not true cracks, although the cracks which lead to eventual failure of the specimen usually nucleate within pre-existing crazes.

Fig. 5.30 *Sample of polycarbonate containing several large crazes. (Kambour, reproduced with permission.)*

Crazing can be demonstrated quite simply by rubbing one's fingers over the surface of a sample of glassy polystyrene and bending it to near its point of failure. The crazes appear as a fine surface mist on the tension surface. The crazes are very small, ~ 1000 Å thick and several microns in lateral dimensions, but they can be seen by the naked eye because they are less dense than the undeformed matrix and so reflect and scatter light. The demonstration also shows how crazing can be made easier by the presence of certain liquids ('crazing agents') such as finger grease. In practice the use of many glassy polymers is often limited by their tendency to undergo crazing at relatively low stresses in the presence of crazing agents. The crazes can mar the appearance of the specimen and lead to eventual catastrophic failure.

As with other mechanical properties, the formation of crazes in polymers depends upon testing temperature and the rate or time-period of loading. This is demonstrated clearly in Fig. 5.31 where the stress required to cause crazing in simple uniaxial tensile specimens of polystyrene is plotted against testing temperature. The craze initiation stress drops as the testing temperature is increased and as the strain-rate is decreased. The data in Fig. 5.31 are for dry crazing in air. The presence of crazing agents (e.g. finger grease or methanol with polystyrene) causes the craze initiation stress to drop further.

5.4.7 *Craze criteria*

As crazing is a yield phenomenon there have been attempts to establish a *craze criterion* in the same way that Tresca and von Mises criteria have been used for shear yielding. Crazing occurs typically about one-half of the yield stress of the polymer. This is in the 'elastic' region (Fig. 5.24), but the occurrence of crazing does not usually cause any detectable change in slope of the stress–strain curve as the volume fraction of crazed material is initially very low. Early suggestions of craze criteria were that a critical uniaxial tensile stress or strain were required for craze formation, but these were inadequate to describe the behaviour in multiaxial stress systems. Observations upon the way in which polymers craze in uniaxial tension under the action of a superimposed hydrostatic pressure have been made in a similar way to the measurement of shear yielding under hydrostatic pressure described earlier (Fig. 5.27). It is found that superimposed

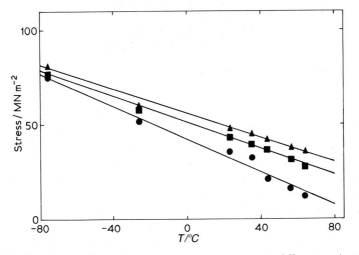

Fig. 5.31 *Dependence of crazing stress upon temperature at different strain-rates for polystyrene. Strain rates,* ●, 0.00067s^{-1}, ■, 0.0267s^{-1}, ▲, 0.267s^{-1}.

hydrostatic compression makes crazing more difficult and increases the tensile stress or strain needed to initiate crazes. Hydrostatic tension on the other hand produces a dilatation which helps to open up voids in the structure and hence aids the formation of crazes.

Since crazing involves the opening up of cavitated regions it would be reasonable to expect that it might take place at a critical strain for crazing, e_c. The effect of pressure outlined above could then be incorporated by writing, when there is an overall hydrostatic tension

$$e_c = C + D/p \tag{5.168}$$

where p is the hydrostatic stress and C and D are time- and temperature-dependent parameters. The maximum tensile strain in an isotropic body subjected to a general state of stress defined by the principal stresses σ_1, σ_2 and σ_3 (where $\sigma_1 > \sigma_2 > \sigma_3$) is given by

$$e_1 = \frac{1}{E}(\sigma_1 - \nu\sigma_2 - \nu\sigma_3) \tag{5.26}$$

where ν is Poisson's ratio and this strain always acts in the direction of the maximum principal stress. The critical strain criterion requires that crazing occurs when e_1 reaches a critical value and so Equation (5.168) can be rewritten to define the criterion in terms of stresses only as

$$\sigma_1 - \nu\sigma_2 - \nu\sigma_3 = X + Y/(\sigma_1 + \sigma_2 + \sigma_3) \tag{5.169}$$

where X and Y are new time- and temperature-dependent constants. This criterion is useful in that it gives both the conditions under which crazes form and the direction in which the crazes grow (i.e. perpendicular to σ_1). As with yield criteria the easiest way to test craze criteria is by making measurements in plane stress ($\sigma_3 = 0$) where Equation (5.169) becomes

$$\sigma_1 - \nu\sigma_2 = X + Y/(\sigma_1 + \sigma_2) \tag{5.170}$$

This equation is plotted in Fig. 5.32 and X and Y have been chosen by fitting the curve to experimental data on crazing in polystyrene under biaxial stress. Crazing cannot take place unless there is the overall hydrostatic tension [i.e. $(\sigma_1 + \sigma_2 + \sigma_3) > 0$] which is necessary to allow cavitation to occur. The curves from Equation (5.170) will then be asymptotic to the line where $\sigma_1 = -\sigma_2$. It is instructive to consider the criteria for shear yielding at the same time as craze criteria are examined and so the pressure-dependent von Mises curve from Fig. 5.28 for polystyrene tested under similar conditions is also given in Fig. 5.32. Having both the yield and craze envelopes on the same plot allows a prediction of the type of yielding that may occur under any general state of stress to be made.

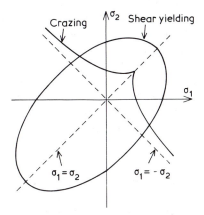

Fig. 5.32 *Envelopes defining crazing and yield for an amorphous polymer undergoing plane stress deformation. (After Sternstein and Ongchin (1969), ACS Polymer Reprints, **19**, 1117.)*

5.5 Deformation mechanisms

The response of a material to mechanical deformation reflects the microscopic deformation processes which are occurring on a molecular or atomic level. There is now a clear understanding of how this happens in metals. Elastic deformation reflects the displacement of the metal atoms out of their low energy positions in the 'potential well' and plastic deformation occurs by the motion of dislocations through the crystal lattice. In addition methods have been developed of strengthening metals by precipitation hardening. The presence of precipitate particles makes dislocation motion more difficult. Our understanding of the deformation processes which occur in polymers is by no means so well developed and this aspect of polymer deformation has been somewhat ignored previously in books at this level. There have been important developments over recent years which have increased our knowledge of the mechanisms by which polymers deform and so it is felt that it may be useful at this stage to review the current ideas of the mechanisms of polymer deformation.

The area has been divided up into elastic deformation and plastic deformation (shear yielding and crazing). They have been considered as essentially time-independent processes. It is clear from Section 5.2 that there will always be an underlying time dependence, but since the discussion is limited to polymers which are generally well away from the viscoelastic region the approximation of time-independent behaviour does not cause too many problems.

5.5.1 Elastic deformation of single-phase polymers

It is useful to consider the deformation of single- and multi-phase polymers separately. Unfilled elastomers, polymer glasses and polymer single crystals

TABLE 5.1 *Typical values of Young's modulus for different types of polymers.*

Types of polymer	Young's modulus/N m^{-2}
Elastomer	$\sim 10^6$
Glassy polymer	$\sim 10^9$
Polymer crystal ($\perp$ to c)	$\sim 10^9$
($\parallel$ to c)	$\sim 10^{11}$

can be considered as being essentially single-phase materials and this makes determination of their moduli somewhat more straightforward than for multi-phase systems such as semi-crystalline polymers which can be thought of as consisting of mixtures of the separate phases.

One of the most useful aspects of polymers is the wide range of moduli that can be obtained. Typical values of Young's modulus are given in Table 5.1. *Elastomers* tend to have very low Young's moduli of the order of 10^6 N m^{-2}. This is because their deformation consists essentially of uncoiling the molecules in a crosslinked network. It is possible to calculate a value of the modulus of an elastomer using the statistical theory and this gives very good agreement with the experimentally-determined values (Section 5.3.2). The moduli of *polymer glasses* are typically of the order of 10^9 N m^{-2}. The deformation of a polymer glass is somewhat more difficult to model theoretically. The structure is thought to be completely random and so elastic deformation will involve the bending and stretching of the strong covalent bonds on the polymer backbone as well as the displacement of adjacent molecules which is opposed by relatively weak secondary (e.g. van der Waals) bonding.

The most striking feature of the elastic properties of *polymer crystals* is that they are very anisotropic. The moduli parallel to the chain direction are $\sim 10^{11}$ N m^{-2} which is similar to the values of moduli found for metals (e.g. steel $E \approx 2.1 \times 10^{11}$ N m^{-2}). However, the moduli of polymer crystals deformed in directions perpendicular to the chain axes are much lower at about 10^9 N m^{-2}. This large difference in modulus reflects the anisotropy in the bonding in polymer crystals. Deformation parallel to the chain direction involves stretching of the strong covalent bonds and changing the bond angles or if the molecule is in a helical conformation distorting the molecular helix. Deformation in the transverse direction is opposed by only relatively weak secondary van der Waals or hydrogen bonding. Because of the high anisotropy of the bonding in polymer crystals it is possible with knowledge of the crystal structure (Section 4.1.3) and the force constants of the chemical bonds to calculate the theoretical moduli of polymer crystals. An exact calculation is rather complex as it requires detailed knowledge of inter- and intramolecular interactions, but Treloar has shown how an estimate can be made of the chain-direction modulus of a polymer which crystallizes with its backbone in the form of a planar zig-zag. This involves a relatively simple calculation which assumes that deformation involves only the bending and stretching of the bonds along the molecular backbone.

The model of the polymer chain used in the calculation is shown schematically in Fig. 5.33. It is treated as consisting of n rods of length, l, which are capable of being stretched along their lengths but not of bending, and are joined together by torsional springs. The bond angles are taken initially to be Θ and the angle between the applied force, f, and the individual bonds as initially α. The original length of the chain, L, is then $nl \cos \alpha$ and so the change in length δL on deforming the chain is given by

$$\delta L = n\delta(l \cos \alpha) = n(\cos \alpha \delta l - l \sin \alpha \delta \alpha) \qquad (5.171)$$

It is possible to obtain expressions for δl and $\delta \Theta$ in terms of the applied force f. Consideration of the bond stretching allows δl to be determined. The component of the force acting along the bond direction is $f \cos \alpha$. This can be related to the extension through the force constant for bond stretching k_l which can be determined from infra-red or Raman spectroscopy. k_l is the constant of proportionality relating the force along the bond to its extension. It follows therefore that

$$\delta l = f \cos \alpha / k_l \qquad (5.172)$$

The force required to cause valence angle deformation can be determined by using the angular deformation force constant k_Θ which relates the change in bond angle $\delta \Theta$ to the torque acting around each of the bond angles. This torque is equal to the moment of the applied force about the angular vertices, $\frac{1}{2} fl \sin \alpha$ and so the change in the angle between the bonds is given by

$$\delta \Theta = (fl \sin \alpha)/2k_\Theta \qquad (5.173)$$

But since it can be shown by a simple geometrical construction that $\alpha = 90°$ $- \Theta/2$ then it follows that

$$\delta \alpha = -\delta \Theta/2$$

and Equation (5.173) becomes

$$\delta \alpha = -(fl \sin \alpha)/4k_\Theta \qquad (5.174)$$

Putting Equations (5.172) and (5.174) into (5.171) gives

$$\frac{\delta L}{f} = n\left[\frac{\cos^2 \alpha}{k_l} + \frac{l^2 \sin^2 \alpha}{4k_\Theta} \right] \qquad (5.175)$$

Fig. 5.33 *Model of a polymer chain undergoing deformation (after Treloar, Polymer,* **1** *(1960)* *95).*

The Young's modulus of the polymer is given by

$$E = (f/A)/(\delta L/L) \tag{5.176}$$

where A is the cross-sectional area supported by each chain and so

$$E = \frac{l \cos \alpha}{A} \left[\frac{\cos^2 \alpha}{k_l} + \frac{l^2 \sin^2 \alpha}{4k_\theta} \right]^{-1}$$

or in terms of the bond angle Θ

$$E = \frac{l \sin(\Theta/2)}{A} \left[\frac{\sin^2(\Theta/2)}{k_l} + \frac{l^2 \cos^2(\Theta/2)}{4k_\theta} \right]^{-1} \tag{5.177}$$

This equation allows the modulus to be calculated with knowledge of the crystal structure, which gives A, l and Θ, and the spectroscopically measured values of the force constants k_l and k_θ. It predicts a value of E for polyethylene of about 180 GN m^{-2} which is generally rather lower than values obtained from more sophisticated calculations. This is because the above analysis does not allow for intramolecular interactions which tend to increase the modulus further. The chain-direction modulus does not completely describe the elastic behaviour of a polymer crystal. A complete set of elastic constants c_{ij} (Equation 5.12) are required in order to do this. The number of constants depends upon the crystal symmetry and for an orthorhombic crystal such as polyethylene nine constants are necessary. The measurement of all the elastic constants is very difficult even for good single crystals like the diacetylenes (Section 4.1.6). A set of elastic constants has been calculated for orthorhombic polyethylene and they are given in Table 5.2. They were obtained by Odajima and Maeda who calculated the forces required to stretch the bonds along the polymer backbone and open up the angles between the bonds as in the Treloar model. In addition, inter- and intramolecular interactions were taken into account.

Table 5.3 gives some calculated and experimentally measured values of chain-direction crystal moduli of several different polymer crystals. Single crystals of polymers are not normally available with dimensions that are suitable to allow them to be tested mechanically. Highly oriented semi-crystalline polymers tend to have moduli of no more than 50% of the theoretical values. This is because even though they may possess

TABLE 5.2 *Calculated values of the nine elastic constants for a polyethylene crystal at 298 K. (After Odajima and Maeda, J. Polym. Sci.* **C** *15 (1966) 55.)*

Stiffness component	$c_{ij}/10^9 \text{N m}^{-2}$
c_{11}	4.83
c_{22}	8.71
c_{33}	257.1
c_{12}	1.16
c_{13}	2.55
c_{23}	5.84
c_{44}	2.83
c_{55}	0.78
c_{66}	2.06

good crystal orientation the presence of an amorphous phase tends to cause a significant reduction in modulus. Estimates of chain-direction moduli can be made experimentally using an X-ray diffraction method. This involves measuring the relative change under an applied stress in the position of any (*hkl*) Bragg reflection which has a component of the chain-direction repeat (usually indexed *c*). The production of 100% crystalline macroscopic polymer single crystals by solid-state polymerization (Section 2.11) has recently made it possible to measure the moduli of these crystals directly using simple mechanical methods. A stress–strain curve for a polydiacetylene single crystal fibre is given in Fig. 5.34. The

TABLE 5.3 *Calculated and measured values of chain-direction moduli for several different polymer crystals. (After Young, Chapter 7 in 'Developments in Polymer Fracture' edited by E.H. Andrews.)*

Polymer	Calculated		Measured	
	$E/\text{GN m}^{-2}$	method	$E/\text{GN m}^{-2}$	method
Polyethylene	182	VFF	240	X-ray
	340	UBFF	358	Raman
	257	UBFF	329	Neutron (CD_2)
Polypropylene	49	UBFF	42	X-ray
Polyoxymethylene	150	UBFF	54	X-ray
Polytetrafluoroethylene	160	UBFF	156	X-ray
			222	Neutron
Substituted (i)	49	VFF	45	Mechanical
polydiacetylenes (ii)	65	VFF	61	Mechanical

VFF—Valence force field (e.g. Treloar's method).
UBFF—Urey—Bradley force field (e.g. used to determine Table 5.2).
(i) Phenyl urethane derivative.
(ii) Ethyl urethane derivative.

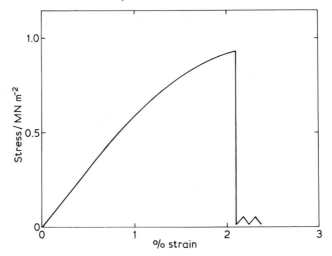

Fig. 5.34 *A stress–strain curve obtained for a polydiacetylene single crystal fibre.*

modulus calculated from the initial slope of the curve is 61 GN m^{-2} which is very close to the value calculated using the simple bond stretching and bending method of Treloar ($\sim$65 GN m^{-2}). The measured and calculated values of moduli for some polydiacetylene single crystals are also given in Table 5.3.

A general picture emerges concerning the values of chain direction moduli of polymer crystals. They tend to be high if the molecule is in the form of a planar zig-zag rather than a helix. For example, polyethylene is stiffer than polyoxymethylene or polytetrafluoroethylene which both have molecules in helical conformations (Table 4.1). The helices can be extended more easily than the polyethylene planar zig-zag. Also the presence of large side groups tend to reduce the modulus because they increase the separation of molecules in the crystal. This causes an increase in the area supported by each chain.

Although polymers are generally considered to be relatively low modulus flexible materials it is clear from Table 5.3 that their potential unidirectional moduli are relatively high. This feature has been utilized to a limited extent for many years in conventional polymer fibres. Recent developments of stiff and inflexible polymer molecules such as poly(p-phenylene terephthalamide) (Table 2.2) have led to very high modulus polymer fibres such as 'kevlar' being produced by the spinning of liquid crystalline solutions. For many purposes it is the *specific modulus* (Young's modulus divided by specific gravity) that is important, especially when light high-stiffness materials are required. Polymers tend to have low specific gravities, typically 0.8–1.5 compared with 7.9 for steel, and so their specific stiffnesses compare even more favourably with metals.

5.5.2 *Elastic deformation of semi-crystalline polymers*

The presence of crystals in a polymer has a profound effect upon its mechanical behaviour. This effect has been known for many years from experience with natural rubber. Fig. 5.35(a) shows the increase in Young's modulus that occurs due to crystallization which can occur during prolonged storage of the material at low temperature. There is a hundred-fold increase in modulus when the degree of crystallinity increases from 0 to 25%. A similar effect is found in other semi-crystalline polymers. Fig. 5.35(b) shows how the modulus of a linear polyethylene increases as the crystallinity is increased through use of different heat treatments. The degree of crystallinity in linear polyethylene is usually quite high (>60%) and the data in Fig. 5.35(b) can be extrapolated to give a modulus of the order of $5\,\text{GN}\,\text{m}^{-2}$ for 100% crystalline material.

It is apparent from considerations of the structure in Section 4.2 that semi-crystalline polymers are essentially two-phase materials and that the increase in modulus is due to the presence of the crystals. Traditional ideas of the stiffening effect due to the presence of crystals were based upon the statistical theory of elastomer deformation (Section 5.3.2). It was thought that the crystals in the amorphous 'rubber' behaved like crosslinks and produced the stiffening through an increase in crosslink density rather than through their own inherent stiffness. Although this mechanism may be relevant at very low degrees of crystallinity it is clear that most semi-

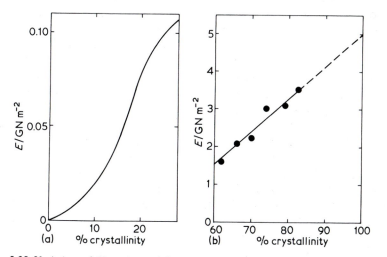

Fig. 5.35 *Variation of Young's modulus, E, with the degree of crystallinity for different polymers.* (a) *Increase in modulus of natural rubber with crystallization (data of Leitner reported by Treloar).* (b) *Dependence of the modulus of polyethylene upon crystallinity (data taken from Wang, J. Appl. Phys.,* **44** *(1973) 4052).*

crystalline polymers behave as composites and the observed modulus is a reflection of the combined modulus of the amorphous and crystalline regions. The exact way in which this combination of moduli can be represented mathematically is not a trivial problem. For example, spherulitic polymers have elastically isotropic elastic properties, but the individual crystals within the spherulites are themselves highly anisotropic. It is therefore necessary to calculate an effective modulus for the particular type of morphology under consideration. The effective crystal modulus depends not only upon the proportion of crystalline material present, given by the degree of crystallinity, but also upon the size, shape and distribution of the crystals within the polymer sample. It is clear, therefore, that reliable estimates of the properties of the semi-crystalline polymer composite can only be made for specific well-defined morphologies.

A model system which can be used is the diacetylene polymer for which the stress–strain curve was given in Fig. 5.34. By controlling the polymerization conditions it is possible to prepare single crystal fibres which contain both monomer and polymer molecules. The monomer has a modulus of only 9 GN m^{-2} along the fibre axis compared with 61 GN m^{-2} for the polymer and the partly polymerized fibres which contain both monomer and polymer molecules are found to have values of modulus between these two extremes as shown in Fig. 5.36. The variation of the modulus with the proportion of polymer (approximately equal to the conversion) can be predicted by two simple models. The first one due to Reuss assumes that the elements in the structure (i.e. the monomer and polymer molecules) are lined up in series and experience the same stress.

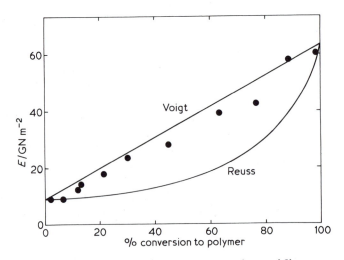

Fig. 5.36 *Dependence of the modulus of polydiacetylene single crystal fibres upon conversion into polymer.*

The modulus in this case, E_R will be given by an equation of the form

$$1/E_R = V_P/E_p + (1 - V_p)/E_m \qquad (5.178)$$

where V_p is the volume fraction of polymer and E_p and E_m are the moduli of the polymer and matrix (monomer) respectively. The second model due to Voigt assumes that all the elements are lined up in parallel and so experience the same strain. The modulus for the Voigt model, E_V, is given by the rule of mixtures as

$$E_V = E_p V_p + E_m(1 - V_p) \qquad (5.179)$$

Lines corresponding to the predictions of Equations (5.178) and (5.179) are also given in Fig. 5.36. It can be seen that the experimental points lie between the two lines and closest to the Voigt line indicating that the elements are experiencing approximately the same strain in this particular structure. Equations similar to (5.178) and (5.179) can be used to predict the moduli of fibre-reinforced composites and it is necessary that in all similar composite structures the measured moduli fall either between or on the Voigt and Reuss lines. The agreement that is found between the measured moduli of the diacetylene polymer fibres and the Voigt prediction is due to the structure of the fibres which consist of long polymer molecules lying parallel to the fibre axis and embedded in a monomer matrix. This morphology ensures that the strain will be uniform in the deformed structure.

The situation is considerably more complicated with more random structures such as spherulitic polymers. The measured modulus of such a structure is an average of that of the crystalline and amorphous regions. Even when it is assumed that the amorphous material is isotropic there are problems in determining how to estimate the effective average modulus of the crystals and how to combine this with modulus of the amorphous material to give an overall modulus for a polymer sample. This latter problem can be overcome by extrapolating measurements of modulus as a function of crystallinity to 100 percent crystallinity as shown in Fig. 5.35(b). The structure of such a material can be thought of as a randomly oriented poly-crystalline structure not unlike a polycrystalline metal. The Reuss and Voigt models can again be applied with their respective assumptions of uniform stress and uniform strain. Using the nine crystal moduli from Table 5.2 the Reuss model predicts a modulus of 4.9 GN m^{-2} compared with 15.6 GN m^{-2} for the Voigt model. Since the extrapolated value is ~5 GN m^{-2} it can be seen that in the case of spherulitic polyethylene the Reuss model gives a better prediction of the observed behaviour.

5.5.3 *Shear yielding in glassy polymers*

Glassy polymers are generally thought of as being rather brittle materials, but they are capable of displaying a considerable amount of ductility below

T_g especially when deformed under the influence of an overall hydrostatic compressive stress. This behaviour is illustrated in Fig. 5.37 where true stress–strain curves are given for an epoxy resin tested in uniaxial tension and compression at room temperature. The T_g of the resin is 100°C and such cross-linked polymers are found to be brittle when tested in tension at room temperature. In contrast they can show considerable ductility in compression and undergo shear yielding. Another important aspect of the deformation is that glassy polymers tend to show 'strain softening'. The true stress drops after yield, not because of necking which cannot occur in compression, but because there is an inherent softening of the material.

The process of shear yielding in epoxy resins is found to be homogeneous. The specimens undergo uniform deformation with no evidence of any localization. The situation is rather similar in other glassy polymers but in certain cases strongly-localized deformation is obtained. Fig. 5.38 shows a polarized-light optical micrograph of a section of polystyrene deformed in plane strain compression. Fine deformation bands are obtained in which the shear is highly localized. It is thought that the formation of shear bands may be associated with the strain softening in the material. Once a small region starts to undergo shear yielding it will continue to do so because it has a lower flow stress than the surrounding relatively undeformed regions.

There have been many attempts over the years to explain the process of shear yielding on a molecular level and some of the approaches are outlined below.

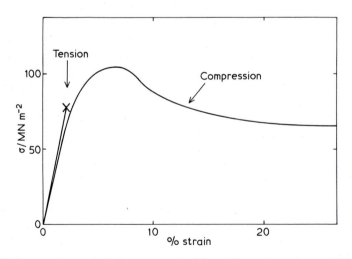

Fig. 5.37 *Stress–strain curves for an epoxy resin deformed in tension or compression at room temperature.*

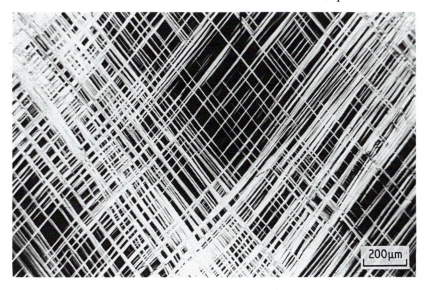

Fig. 5.38 *Shear bands in polystyrene deformed in plane strain compression (courtesy Dr. R.J. Oxborough).*

(i) *Stress-induced increase in free volume*

It has been suggested that the one effect of an applied strain upon a polymer is to increase the free volume (Section 4.4.2) and hence increase the mobility of the molecular segments. It follows that this would also have the effect of reducing T_g. Developments of this idea predict that during tensile deformation the overall hydrostatic tension should cause an increase in free volume and so aid plastic deformation. However, there are problems in the interpretation of compression tests where plastic deformation can still take place even though the hydrostatic stress is compressive. Moreover, careful measurements have shown that there is generally a slight reduction in overall specimen volume during plastic deformation in both tension and compression. It would seem, therefore, that the concept of free volume is not very useful in the explanation of plastic deformation of glassy polymers.

(ii) *Application of the Eyring theory to yield in polymers*

It is possible to think of yield and plastic deformation in polymers as a type of viscous flow, especially since glassy polymers are basically frozen liquids that have failed to crystallize. Eyring developed a theory to describe viscous flow in liquids and it can be readily adapted to describe the behaviour of glassy polymers. The segments of the polymer chain can be thought of as being in a pseudo-lattice and for flow to occur a segment must move to an adjacent site. There will be a potential barrier to overcome,

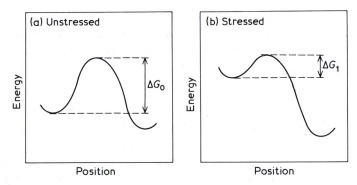

Fig. 5.39 *Schematic illustration of the potential barrier for flow in a glassy polymer.* (a) *Unstressed state with barrier of* ΔG_0. (b) *Stressed state with barrier reduced to* $\Delta G_1 = (\Delta G_0 - \frac{1}{2}\sigma A x)$.

because of the presence of adjacent molecular segments, and the situation is shown schematically in Fig. 5.39(a). The height of the barrier is given by ΔG_0 and the frequency ν_0 at which the segments jump the barrier and move to the new position can be represented by an Arrhenius type equation

$$\nu_0 = B\exp(-\Delta G_0/\mathbf{k}T) \tag{5.180}$$

where B is a temperature-independent constant and $\mathbf{k}$ is Boltzmann's constant. The rate at which segments jump forward over the barrier will be increased by the application of a shear stress σ. This has the effect of reducing the energy barrier by an amount $\frac{1}{2}\sigma A x$ which is the work done in moving the segment a distance, x, through the lattice of unit cross-sectional area A. The new situation is illustrated schematically in Fig. 5.39(b) and the frequency of forward motion increases to become

$$\nu_f = B\exp[-(\Delta G_0 - \frac{1}{2}\sigma A x)/\mathbf{k}T]$$
$$\text{or} \quad \nu_f = \nu_0\exp(\sigma A x/2\mathbf{k}T] \tag{5.181}$$

There is also the possibility of backward motion occurring. The frequency of backward motion ν_b is given by

$$\nu_b = \nu_0\exp(-\sigma A x/2\mathbf{k}T) \tag{5.182}$$

and the application of stress makes backward motion more difficult. The resultant strain rate $\dot{e}$ of the polymer will be proportional to the difference between the frequencies of forward and backward motion

$$\text{i.e. } \dot{e} \propto (\nu_f - \nu_b) = \nu_0\exp(\sigma A x/2\mathbf{k}T) - \nu_0\exp(-\sigma A x/2\mathbf{k}T)$$

and this can be written as

$$\dot{e} = K\sinh(\sigma V/2\mathbf{k}T) \tag{5.183}$$

where K is a constant and V (= Ax) is called the 'activation volume' or 'Eyring volume' and is the volume of the polymer segment which must move to cause plastic deformation. If the stress is high and the frequency of backward motion is small then Equation (5.183) can be approximated to

$$\dot{e} = (K/2)\exp(\sigma V/2\mathbf{k}T) \tag{5.184}$$

This approach has met with considerable success in the prediction of the strain-rate dependence of the yield stress of polymers. It is tempting to relate V to structural elements in the polymer. For most glassy polymers it comes out to be the volume of several polymer segments (typically between 2 and 10) although there is no clear correlation between the size of the activation volume and any particular aspect of the polymer structure.

(iii) *Molecular theories of yield*
Recent developments in the interpretation of the yield behaviour of glassy polymers have involved attempts to relate plastic deformation to specific kinds of molecular motion. One particularly successful approach was suggested by Robertson for the case of a planar zig-zag polymer chain. He assumed that at any time the molecular chain segments would be distributed between low-energy *trans* and high-energy *cis* conformations and that the effect of an applied stress would be to cause certain segments to change over from *trans* to *cis* conformations as illustrated in Fig. 5.40. This and other similar molecular theories have led to accurate predictions of the variation of yield stress with strain-rate and temperature for glassy polymers. The reader is referred to more advanced texts for further details.

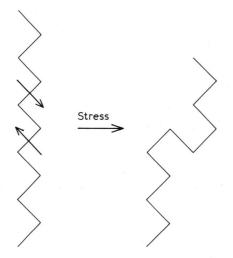

Stress

Fig. 5.40 *Schematic illustration of the conformational changes involved in the deformation of a planar zig-zag polymer molecule (after Robertson and Kambour).*

5.5.4 *Crazing in glassy polymers*

The phenomenology of crazing was discussed in Section 5.4, but as yet the mechanisms by which crazing occurs have not been considered. Crazes normally nucleate on the surface of a specimen, probably at the site of flaws such as scratches or other imperfections. They tend to be lamellar in shape and grow into the bulk of the specimen from the surface as shown in Fig. 5.41. The flaws on the surface tend to raise the magnitude of the applied stress locally and allow craze initiation to occur.

The basic difference between crazes and cracks is that crazes contain polymer, typically about 50% by volume, within their bulk whereas cracks do not. The presence of polymer within crazes was originally deduced from measurements of the refractive index of crazed material, but it can be demonstrated more directly by using electron microscopy. Fig. 5.42 shows a section through a craze in polystyrene viewed at high magnification. It can be clearly seen that there is a sharp boundary between the craze and the polymer matrix and that the craze itself contains many cavities and is bridged by a dense network of fine fibrils of the order of 20 nm in diameter. If a craze is followed along its length to the tip it is found to taper gradually until it eventually consists of a row of voids and microfibrils on the 5 nm scale.

Microscopic examination has given the most important clues as to how crazes nucleate and grow in a glassy polymer. The basic mechanism is thought to be a local-yielding and cold-drawing process which takes place in a constrained zone of material. It is known that the local stresses at the tip of a surface flaw or a growing craze are higher than the overall tensile

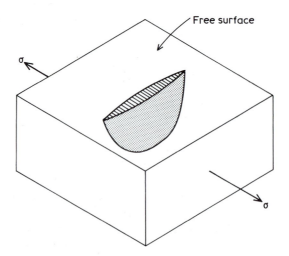

Fig. 5.41 *Schematic diagram of a craze nucleating at the surface of a polymer.*

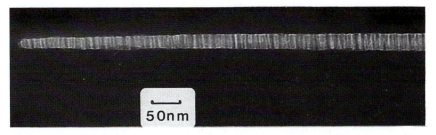

50nm

Fig. 5.42 *Transmission electron micrograph of a craze in a thin film of polystyrene (courtesy of Professor E. Kramer).*

stress applied to the specimen. It is envisaged that yielding takes place locally at these stress concentrations at a stress below that which is required to cause general yielding in the rest of the specimen. If the bulk tensile uniaxial stress is σ_0 then there will be an overall bulk tensile hydrostatic stress of $p = \sigma_0/3$ but this hydrostatic component will be very much higher at the tip of the flaw or growing craze. These conditions of stress then favour the initiation of a craze at a flaw or the propagation of a preexisting craze. It is envisaged that the fibrils within the craze form by a local yielding and cold-drawing process and the fibrils remain intact because they contain oriented molecules and are consequently stronger than the surrounding uncrazed polymer. The thickening of the craze is thought to take place by uncrazed material at the craze/matrix boundary being deformed into fibrils and so transformed into craze matter. The crazes form as very thin lamellae in a plane perpendicular to the tensile axis because the tensile stress is highest across this plane and the elastic constraint of the surrounding uncrazed material allows the craze to grow only laterally and thicken.

5.5.5 *Plastic deformation of polymer crystals*

There is now clear evidence to suggest that polymer crystals are capable of undergoing plastic deformation in a similar way to other crystalline solids through processes such as slip, twinning and martensitic transformations. It is thought that similar deformation processes occur in both isolated single crystals and in the crystalline regions of semi-crystalline polymers. The main difference between the deformation of polymer crystals and crystals of atomic solids is that the deformation occurs in the polymer such that molecules are not broken and, as far as possible, remain relatively undistorted. This problem of molecular integrity does not arise in atomic solids. Deformation, therefore, generally involves only the sliding of molecules in directions parallel or perpendicular to the chain axis which breaks only the secondary van der Waals bonds. The specific deformation mechanisms will be discussed separately.

(i) *Slip*

The deformation of metal crystals takes place predominantly through slip processes which involve the sliding of one layer of atoms over the other through the motion of dislocations. In fact the ductility of metals stems from there normally being a large number of slip systems (i.e. slip planes and directions) available in metal crystals. Slip is also important in the deformation of polymer crystals, but the need to avoid molecular fracture limits the number of slip systems available. The slip planes are limited to those which contain the chain direction (i.e. $(hk0)$ planes when [001] is the chain direction). The slip directions can in general be any direction which lies in these planes, but normally only slip which takes place in directions parallel or perpendicular to the chain direction is considered in detail. Slip in any general direction in a $(hk0)$ plane will have components of these two specific types. *Chain direction slip* is illustrated schematically in Fig. 5.43. The molecules slide over each other in the direction parallel to the molecular axis. It is known to occur during the deformation of both bulk semi-crystalline polymers and of solution-grown polymer single crystals. It can be readily detected from the change of orientation of the molecules within the crystals as shown in Fig. 5.44 for polytetrafluoroethylene. Initially the molecules are approximately normal to the lamellar surface (Fig. 4.21), but during deformation they slide past each other and become tilted within the crystal. Slip in directions perpendicular to the molecular axes is sometimes called *transverse slip*. It has been shown to occur during the deformation of both melt-crystallized polymers and solution-grown single crystals, but is rather more difficult to detect because it involves no change in molecular orientation within the crystals.

(ii) *Dislocation motion*

It is well established that slip processes in atomic solids are associated with the motion of dislocations. The theoretical shear strength of a metal lattice has been shown (Section 5.1.4) to be of the order of $G/10$ to $G/30$ where G

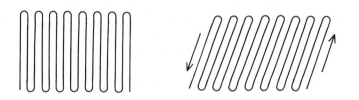

Fig. 5.43 *Schematic representation of chain direction slip in a chain-folded polymer crystal.*

Fig. 5.44 *A replica of a fracture surface of polytetrafluoroethylene after compressive deformation in the vertical direction.*

is the shear modulus of the crystal. The measured shear yield strengths of many crystalline solids are typically between 10^{-4} G and 10^{-5} G. The difference between the theoretical and experimentally determined values is due to deformation taking place by the glide of dislocations which occurs at stresses well below the theoretical strength. The presence of dislocations in polymer crystals has been discussed previously (Section 4.2.5) and there is no fundamental reason why slip in polymer crystals cannot occur through dislocation motion.

The energy per unit length, $E(l)$, of a dislocation in an isotropic medium can be shown* to be given approximately by

$$E(l) = \frac{Gb^2}{4\pi} \ln\left(\frac{r}{r_0}\right) \qquad (5.185)$$

where b is the Burgers vector of the dislocation, r the radius of the crystal and r_0 is the core radius of the dislocation (normally taken $\approx b$). The shear modulus G in Equation (5.185) is that for an isotropic medium. Since the bonding in polymer crystals is highly anisotropic G must be replaced by an effective crystal modulus when the equation is applied to polymer crystals. Inspection of Equation (5.185) shows that for a crystal of a given size, dislocations will have low energy when the shear modulus and b are both

*J. Friedel, *Dislocations*, Pergamon Press (1964).

small. The total energy of a dislocation will also be reduced if the length of the dislocation line is as short as possible. Since dislocations cannot terminate within a crystal then the dislocations with the shortest length can be obtained in lamellar polymer crystals by having the dislocation line perpendicular to lamellar surface and so parallel to the chain direction. The need to avoid excess deformation of the polymer chains means that b should also be parallel to the chain direction. The result of these considerations is that a particularly favourable type of dislocation in lamellar crystals would be expected to be a screw dislocation with its Burgers vector parallel to the chain direction as illustrated schematically in Fig. 4.24.

Such dislocations have been shown to be present in solution-grown polyethylene single crystals and their number is found to increase following deformation (Fig. 5.45). The Burgers vector of such dislocations is the same as the chain direction repeat 2.54 Å (Table 4.1) and the shear modulus for slip in the chain direction will be either $c_{44} = 2.83$ GN m^{-2} or $c_{55} = 0.78$ GN m^{-2} (Table 5.2) depending upon whether the slip plane is (010) or (100). It is found that at room temperature [001] chain direction slip in bulk polyethylene occurs at a stress of the order of 0.015 GN m^{-2} (i.e. ~1/50 to 1/200 of the shear modulus) and calculations have shown that chain direction slip could occur through the thermal activation of the type of dislocations shown in Fig. 4.24 and 5.45. This mechanism is, however, only favourable when the crystals are relatively thin where the length of the dislocation line is small. It is expected that such a dislocation would be favourable in thin crystals of other polymers since they generally have low shear moduli. Most other polymer crystals have larger chain-direction repeats than polyethylene (Table 4.1) especially when the molecule is in a helical conformation. This would give rise to larger values of $E(l)$ through higher Burgers vectors, but it would be expected that in this case the dislocations might split into partials in order to reduce their energy.

(iii) *Twinning*

Twinned crystals of many materials are produced during both growth and deformation. Many mineral crystals are found in a twinned form in nature and deformation twins are often obtained when metals are deformed, especially at low temperatures. The most commonly obtained types of twinned crystals are those in which one part of the crystal is a mirror image of another part. The boundary between the two regions is called the *twinning plane*. The particular types of twins that form in a crystal depend upon the structure of the crystal lattice and deformation twins can be explained in terms of a simple shear of the crystal lattice.

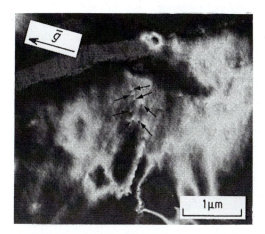

Fig. 5.45 *Dark-field electron micrograph of a polyethylene crystal containing screw dislocations with Burgers vectors parallel to the chain direction (normal to the crystal surface). The dislocations can be seen arrowed as dots with black and white contrast. (Courtesy of Dr J. Petermann.)*

As in the case of slip, there is no fundamental reason why polymer crystals cannot undergo twinning and it is found that twinning frequently occurs when crystalline polymers are deformed. The types of twins that are obtained are formed in such a way that the lattice is sheared without either breaking or seriously distorting the polymer molecules. A good example of the formation of twins in polymer crystals is obtained by deforming solid-state polymerized polydiacetylene single crystals (Section 2.11). Fig. 5.46 shows a micrograph of such a crystal viewed at 90° to the chain direction. The striations on the crystal define the molecular axis and it can be seen that the molecules kink over sharply at the twin boundary. Accurate measurements have shown that there is a 'mirror image' orientation of the crystal on either side of the boundary. This clearly indicates that the crystal contains a twin. A schematic diagram of the molecular arrangements on either side of the twin boundary is also given in Fig. 5.46. The angle that the molecules can bend over and still retain the molecular structure can be calculated from a knowledge of the crystal structure and exact agreement is found between the measured and calculated angles. The result of the twinning process is that the molecules on every successive plane in the twin are displaced by one unit of c parallel to the chain direction. This is clearly identical to the result of chain direction slip taking place and it has now become recognized that this type of twinning could be important in the deformation of many types of polymer crystals.

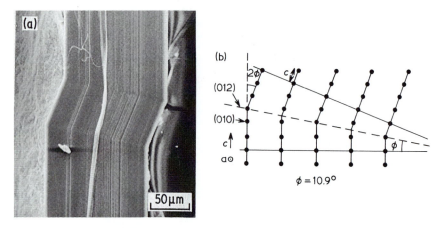

Fig. 5.46 *A large twin in a polydiacetylene single crystal. (a) Scanning electron micrograph of twin; (b) schematic diagram of twin on a molecular level. (Young, Bloor, Batchelder and Hubble, J. Mater, Sci., **13** (1978) 62, reproduced with permission.)*

Another type of twinning that has been found to occur in polymer crystals involves a simple shear in directions perpendicular to the chain axis and in this case does not bend the molecules. An example of this type of twinning is shown schematically in Fig. 5.47 for the polyethylene unit cell projected parallel to the chain direction. Two particular twinning planes (110) and (310) have been found and the twinning has been reported after deformation of both solution-grown single crystals and melt-crystallized polymer. The occurrence of twinning can be detected from measurements of the rotation of the *a* and *b* crystal axes about the molecular *c* axis since the twinning causes rotation of the crystal lattice.

(iv) *Martensitic transformations*
This is a general term which covers transformations which bear some similarity to twinning, but involve a change of crystal structure. The name comes from the well-known, important transformation which occurs when austenitic steel is rapidly cooled and forms the hard tetragonal metastable phase called 'martensite'. These transformations can also occur on the deformation or cooling of many crystalline materials and all of these martensitic transformations share the similar feature that they take place with no long-range diffusion.

Martensitic transformations have not been widely reported in polymers, but an important one occurs on the deformation of polyethylene. The normal crystal structure of both solution-grown single crystals and melt-crystallized polymer is orthorhombic, but after deformation extra Bragg reflections are found in diffraction patterns and these have been

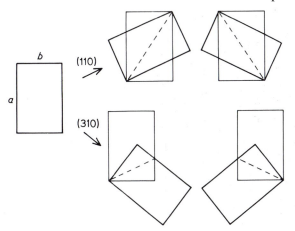

Fig. 5.47 *Schematic representation of (110) and (310) twinning in orthorhombic polyethylene. The unit cells are viewed projected parallel to the chain direction. (After Frank, Keller and O'Conner, Phil. Mag.* **3** *(1958) 64.)*

explained in terms of the formation of a monoclinic form (Table 4.1). The transformation takes place by means of a simple two-dimensional shear of the orthorhombic crystal structure in a direction perpendicular to c. The chain axis remains virtually unchanged by the transformation at about 2.54 Å, but in the monoclinic cell (Table 4.1) it is conventionally indexed as b.

5.5.6 *Plastic deformation of semi-crystalline polymers*

Melt-crystallized polymers have a complex structure consisting of amorphous and crystalline regions in a spherulitic microstructure (Fig. 4.14). They display their most useful properties in the temperature range $T_g < T < T_m$ and they are normally designed such that the temperature of operation comes within this range. Below the T_g they are brittle and above T_m they behave as viscous liquids or rubbers. The deformation of semi-crystalline polymers can be considered on several levels of structure. Fig. 5.48 shows how the spherulitic microstructure changes with deformation. The spherulites become elongated parallel to the stretching direction as the strain is increased and at high strains the spherulitic microstructure breaks up and a fibre-like morphology is obtained. Clearly the change in appearance of the spherulitic structure must be reflecting changes which are occurring on a finer structural level within the crystalline and amorphous areas. It is thought that when the spherulites are deforming homogeneously, during the early stages of deformation, the crystalline regions deform by a combination of slip, twinning and martensitic transformations as outlined in the last section. In addition it is thought that

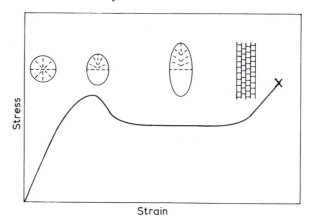

Fig. 5.48 *Schematic illustration of the change in spherulite morphology during the cold-drawing of a semi-crystalline polymer (after Magill).*

because the amorphous phase is in a rubbery state it allows deformation to take place by the shear of crystals relative to each other. During these early and intermediate stages of deformation the crystals become distorted but remain essentially intact. Towards the later stages when the spherulitic morphology is lost there appears to be a complete break-up of the original crystalline microstructure and a new fibre-like structure is formed. The mechanisms whereby this happens are not fully understood but one possible method by which this could take place is illustrated in Fig. 5.49. It is assumed for simplicity that the undeformed polymer has crystals which are stacked with the molecules regularly folded. Deformation then takes place by slip and twinning etc., until eventually the crystals start to crack and chains are pulled out. At sufficiently high strains the molecules and crystal blocks become aligned parallel to the stretching direction and a fibrillar structure is formed. Earlier theories of cold-drawing suggested that the structural changes occurred because the heat generated during deformation caused melting to occur and the material subsequently recrystallized on cooling. Although it is clear that there is a small rise in temperature during deformation this is not sufficient to cause melting, and at normal strain rates the heat is dissipated relatively quickly.

Polymers oriented by drawing are of considerable practical importance because they can display up to 50% of the theoretical chain direction modulus of the crystals. This short-fall in modulus is caused by imperfections within the crystals and the presence of folds and non-crystalline material. The value of modulus obtained depends principally upon the crystal structure of the polymer and the extent to which a sample is drawn (i.e. the draw ratio). The maximum draw ratio that can be obtained in turn

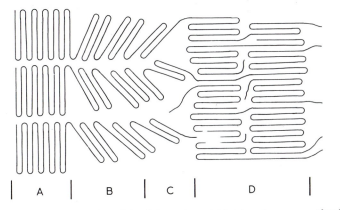

Fig. 5.49 *Schematic representation of the deformation which takes place on a molecular level during the formation of a neck in a semi-crystalline polymer. (A) Idealized representation of undeformed structure; (B) crystals deforming by slip, twinning and martensitic transformations; (C) crystals break up and molecules are pulled out; (D) fibrillar structure forms.*

depends upon the temperature of drawing and the average molar mass and molar mass distribution of the polymer. There is now considerable interest is optimizing these variables to obtain ultra-high modulus, oriented polymers.

5.6 Fracture

Fracture can be defined simply as the creation of new surfaces within a body through the application of external forces. The failure of engineering components through fracture can have catastrophic consequences and much effort has been put in by materials scientists and engineers in developing materials and designing structures which are resistant to premature failure through fracture. It is possible to classify materials as being either 'brittle' like glass which shatters readily or 'ductile' like pure metals such as copper or aluminium which can be deformed to high strains before they fail. Polymers are found to display both of these two types of behaviour depending upon their structure and the conditions of testing. In fact, the same polymer can sometimes be made either brittle or ductile by changing the temperature of rate of testing. Over the years some polymers have been advertised as being 'unbreakable'. Of course, this is very misleading as every material can only stand limited stresses. On the other hand, domestic containers made from polymers such as polypropylene or low-density polyethylene are remarkably tough and are virtually indestructible under normal conditions of use.

Polymers are capable of undergoing fracture in many different ways and the stress–strain curves of several different polymers tested to failure are

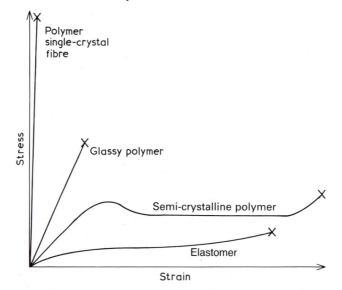

Fig. 5.50 *Schematic stress–strain curves of different types of polymers, drawn approximately to scale.*

given schematically in Fig. 5.50. Glassy polymers tend to be rather brittle, failing at relatively low strains with very little plastic deformation. On the other hand, semi-crystalline polymers are normally more ductile, especially between T_g and T_m, and undergo cold-drawing before ultimate failure. Crosslinked rubbers (elastomers) are capable of being stretched elastically to high extensions, but a tear will propagate at lower strains if a pre-existing cut is present in the sample. Polymer single-crystal fibres break at very high stresses, but because their moduli are high the stresses correspond to strains of only a few percent. Because of this wide variation in fracture behaviour it is convenient to consider the different fracture modes separately.

5.6.1 *Ductile–brittle transitions*

Many materials are ductile under certain testing conditions, but when these conditions are changed such as, for example, by reducing the temperature they become brittle. This type of behaviour is also encountered in steels and is very common with polymers. Fig. 5.51(a) shows how the strength of poly(methyl methacrylate) varies with testing temperature. At low temperatures the material fails in a brittle manner, but when the testing temperature is raised to just above room temperature the polymer undergoes general yielding and brittle fracture is suppressed. This type of brittle–ductile transition has been explained in terms of brittle fracture

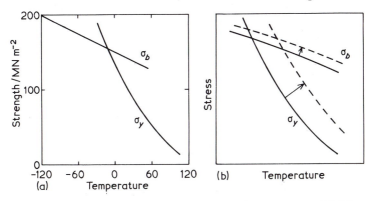

Fig. 5.51 *Ductile-to-brittle transitions.* (a) *Variation with temperature of brittle strength measured in flexure,* σ_b *and yield strength measured in tension,* σ_y *for poly(methyl methacrylate).* (b) *Effect of strain-rate upon the transition. Curves for a higher strain-rate are drawn as dashed lines (After Ward).*

and plastic deformation being independent processes which have a different dependence on temperature as shown schematically in Fig. 5.51(b). Since the yield stress increases more rapidly as the temperature is reduced, at a critical temperature it becomes higher than the stress required to cause brittle fracture. The temperature at which this occurs corresponds to that of the *brittle–ductile transition*, T_b, and it is envisaged that the process which occurs during deformation is that which can take place at the lowest stress, i.e. brittle fracture at $T < T_b$ and general yielding at $T > T_b$.

With polymers there is the added complication of there being a strong dependence of the mechanical properties upon strain rate as well as upon temperature. For example, nylons can be cold-drawn at room temperature when relatively low strain-rates are used, but become brittle as the strain-rate is increased. It is thought that this is because the curves of yield stress and brittle stress against temperature are moved to the right by increasing the strain-rate and so the temperature of the brittle–ductile transition is increased (Fig. 5.51(b)).

It is known that the temperature at which the brittle–ductile transition occurs for a particular polymer is sensitive to both the structure of the polymer and the presence of surface flaws and notches. The temperature of the transition is generally raised by crosslinking and increasing the crystallinity. Both of these factors tend to increase the yield stress without affecting the brittle stress significantly. The addition of plasticizers to a polymer has the effect of reducing yield stress and so lowers T_b. Plasticization is widely used in order to toughen polymers such as poly(vinyl chloride) which would otherwise have a tendency to be brittle at room temperature.

The presence of surface scratches, cracks or notches can have the effect of both reducing the tensile strength of the material and also increasing the temperature of the brittle–ductile transition. In this way the presence of a notch can cause a material to fail in a brittle manner at a temperature at which it would otherwise be ductile. As one might expect this phenomenon can play havoc with any design calculations which assume the material to be ductile. This type of 'notch sensitivity' is normally explained by examining the state of stress that exists at the root of the notch. When an overall uniaxial tensile stress is applied to the body there will be a complex state of triaxial stress at the tip of the crack or notch. This can cause a large increase (up to three times) in the yield stress of the material at the crack tip as shown in Fig. 5.52. The brittle fracture stress will be unaffected and so the overall effect is to raise T_b. If the material is used between T_b for the bulk and T_b for the notched specimen it will be prone to suffer from 'notch brittleness'.

It is tempting to relate the temperature at which the ductile–brittle transition takes place to either the glass transition or secondary transitions (Section 5.2.6) occurring within the polymer. In some polymers such as natural rubber or polystyrene T_b and T_g occur at approximately the same temperature. Many other polymers are ductile below the glass transition temperature (i.e. $T_b < T_g$). In this case it is sometimes possible to relate T_b to the occurrence of secondary low-temperature relaxations. However, more extensive investigations have shown that there is no general correlation between the brittle–ductile transition and molecular relaxations. This may not be too unexpected since these relaxations are detected at low strains whereas T_b is measured at high strains and depends upon factors such as the presence of notches which do not affect molecular relaxations.

5.6.2 *Brittle fracture and flaws*

One of the most useful approaches that can be used to explain the fracture of brittle polymers is the theory of brittle fracture developed by Griffith to

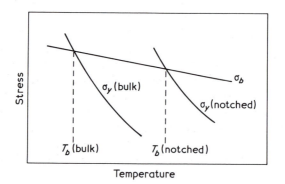

Fig. 5.52 *Schematic representation of the effect of notching upon the ductile-to-brittle transition (after Andrews).*

account for the fracture behaviour of glass. The theory is concerned with the thermodynamics of initiation of cracks from pre-existing flaws in a perfectly elastic brittle solid. These flaws could be either scratches or cracks which both have the effect of causing a stress concentration. This means that the local stress in the vicinity of the crack tip is higher than that applied to the body as a whole. The theory envisages that fracture starts at this point and the crack then continues to propagate catastrophically.

The effect of the presence of a crack in a body can be determined by considering an elliptical crack of major and minor axes a and b in a uniformly loaded infinite plate as illustrated in Fig. 5.53. If the stress at a great distance from the crack is σ_0 then the stress, σ, at the tip of the crack is given by*

$$\sigma = \sigma_0(1 + 2a/b) \tag{5.186}$$

The ratio σ/σ_0 defines the stress-concentrating effect of the crack and is three for a circular hole $(a = b)$. It will be even larger when $a > b$. The equation can be put in a more convenient form using the radius of curvature ρ of the sharp end of the ellipse which is given by

$$\rho = b^2/a$$

Equation (5.186) then becomes

$$\sigma = \sigma_0[1 + 2(a/\rho)^{1/2}] \tag{5.187}$$

If the crack is long and sharp $(a \gg \rho)$, σ will be given approximately by

$$\sigma = \sigma_0 2(a/\rho)^{1/2} \tag{5.188}$$

Inspection of these equations shows that the effect of a sharp crack will be to cause a large concentration of stress which is a maximum at the tip of the crack, the point from which the crack propagates.

Griffith's contribution to the argument concerned a calculation of the energy released by putting the sharp crack into the plate and relaxing the surrounding material, and relating this to the energy required to create new surface. The full calculation requires an integration over the stress field around the crack but an estimate of the energy released in the creation of new surface can be made by considering that the crack completely relieves the stresses in the shaded circular zone of material of radius a around the crack as shown in Fig. 5.53. If the plate has unit thickness then the volume of this region is $\sim\pi a^2$. Since the material is assumed to be linearly elastic then the strain energy per unit volume in an area deformed to a stress of σ_0 is $\frac{1}{2}\sigma_0^2/E$, where E is the modulus. If it is assumed that the circular area is completely relieved of stress when the crack is present then

*See J.G. Williams, *Stress Analysis of Polymers.*

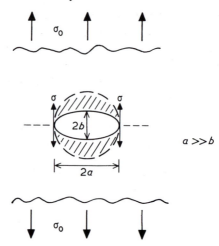

Fig. 5.53 *Model of an elliptical crack of length 2a in a uniformly loaded infinite plate.*

the energy released is $\frac{1}{2}\pi\, a^2\sigma_0^2/E$ per unit thickness. The surface energy of the crack is $4a\gamma$ per unit thickness, where γ is the surface energy of the material. The extra factor of 2 appears because the crack of length $2a$ has two surfaces. It is postulated that the crack will propagate if the energy gained by relieving stress is greater than that needed to create new surface and this can be expressed mathematically as

$$\frac{\partial}{\partial a}\left(-\frac{\pi\sigma_0^2 a^2}{2E} + 4a\gamma\right) > 0 \tag{5.189}$$

or at the point of fracture

$$\sigma_0 = \sigma_f = \left(\frac{4E\gamma}{\pi a}\right)^{1/2} \tag{5.190}$$

where σ_f is the fracture stress. This equation is similar to that obtained by Griffith who showed by a better calculation of the stresses around the crack that

$$\sigma_f = \left(\frac{2E\gamma}{\pi a}\right)^{1/2} \tag{5.191}$$

This equation is the usual form in which the Griffith criterion is expressed, but it is only strictly applicable when plane stress conditions prevail. In plane strain it becomes

$$\sigma_f = \left[\frac{2E\gamma}{\pi(1 - \nu^2)a}\right]^{1/2} \tag{5.192}$$

where ν is Poisson's ratio.

The Griffith theory was tested initially for glass and it has since been applied to the fracture of other brittle solids. The main predictions are borne out in practice. The theory predicts that there should be a proportional relationship between the fracture strength and $a^{-1/2}$ and so in a uniform stress field once a crack starts to propagate it should continue to do so since the stress required to drive it drops as the crack becomes longer.

The theory provides a good qualitative prediction of the fracture behaviour of brittle polymers such as polystyrene and poly(methyl methacrylate). Fig. 5.54 shows how the strength of uniaxial tensile specimens of these two polymers varies with the length of artificially induced cracks. There is a clear dependence of σ_f upon $a^{-1/2}$ as predicted by the Griffith theory but detailed investigations have shown that there are significant discrepancies. For example, since the Young's moduli of the polymers are known, the values of γ can be calculated from the curves in Fig. 5.54 for the two polymers using Equation (5.191). The value of γ for poly(methyl methacrylate) is found to be $\sim 210 \, \mathrm{J \, m^{-2}}$ and that for polystyrene $\sim 1700 \, \mathrm{J \, m^{-2}}$. These values are very much larger than the surface energies of these polymers which are thought to be of the order of $1 \, \mathrm{J \, m^{-2}}$. The values of γ determined for other materials are also larger than the theoretical surface energies (see for example Table 5.4). The discrepancy arises because the Griffith approach assumes that the material behaves elastically and does not undergo plastic deformation. It is known that even if a material appears to behave in a brittle manner and does not undergo general yielding there is invariably a small amount of plastic deformation, perhaps only on a very local level, at the tip of the crack. The energy absorbed during plastic deformation is very much higher than the surface energy and this is reflected in the measured values of γ for most materials being much larger than the theoretical ones.

The Griffith approach assumes that for any particular material the fracture stress is controlled by the size of the flaws present in the structure. The strength of any particular sample can be increased by reducing the size of these flaws. This can be easily demonstrated for glass in which flaws are introduced during normal handling and so it normally has a low fracture stress. The strength can be greatly increased by etching the glass in hydrofluoric acid which removes most of the flaws. The reduction of the size of artificially induced flaws clearly also causes an increase in the strength of brittle polymers (Fig. 5.54). However, this rise in strength does not go on indefinitely and when the flaw size is reduced below a critical level, about 1 mm for polystyrene and 0.07 mm for poly(methyl methacrylate) at room temperature, σ_f becomes independent of flaw size. The materials, therefore, behave as if they contain natural flaws of these critical sizes and so the introduction of smaller artificial flaws does not

TABLE 5.4 *Approximate values of fracture surface energy γ determined using the Griffith theory.*

Material	$\gamma/\text{J m}^{-2}$
Inorganic glass	7–10
Poly(methyl methacrylate)	200–400
Polystyrene	1000–2000
Epoxy resins	50–200
MgO	8–10
High-strength aluminium	~8000
High-strength steel	~25 000

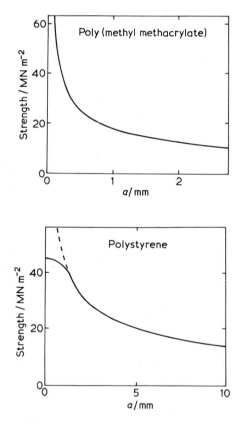

Fig. 5.54 *Dependence of fracture strength of poly(methyl methacrylate) and polystyrene upon the crack length, a. (After Berry, Chapter 2 in Fracture VII, edited by H. Liebowitz, Academic Press, 1972.)*

affect the strength. The exact size of the natural or inherent flaws can be determined from the fracture strength of an unnotched specimen, the modulus and the value of γ measured from the curves in Fig. 5.54. The relatively low inherent flaw size of poly(methyl methacrylate) accounts for the observation that in the unnotched state it has a higher strength than polystyrene even though polystyrene has a higher value of γ (Table 5.4).

Since polystyrene behaves as if it contains natural flaws of the order of 1 mm it might be expected that they could be detected in undeformed material but this is not the case. The inherent flaws are in fact crazes which form during deformation. When an unnotched specimen of polystyrene is deformed in tension large crazes of the order of 1 mm in length are seen to form whereas in the case of poly(methyl methacrylate) the crazes are much smaller. Fracture then takes place through the breakdown of one of these crazes which becomes transformed into a crack. On the other hand, when specimens containing cracks of greater than the inherent flaw size are deformed general crazing does not occur and failure takes place at the artificial flaw.

5.6.3 *Linear elastic fracture mechanics*

The derived value of γ is usually greater than the true surface energy because other energy-absorbing processes occur and so it is normal to replace 2γ in Equation (5.191) with $\mathcal{G}_c$ which then represents the total work of fracture i.e. the *fracture energy*. The equation then becomes

$$\sigma_f = \left(\frac{E\mathcal{G}_c}{\pi a}\right)^{1/2} \qquad (5.193)$$

and this equation can be thought of as a prediction of the critical stress and crack length at which a crack will start to propagate in an unstable manner. Failure would be expected to occur when the term $\sigma\sqrt{\pi a}$ reaches a value of $\sqrt{E\mathcal{G}_c}$. The term $\sigma\sqrt{\pi a}$ can be considered as the driving force of crack propagation and so the stress intensity factor, K, is defined as

$$K = \sigma\sqrt{\pi a} \qquad (5.194)$$

and the condition for crack propagation to occur is that K reaches a critical value, K_c given by

$$K_c = \sqrt{E\mathcal{G}_c} \qquad (5.195)$$

The critical stress intensity factor, K_c, is sometimes called the *fracture toughness* since it characterizes the resistance of a material to crack propagation.

K_c and $\mathcal{G}_c$ are the two main parameters used in *linear elastic fracture mechanics*. Equations (5.194) and (5.195) can only be strictly applied if the material is linearly elastic and any yielding is confined to a small region in the vicinity of the crack tip. These restrictions can then only normally be applied to glassy polymers undergoing brittle fracture and the application of linear elastic fracture mechanics to the failure of brittle polymers has allowed significant progress to be made in the understanding of polymer fracture processes. In the case of more ductile polymers and elastomers generalized versions of the Griffith equation must be applied.

The understanding of brittle fracture is also greatly improved by using more sophisticated test pieces than the simple uniaxial tensile specimens. It is possible to design test pieces such that cracks can be grown in a stable manner at controlled velocities. In the case of the uniaxial tensile specimen once a crack starts to propagate the crack runs away very rapidly. This can be overcome by having a specimen in which the effective stress driving the

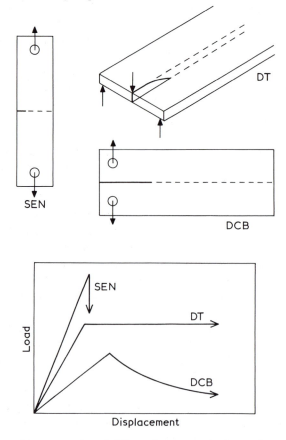

Fig. 5.55 *Schematic representation of different fracture mechanics specimens used with brittle polymers and load–displacement curves for each specimen for a constant displacement rate. SEN, single-edge notched; DT, double torsion; DCB, double-cantilever beam.*

crack drops as the crack grows longer. Fig. 5.55 shows typical fracture mechanics test pieces. In the double-torsion specimen the crack is driven by twisting the two halves of one end of the specimen in opposite directions and the stress distribution is such that for a given load applied at the end of the specimen the stress-intensity factor, K, is independent of the crack length. This must be contrasted with the unaxial tensile specimen where, for a given applied stress, K increases rapidly with increasing crack length. It is possible to have specimens where K actually drops as the crack length increases such as in the case of the double cantilever beam which is also shown in Fig. 5.55. The effect of this variation of K with crack length upon load–displacement curves obtained from poly(methyl methacrylate) specimens when the specimen ends are displaced at a constant rate are also given in Fig. 5.55. In all three cases the load is initially approximately proportional to the displacement. For the simple tensile specimen the load drops immediately the crack starts to grow and the crack runs rapidly in an uncontrolled manner to the end of the specimen. When the crack starts to grow in the double-torsion specimen it propagates at a constant load (and also constant velocity) until it reaches the end of the specimen. Finally, in the double cantilever beam the crack propagates at a decreasing load and velocity when the specimen ends are displaced at a constant rate.

The value of K_c at which crack propagation occurs in a specimen of finite size depends upon the applied load, crack length and specimen dimensions. It is only given by Equation (5.194) for a crack of length $2a$ in an infinite plate. At other times it must be determined from an elastic analysis of the particular specimen used and this so-called 'K-calibration' has only been done for certain specific specimens. The formulae for K which have been determined for the three specimens in Fig. 5.55 are given in Table 5.5. It should be noted that for the double-torsion specimen the crack length, a does not appear in the formula.

It is sometimes more convenient to use $\mathcal{G}_c$ rather than K_c to characterize the fracture behaviour of a material especially since $\mathcal{G}_c$ is an energy parameter and so can be related more easily to particular energy-absorbing fracture processes. If the value of K_c is known then $\mathcal{G}_c$ can be obtained through Equation (5.195) but care is required with polymers since E depends upon the rate of testing and it may not be obvious as to what is the appropriate strain-rate for the fracture experiment. It is possible to calculate $\mathcal{G}_c$ without any prior knowledge of K_c or E for any general type of fracture specimen using a compliance calibration technique. This involves measurement of $(\partial C/\partial a)$, the variation of specimen compliance C with crack length, the compliance in this case being defined as the ratio of the displacement of the ends of the specimens to the applied load, L. When crack propagation occurs $\mathcal{G}_c$

TABLE 5.5 *Expression for K_c for different fracture mechanics specimens used in the study of the fracture of brittle polymers. (After Young, Chapter 6 in 'Developments in Polymer Fracture' ed. E.H. Andrews.)*

	Geometry of specimen	K_c
	Single-edge notched (SEN)	$\dfrac{La^{1/2}}{bt}[1.99 - 0.41(a/b) + 18.7(a/b)^2$ $-38.48(a/b)^3 + 53.85(a/b)^4]$
	Double torsion (DT)	$LW_m\left[\dfrac{3(1+\nu)}{Wt^3t_n}\right]^{1/2}$ $(W/2 \gg t)$
	Double cantilever beam (DCB)	$\dfrac{2L}{t}\left[\dfrac{3a^2}{h^3} + \dfrac{1}{h}\right]^{1/2}$

L = applied load
b = specimen breadth
t = sheet thickness
a = crack length
h = distance of specimen edge from fracture plane
t_n = thickness of sheet in plane of crack
W = specimen width
W_m = length of moment arm
ν = Poisson's ratio

is given by*

$$\mathcal{G}_c = \frac{L^2}{2t}\left(\frac{\partial C}{\partial a}\right) \tag{5.196}$$

where t is the specimen thickness.

5.6.4 *Crack propagation in poly(methyl methacrylate)*

The slow growth of cracks in poly(methyl methacrylate) is an ideal application of linear elastic fracture mechanics to the failure of brittle polymers. Cracks grow in a very well-controlled manner when stable test pieces such as the double-torsion specimen are used. In this case the crack will grow steadily at a constant speed if the ends of the specimen are displaced at a constant rate. The values of K_c or $\mathcal{G}_c$ at which a crack propagates depends upon both the crack velocity and the temperature of testing, another result of the rate- and temperature-dependence of the mechanical properties of polymers. This behaviour is demonstrated clearly

*See for example, J.F. Knott, *Fundamentals of Fracture Mechanics*, Butterworths, London (1973).

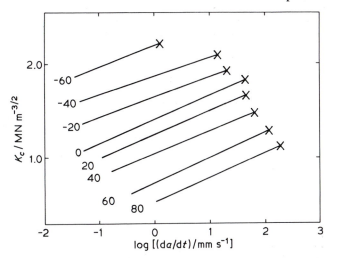

Fig. 5.56 *Dependence of K_c upon crack velocity for crack growth in poly(methyl methacrylate) at different temperatures (in °C). The values of K_c^* in each case are indicated by a cross. (Data taken from Marshall, Coutts and Williams, J. Mater, Sci., 9 (1974) 1409.)*

in Fig. 5.56. The value of K_c is a unique function of crack velocity at a given temperature with K_c increasing as the crack velocity increases and the temperature of testing is reduced. The data in Fig. 5.56 are presented in the form of a log–log plot and they can be fitted to an empirical equation of the form

$$K_c = A(\mathrm{d}a/\mathrm{d}t)^n \tag{5.197}$$

where A is a temperature-dependent constant and n is an exponent equal to the slope of the lines (~ 0.07). At a given temperature the value of K_c does not continue to rise indefinitely but reaches a maximum value K_c^*. This corresponds to a point of instability above which the crack grows rapidly at a lower value of K_c. This instability has been explained in terms of an adiabatic-isothermal transition. It is thought that at velocities below the instability, cracks grow isothermally because the heat can be dissipated from the tip of the growing crack. Whereas at velocities above the transition the crack grows too rapidly for heat to be dissipated and adiabatic conditions prevail at the crack tip. The consequent rise in temperature softens the polymer and so reduces the value of K_c for propagation.

A consequence of the dependence of K_c upon crack velocity for poly(methyl methacrylate) is that the polymer is prone to time-dependent failure during periods of prolonged loading. Since polymers are often used under stress for long periods of time this can be a serious problem. A typical use might be as windows in the pressurized cabins of air-craft. The time-to-failure, t_f, for a specimen under stress can be calculated by integrating Equation (5.197). If the specimen contains flaws of size a_0 and is subjected to a constant stress σ_0 then a crack will start to grow slowly at a velocity governed by Equation (5.197). If it is postulated that failure occurs when the crack reaches a critical length a^* then the equation can be rearranged to give

$$\int_0^{t_f} dt = \int_{a_0}^{a^*} \frac{A^{1/n}}{K_c^{1/n}} da \tag{5.198}$$

If the flaw is small and so can be considered to be in an infinite plate then since K is given by $\sigma \sqrt{\pi a}$ Equation (5.198) can be written as

$$t_f = \int_{a_0}^{a^*} A^{1/n} \sigma_0^{-1/n} \pi^{-1/2n} a^{-1/2n} da \tag{5.199}$$

and integrating gives

$$t_f = \frac{A^{1/n}(\sigma_0^2 \pi)^{-1/2n}}{(1 - 1/2n)} \left[a^{*(1-1/2n)} - a_0^{(1-1/2n)} \right]$$

Since $a^* \gg a_0$ and $n \ll 1$ (Fig. 5.56) then the term in a^* can be neglected and so the time-to-failure will be given approximately by

$$t_f = \frac{A^{1/n} a_0^{(1-1/2n)}}{(\sigma_0^2 \pi)^{1/2n}(1/2n - 1)} \tag{5.200}$$

The time-to-failure can therefore be determined from a knowledge of σ_0, a_0 and the variation of K_c with crack velocity. This particular equation is for a small crack in an infinite plate. Similar equations can be derived for other specimen geometries.

So far the mechanisms of crack propagation have not been considered but they are important since the observed fracture behaviour is a reflection of deformation processes taking place at the crack tip. There is accumulated evidence to show that in poly(methyl methacrylate), and probably many other glassy polymers, crack propagation takes place through the breakdown of a craze at the tip of the growing crack. One way that this can be demonstrated indirectly is from the presence of craze debris on the polymer fracture surfaces which gives rise to interference colours when the surfaces are viewed in reflected light. Fig. 5.57 shows the geometry of the crack/craze entity. It is envisaged that propagation occurs through the growth of a crack with a single craze at its tip and a steady-state situation is reached whereby the rate of craze growth equals that of craze breakdown and the crack/craze entity propagates gradually through the specimen. The formation of craze and the subsequent breakdown absorbs

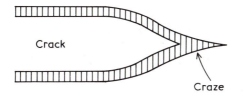

Fig. 5.57 *Schematic representation of a crack growing through a single craze in poly(methyl methacrylate).*

a considerable amount of energy and it is these processes which account for the fracture energies being much higher than the energy required just to create new surface (Table 5.4).

The mechanisms of crack propagation in poly(methyl methacrylate) are particularly amenable to analysis. The situation is not so simple for other polymers. For example, in polystyrene there is usually multiple crazing in the vicinity of the crack tip. In tougher polymers such as polycarbonate shear yielding as well as crazing often takes place at the crack tip. In these cases crack propagation does not occur in such a well-controlled manner as in poly(methyl methacrylate) and it is more difficult to analyse.

5.6.5 *Tearing of elastomers*

It is impossible to apply linear elastic fracture mechanics to polymers which fail at high strains and when the deformation is not linear elastic. There are, however, certain cases where the Griffith approach (Section 5.6.2) can be generalized and failure criteria established. Such a case is in the tearing of elastomers where the crack growth can be considered as being a balance between strain energy in the material and the tearing energy $\mathcal{T}$ of the material. This case is particularly easy to analyse because the material behaves in a reasonably (non-linear) elastic manner and the energy losses are confined to regions in the vicinity of the crack tip. The situation can best be analysed by taking a specific example such as the pure shear test piece of Rivlin and Thomas which is shown in Fig. 5.58. It is possible to calculate for this type of specimen the 'energy release rate', $-d\mathcal{E}/dA$, which is the amount of energy released per unit increase in crack area, A.

When the specimen is deformed the strain energy will not be the same in the different zones of the specimen and so they must be considered separately. Region A is unstressed and so the strain energy $\mathcal{E}_A$ will be zero. In region B around the crack tip the stress field is complex and the strain energy is difficult to determine. Region C will be in pure shear and so the energy in this region will be given by $\mathcal{E}_c = WV_C$ where W is the energy density (i.e. strain energy per unit volume) of the material in pure shear at the relevant strain and V_C is the volume of region C. Region D will not be in

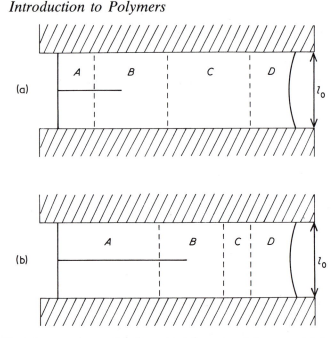

Fig. 5.58 *Pure shear test piece used in the study of the tearing of elastomers.* (a) *Specimen before crack propagation;* (b) *after crack propagation. (After Andrews, Chapter 9, in Polymer Science, ed. Jenkins, North-Holland, 1972.)*

pure shear because of its proximity to the ends of the specimen and so $\mathscr{E}_D$ will be difficult to calculate. The strain energy in the different regions can be summarized as follows

$$\mathscr{E}_A = 0, \qquad \mathscr{E}_B = ?$$
$$\mathscr{E}_C = WV_C, \quad \mathscr{E}_D = ?$$

The lack of knowledge of $\mathscr{E}_B$ and $\mathscr{E}_D$ can be overcome by calculating only the *change* in energy when crack propagation occurs. If the specimen grips are held at constant displacement and the crack propagates a distance Δa then the net result is to increase region A at the expense of region C and so since region A is unstrained the change in specimen strain energy $\Delta\mathscr{E}$ is

$$\Delta\mathscr{E} = -W\Delta V_C \qquad (5.201)$$

where ΔV_C is the reduction in the volume of C due to crack growth. The size of regions B and D are unchanged and so the fact that their strain energies are not known does not matter. Holding the grips at constant displacement ensures that no work is done by or on the specimen and so the overall change in stored energy is given by Equation (5.201). Since elastomers shear at approximately constant volume ΔV_C can be determined from either the unstrained or strained specimen dimensions and is

$$\Delta V_C = l_0 t_0 \Delta a \tag{5.202}$$

where l_0 and t_0 are the initial specimen length and thickness. The energy change is then given by

$$\Delta \mathscr{E} = -W l_0 t_0 \Delta a \tag{5.203}$$

and so

$$\frac{-\partial \mathscr{E}}{\partial a} = W l_0 t_0 \tag{5.204}$$

Since the crack has two surfaces then the change in crack area ΔA is given by

$$\Delta A = 2 t_0 \Delta a$$

and so

$$\frac{-\partial \mathscr{E}}{\partial A} = \tfrac{1}{2} W l_0 \tag{5.205}$$

The Griffith criterion for brittle solids assumes that crack propagation occurs when the release of elastic strain energy during an increment of crack growth is greater than the corresponding increase in surface energy due to the creation of new surface. This can be expressed in a similar way for the case of an elastomer as

$$\frac{-\partial \mathscr{E}}{\partial A} \geq \mathscr{T} \tag{5.206}$$

where $\mathscr{T}$ is the tearing energy or the characteristic energy for unit area of crack propagation in the material (analogous to γ in the Griffith theory or $\mathscr{G}_c/2$ in linear elastic fracture mechanics). In terms of a critical stored energy density W_c the criterion for crack growth in the pure shear test piece is

$$W_c = 2\mathscr{T}/l_0 \tag{5.207}$$

It is possible to determine W_c from the state of strain in the sample and the stress/strain characteristics of the material.

The quantity $-\partial \mathscr{E}/\partial A$ can be determined for testing geometries other than the pure shear specimen. The values of $-\partial \mathscr{E}/\partial A$ for a variety of other specimens commonly used to study the tearing of elastomers are given with schematic diagrams of the specimens in Fig. 5.59. It is found that cracks will propagate at a given crack velocity and temperature at the same value of $-\partial \mathscr{E}/\partial A$ in all of these types of specimen. This is strong evidence that $\mathscr{T}$ can be thought of as a true material property. It is also found that $\mathscr{T}$ varies

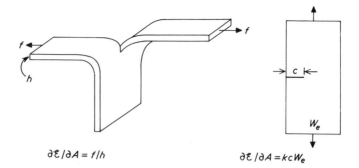

$$\partial \mathscr{E}/\partial A = f/h \qquad\qquad\qquad \partial \mathscr{E}/\partial A = kcW_e$$

Fig. 5.59 *Trouser and edge-crack test pieces used in the study of the tearing of elastomers. Expressions for* $-\partial\mathscr{E}/\partial A$ *are given in each case. (After Andrews, Chapter 9 in Polymer Science, ed. Jenkins, North-Holland, 1972.)*

with crack velocity and temperature for elastomers in a similar way that K_c depends upon these variables for glassy polymers such as poly(methyl methacrylate) (Fig. 5.56). This is neatly summarized in the three-dimensional plot in Fig. 5.60 for the tearing of a vulcanized styrene–butadiene rubber (SBR). The value of $\mathscr{T}$ increases with increasing crack velocity and decreasing temperature. This variation with rate and temperature can be represented in the form of a single WLF plot (Section 5.2.7) which enables the value of $\mathscr{T}$ to be determined for any velocity or temperature. At any given temperature the variation of $\mathscr{T}$ with crack velocity can be represented by an equation of the form

$$\mathscr{T} = B(da/dt)^n \tag{5.208}$$

where n has a value of about 0.25 for the SBR used and B is a temperature-dependent constant. This equation can be used to predict the time-dependent failure of the SBR in the same way that Equation (5.197) is used with poly(methyl methacrylate).

The values of $\mathscr{T}$ for elastomers are typically between 10^3 and $10^5\,\mathrm{J\,m^{-2}}$ which is very much higher than the energies required to break bonds. The extra energy is not used up in crazing or plastic deformation but is dissipated when material in the vicinity of the crack tip is deformed and relaxed viscoelastically as the crack propagates. If the cracks are propagated very slowly at high temperatures these losses can be reduced and the values of $\mathscr{T}$ become closer to those expected for bond breaking. The behaviour shown in Fig. 5.60 is only encountered if the elastomer is unfilled and does not crystallize. In these more complex materials crack propagation tends to be unstable with the crack moving in a series of jumps and the plots of $\mathscr{T}$ against crack velocity and temperature tend to be more complicated than that in Fig. 5.60. This difference in behaviour is due to the presence of filler

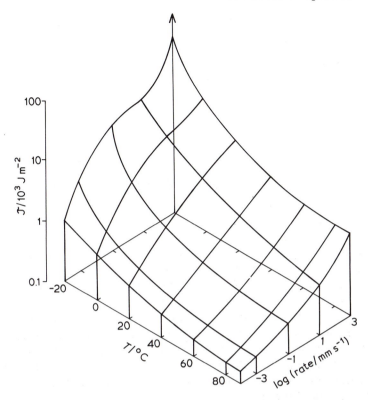

Fig. 5.60 *Dependence of tearing energy, $\mathcal{T}$ upon rate and temperature for a gum styrene–butadiene rubber. (After Greensmith and Thomas, J. Polym. Sci., **18**, (1955) 189.)*

particles and crystallization modifying the viscoelastic properties of the elastomer.

5.6.6 *Fatigue*

Many materials suffer failure under cyclic loading at stresses well below those they can sustain during static loading. This phenomenon is known as *fatigue* and the development of fatigue cracks in the metallic components of aircraft and other structures has lead to disastrous consequences. It is perhaps less well known that polymers can also undergo fatigue fracture. Fig. 5.61 gives the dependence of the failure stress of a variety of polymers upon the number, N, of cycles during fatigue loading. The polymers had been cycled at a frequency of 0.5 Hz between zero stress and the stress indicated. The curves have a characteristic sigmoidal shape and tend to flatten out when N becomes sufficiently large, suggesting a fatigue limit. This is stress to which the material can be cycled without causing failure. Similar so-called *S–N* curves have been found for metals, but one

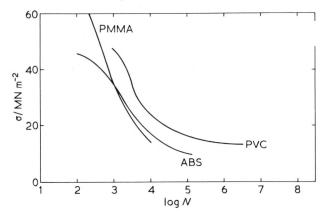

Fig. 5.61 *Dependence of failure stress of a variety of engineering polymers upon the number of cycles accumulated during a dynamic fatigue test at 20°C. The polymers used were poly(methyl methacrylate) (PMMA), rigid poly(vinyl chloride) (PVC) and an acrylonitrile–butadiene–styrene copolymer (ABS). (After Bucknall, Gotham and Vincent, Chapter 10 in Polymer Science, ed. Jenkins, North-Holland, 1972.)*

important difference with polymers is the strong dependence of the fatigue behaviour upon the testing frequency. There is often a large rise in specimen temperature, especially at high stresses and high testing frequencies, which causes failure to take place through thermal softening. The temperature rise is due to the energy dissipated by the viscoelastic energy loss processes not being conducted away sufficiently rapidly because of the low thermal conductivity of the polymer.

Fatigue failure occurs through a process of initiation and propagation of cracks. In the propagation phase the fatigue crack grows by a small amount during each cycle. This behaviour can be monitored by measuring the amount of crack growth per cycle, da/dN as a function of the applied stress. It is found that for many brittle polymers the propagation rate is related to the range of stress-intensity factor ΔK by an equation of the form

$$da/dN = C(\Delta K)^m \qquad (5.209)$$

where m is a constant of the order of 4 and C is a temperature-dependent constant for the particular polymer. This type of relationship is also found for many metals. The exact form of behaviour for a particular polymer is found to depend upon the testing frequency and the mean value of K used. Analogous behaviour is also found for non-crystallizing elastomers where the crack growth rate can be related to $\mathcal{T}$ through the equation

$$da/dN = D\mathcal{T}^m \qquad (5.210)$$

where D and m are constants which depend on the type of material and testing conditions used.

5.6.7 *Environmental fracture*

The Achilles heel of many polymers is their tendency to fail at relatively low stress levels through the action of certain hostile environments. This type of failure falls into several different categories.

(a) One of the most widely-studied forms of environmental failure is that of vulcanized natural rubber in the presence of *ozone*. The ozone reacts at the surface of unsaturated hydrocarbon elastomers and even when the materials are subjected to relatively low stresses cracks can nucleate and grow catastrophically through the material. It is found that ozone cracking takes place above a critical value of tearing energy $\mathcal{T}_0$. This can be of the order of $0.1 \, \mathrm{J\,m^{-2}}$ which is close to the true surface energy of the polymer and well below the tearing energies of elastomers in the absence of a hostile environment.

(b) Another type of environmental failure which occurs through surface interactions is the failure of polyolefins such as high density polyethylene in the presence of *surfactants* such as certain detergents. It is found that when this material is held under constant stress in these environments its behaviour changes from ductile failure at high stresses over short time periods to brittle fracture at low stresses after longer periods of time.

(c) Glassy polymers in particular are prone to failure through the action of certain organic liquids which act as *crazing and cracking* agents. Failure can occur at stresses and strains well below those required in the absence of the agents. The detailed behaviour is rather complex and it is difficult to generalize from one system to another. It is thought that in most cases the failure occurs through the initial formation of crazes although the crazes may not be particularly stable and may transform directly into cracks. Two particular mechanisms have been proposed to account for the action of crazing and cracking agents in glassy polymers and in many cases it is thought that both mechanisms may apply. One suggestion is that the presence of the liquid lowers the surface energy of the polymer and makes the formation of new surface during crazing easier. Another is that the organic liquid swells the polymer and lowers its T_g allowing deformation and crazing to take place at lower stresses and strains.

The effect of certain liquids with different solubility parameters, δ, upon the critical strain for crazing and cracking poly(2,6-dimethyl-1,4-phenylene oxide) is shown in Fig. 5.62. In certain cases the crazes are not stable and they transform directly into cracks. The minimum in the curve corresponds to liquids which have the highest solubilities in the polymer, when the values of δ are matched for the polymer and solvent (Section 3.2.5). It is clear, therefore, that the crazing and cracking in this particular system is due principally to a reduction in the T_g by swelling. However, this is not the

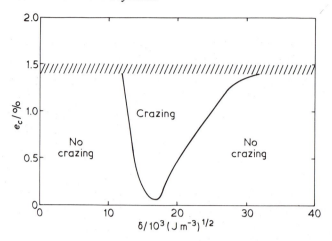

Fig. 5.62 *Dependence of critical strain for crazing, e_c, upon solubility parameter δ for poly(2,6-dimethyl-1,4-phenylene oxide) in various liquids. The shaded area gives e_c for the polymer in air. (After Kambour and Robertson, Chapter 11 in Polymer Science, ed. Jenkins, North-Holland, 1972.)*

whole story since in the liquids with high or low values of δ which produce no swelling at all, the polymer fails by crazing whereas when tested at constant strain-rate in air it undergoes cold-drawing with very little crazing. It is thought that these liquids tend to reduce the surface energy and make the formation of holes and voids during crazing easier.

5.6.8 *Molecular failure processes*

In this section the processes which take place on a molecular level when a polymer is fractured will be examined. It is clear that, in general, failure will occur through a combination of molecular rupture and the slippage of molecules past each other, but the extent to which these two processes occur is not easy to determine except for particular specific polymer structures and morphologies.

The rupture of primary bonds is likely to be the principal mechanism of failure during the fracture of polymer single crystals deformed in tension parallel to the chain direction. Fig. 5.63 shows a side view of a polydiacetylene single crystal fibre that has been fractured in tension. The stress–strain curve of a similar crystal is given in Fig. 5.34. Examination of the fractured end of the crystal shows clearly that molecular fracture has taken place. Since the covalent bonds along the polymer chain are very strong the material has a very high fracture strength. Fig. 5.64 shows a plot of the fracture strength of the polymer as a function of fibre diameter. The fracture strength increases as the diameter is reduced. This is because the

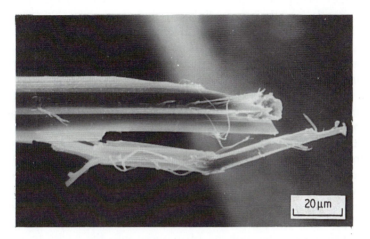

Fig. 5.63 *Scanning electron micrograph of a fractured polydiacetylene single crystal fibre.*

thinner fibres contain less strength-reducing flaws (e.g. Griffith theory). The narrowest fibres tested had a strength of the order of $1.5 \, \text{GN m}^{-2}$. The modulus of the polymer is about $60 \, \text{GN m}^{-2}$ and so the maximum fracture strength is of the order of $E/40$. Theoretical considerations (Section 5.1.5) show that the highest fracture strength might be expected to be about $E/10$ and most materials are, because of flaws, well below this value. The values of fracture strengths of the polydiacetylene single crystals are therefore close to the theoretical estimates and compare favourably with those of other high-strength materials.

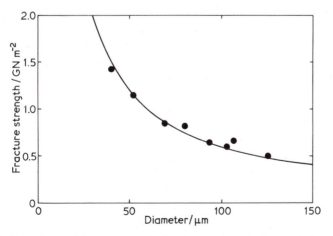

Fig. 5.64 *Dependence of fracture strength of polydiacetylene single crystal fibres upon fibre diameter.*

Another group of materials in which molecular fracture must occur during crack growth is cross-linked polymers such as elastomers. The crosslinking points do not allow molecular pull-out to occur and bond breakage must then take place. It can be demonstrated that this is indeed the failure mechanism by determining the tearing energy $\mathcal{T}_0$ of an elastomer in the absence of any viscoelastic energy losses. This is found to be close to that expected if failure takes place through bond breakage.

There is now a good deal of interest in using techniques which are capable of monitoring directly the occurrence of molecular fracture during the deformation of polymer. The most important method in this respect is electron spin resonance spectroscopy (ESR) (Section 3.21) which detects the production of free radicals produced through the fracture of covalent bonds. Specimens can be fractured within the cavity of the spectrometer and molecular fracture is manifest be the generation of an ESR signal. The technique suffers from certain limitations such as difficulty in being able to get a sufficiently large volume of material into the cavity to produce a strong enough signal and the drop in signal strength due to the decay of radicals through combination and other reactions. The radicals are found to be more stable at low temperature where the mobility of the chains is reduced. It is found that for many polymers only weak signals are obtained because very little molecular fracture takes place and failure occurs primarily through molecular slippage. The strongest signals are obtained from crosslinked polymers and crystalline polymer fibres where the crosslinks and crystals act as anchorage points and reduce the amount of slippage that can occur.

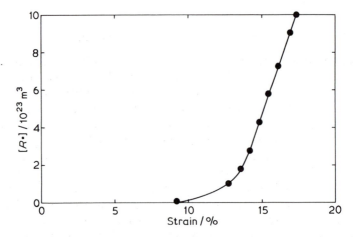

Fig. 5.65 *Dependence of radical concentration, $[R\cdot]$ upon the strain for nylon 6 fibres deformed at room temperature. (Data taken from Becht and Fischer, Kolloid-Z.u.Z. Polymere,* **229**, *(1969) 167.)*

Fig. 5.65 shows how the radical concentration increases with strain as nylon 6 fibres are deformed in tension. An appreciable signal starts to build up above a strain of about 10 percent, well before the specimen fractures. The molecules in the fibre are aligned and strained to a certain extent before deformation. The build up in ESR signal is thought to be due to the failure of the most highly-stressed molecules within the structure before overall fracture of the specimen takes place.

5.7 Toughened polymers

Glassy polymers have properties which are adequate for many applications but they are prone to brittle fracture particularly when notched or subjected to high strain-rate impact loading (Section 5.6.1). Because of such problems toughened polymers have been developed in which a glassy polymer matrix is toughened through the incorporation of a second phase consisting of particles which are usually spherical and of a rubbery polymer above its glass transition polymer, T_g. This can lead to significant improvements in the mechanical behaviour of the matrix polymer and is often referred to as *rubber toughening*. There are two general ways of adding the second phase. The first is through blending an incompatible rubbery polymer with the matrix and the second is through copolymerization (Sections 2.15 and 2.16). Block copolymers are sometimes employed for which one part of the molecule provides the glassy matrix and the other part is an incompatible rubber which produces the dispersed phase. The covalent bonding between the blocks ensures a good interface between the two phases.

5.7.1 Mechanical behaviour of rubber-toughened polymers

The incorporation of rubber particles into a brittle polymer has a profound effect upon the mechanical properties as shown from the stress–strain curves in Fig. 5.66. This can be seen in Fig. 5.66(a) for high-impact polystyrene (HIPS) which is a blend of polystyrene and polybutadiene. The stress–strain curve for polystyrene shows brittle behaviour, whereas the inclusion of the rubbery phase causes the material to undergo yield and the sample to deform plastically to about 40% strain before eventually fracturing. The plastic deformation is accompanied by 'stress-whitening' whereby the necked region becomes white in appearance during deformation. As will be explained later, this is due to the formation of a large number of crazes around the rubber particles in the material.

Fig. 5.66(b) shows a series of stress–strain curves for samples of rubber-toughened poly(methyl methacrylate) (RTPMMA) containing different weight fractions w_p of toughening particles. It can be seen that

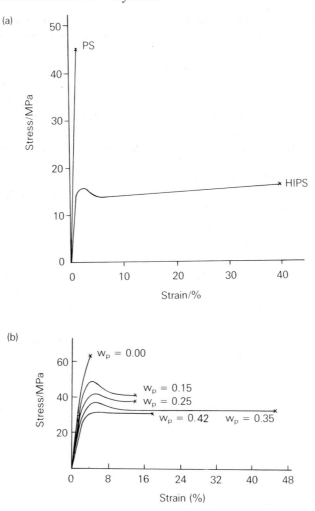

Fig. 5.66 *Stress-strain curves for rubber-toughened polymers. (a) Polystyrene (PS) and high-impact polystyrene (HIPS) (after Bucknall). (b) Rubber-toughened poly(methyl methacrylate) showing the effect of the weight fraction of rubber particles, w_p (courtesy D.E.J. Saunders).*

PMMA is relatively brittle but that following the addition of rubbery particles the material is able to undergo yield and deform plastically. A maximum elongation of over 45% is obtained for $w_p = 0.35$. It is found that as with HIPS, stress-whitening is obtained following yield, although in the case of RTPMMA this is thought to be due to the particles undergoing voiding rather than crazing taking place. The reduction in elongation for

$w_p > 0.35$ is because at high values of w_p the particles become so close that they touch and interact with each other and thereby reduce the efficiency of toughening.

As well as toughening thermoplastics such as polystyrene and PMMA, it is also possible to toughen brittle network polymers such as epoxy resins (Section 2.3.1) by the addition of a rubbery second phase. Fig. 5.67 shows the effect of the volume fraction of rubber particles, v_p upon $\mathcal{G}_c$, the fracture energy for a rubber-toughened epoxy resin. The samples were prepared by the addition of 8.7% by weight of a carboxyl-terminated poly(butadiene-*co*-acrylonitrile) (CTBN) rubber to the epoxy prepolymer and curing agent. The rubber is soluble in the prepolymer but precipitates in the form of particles as the resin increases in molar mass during curing. The variation of v_p in Fig. 5.67 was obtained by varying the cure conditions employed. Although the weight of rubber used in each case was constant it can be seen from Fig. 5.67 that it is the volume fraction of rubber precipitated in the form of particles that controls the toughness of the material. It should be noted that the rubber employed had reactive carboxylic acid end-groups which are capable of reacting with the epoxy prepolymer and so ensuring a strong interface between the rubber particles and the epoxy matrix.

Table 5.6. gives typical values of K_c and $\mathcal{G}_c$ for three different rubber-toughened polymer systems.

5.7.2 *Mechanisms of rubber-toughening*

Both crazing and shear yielding involve the absorption of energy and most methods of toughening brittle polymers involve modifying the polymer

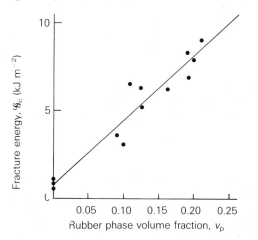

Fig. 5.67 *Dependence of $\mathcal{G}_c$ upon rubber phase volume fraction for a rubber-toughened epoxy resin (after Bucknall and Yoshi (1978) Brit. Polym. J.* **10**, *53).*

Table 5.6 *Typical values of Young's modulus, E, fracture toughness, K_c and fracture energy, $\mathcal{G}_c$, for different polymers*

Polymer	E(GPa)	K_c(MN m$^{-3/2}$)	$\mathcal{G}_c$(kJ m^{-2})
Polystyrene	3.0	1.1	0.4
HIPS	2.1	5.8	16
Poly(methyl methacrylate)	3.0	1.2	0.5
RTPMMA	2.1	2.4	2.6
Epoxy resin	2.8	0.5	0.1
Rubber-toughened epoxy resin	2.4	2.2	2.0
Silica-filled epoxy resin	7.5	1.4	0.3

such that controlled high levels of crazing or shear yielding are able to take place. The second-phase spherical particles of the rubber have a Young's modulus about 3 orders of magnitude lower than that of the glassy matrix (Table 5.1). This leads to a stress concentration at the equators of the particles during mechanical deformation which is similar to the stress concentration found around holes and notches in plates (Section 5.6.2). The presence of the stress concentration can lead to shear yielding or crazing around every particle and hence throughout a large volume of material rather than just at the crack tip. Hence, the polymer absorbs a large amount of energy during deformation and is toughened.

The exact mechanisms of deformation around the rubber particles depends upon the type of polymer and the conditions of testing (rate and temperature). The structure of rubber-toughened polymers is generally visualized by transmission electron microscopic examination of thin microtomed sections such as those shown in Fig. 5.68. It is sometimes possible to follow the deformation processes by examination of such sections. Fig. 5.68(a) shows an OsO$_4$-stained section of high-impact polystyrene (HIPS), after tensile deformation. The OsO$_4$ stains the rubbery regions dark by depositing osmium in areas containing the unsaturated molecules (i.e. the polybutadiene). It can be seen that relatively large crazes emanate from the equators of the rubber particles. It should be noted that the rubber particles are not simple rubber spheres but have inclusions which have not been stained and so must be polystyrene. It has been found that this complex rubber particle morphology can lead to a given volume fraction of rubber v_r giving rise to a larger number of effective toughening particles than when there are no inclusions. Hence it makes the toughening process significantly more efficient.

The behaviour of HIPS can be contrasted with that of rubber-toughened poly(methyl methacrylate) (RTPMMA) shown in Fig. 5.68(b). In this case

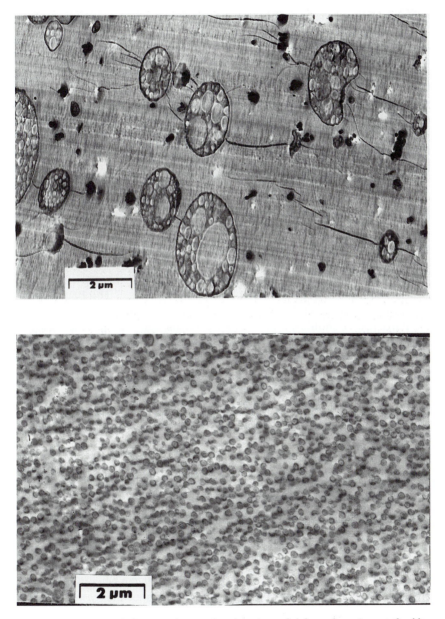

Fig. 5.68 *Transmission electron micrographs of sections of deformed specimens of rubber-toughened polymers. (a) Stained section of a commercial HIPS. The small black particles are pigment (courtesy of Hu Xiao). (b) RTPMMA (courtesy, D. E. J. Saunders). The tensile axis is approximately vertical in each case.*

a large number of smaller particles generally are employed. The particles are typically of the order of $0.2-0.3\mu m$ in diameter and again have a complex morphology. They are made by emulsion polymerization (Section 2.4.15) and have a 'core-shell' structure with up to four alternating layers of rubbery and glassy polymers. The glassy layers are essentially identical to the matrix polymer whereas the rubber is poly[(n-butyl acrylate)-*co*-styrene] with a composition chosen to give a refractive index identical to that of the PMMA matrix at room temperature. Since the particles are also smaller than the wavelength of light the toughened polymer remains transparent and the excellent optical properties of the matrix PMMA are virtually unaltered. The presence of high modulus glassy material within the particles again ensures more efficient toughening than for pure rubber spheres. The micrograph of the RTPMMA in Fig. 5.68(b) shows the core-shell structure of the 4-layer particles employed in this material. In this case no staining was needed since the rubbery phase has a higher electron density than the glassy phase and hence appears dark. It should be noted that only three layers can be visualized since the outer layer of the particles is PMMA and blends with the matrix. There is chemical graftlinking between the different layers in the particles and hence strong interfaces are obtained both between the particles and matrix and within the particles. It should be noted that even though the specimen used to obtain the section in Fig. 5.68(b) had been deformed to over 25% strain no crazes can be seen around the particles. It is thought that the particles in the RTPMMA toughen the material by inducing the formation of shear bands around the particles. Since they do not lead to cavitation in the matrix or a change in polymer density, unlike crazes they cannot be seen by electron microscopy. However, the material does undergo some stress-whitening which is thought to be due to some limited cavitation within the particles.

A further insight into the deformation mechanisms which take place in rubber-toughened polymers can be obtained by measuring the change in specimen volume (ΔV) during deformation as shown in Fig. 5.69 for a variety of materials. It can be seen that in the case of HIPS there is a significant increase in specimen volume following yield. Plastic deformation normally occurs at constant volume and so this large increase in volume shows that crazing takes place on a large scale during deformation of the material. The behaviour of HIPS in Fig. 5.69 can be contrasted with that of RTPMMA for which there is no change in specimen volume during deformation. Since shear yielding takes place at approximately constant volume this is further confirmation that the toughening mechanism in the RTPMMA is shear yielding. The volume-strain behaviour of other polymers in Fig. 5.69 is between that of HIPS and RTPMMA showing that for these materials both crazing and shear yielding are probably taking

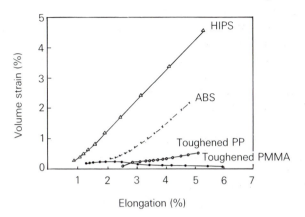

Fig. 5.69 *Volume-strain measurements as a function of axial strain for a series of different rubber-toughened polymers, HIPS, ABS (acrylonitrile–butadiene–styrene copolymer), toughened PP (polypropylene) and RTPMMA. (After Bucknall, Partridge and Ward (1984) J. Mater. Sci.* **19**, *2064).*

place. Hence it can be seen that the deformation mechanisms in rubber-toughened polymers can be readily monitored using volume-strain measurements.

5.7.3 *Particulate reinforcement and composites*

It can be seen from Table 5.6 that the incorporation of a soft rubbery phase into a brittle polymer can lead to significant degrees of toughening by inducing high levels of crazing and/or shear yielding. However, it is also found that an increase in toughness can be achieved in brittle polymers through the addition of a rigid filler such as glass beads or silica particles which have a higher modulus than the matrix (Table 5.6). In this case the presence of the particles impedes the propagation of cracks, increases fracture surface roughness and hence makes crack propagation more difficult resulting in higher K_c or $\mathscr{G}_c$ values. However, the increase in toughness is not as significant as in the case of rubber-toughening and the presence of rigid fillers can give rise to flaws which can reduce both the fracture strength of the polymer and the elongation to failure.

The degree of toughening in the case of rigid particles is found to depend upon the particle shape and in particular the 'aspect ratio' of the reinforcement. Significantly higher levels of toughening are obtained with particles which have high aspect ratios, such as whiskers or fibres. This is because the surface-to-volume ratio is increased and also is due to the possibility of extra energy-dissipating processes such as fibre pull-out. The highest levels of reinforcement are obtained for uniaxially-aligned long fibres, i.e. *fibre-reinforced composites*. However, such materials are highly anisotropic with relatively poor transverse mechanical properties. The subject of fibre-reinforced composites is extremely large and the reader is referred to the many specialized texts on the subject for more details.

Further reading

Andrews, E.H. (1968), *Fracture in Polymers,* Oliver and Boyd Ltd, London.
Arridge, R.G.C. (1975), *Mechanics of Polymers,* Clarendon Press, Oxford.
Booth, C. and Price, C. (Eds) (1989), *Comprehensive Polymer Science*, Vol. 2, Pergamon Press, Oxford.
Bucknall, C.B. (1977), *Toughened Plastics*, Applied Science Publishers, London.
Haward, R.N. (Ed.) (1973), *The Physics of Glassy Polymers,* Applied Science Publishers, London.
Hull, D. (1981), *An Introduction to Composite Materials*, Cambridge University Press, Cambridge.
Jenkins, A.D. (Ed.) (1972), *Polymer Science,* North-Holland Publishing Co., London.
Kambour, R.P. (1973), *Crazing and Fracture in Thermoplastics,* Journal of Polymer Science, Macromolecular Reviews, Vol. 7.
Kausch, H.H. (1988), *Polymer Fracture*, Springer-Verlag, Berlin.
Kelly, A. and Macmillan, N.H. (1986), *Strong Solids*, 3rd Edn, Clarendon Press, Oxford.
Kelly, A. and Groves, G.W. (1970), *Crystallography and Crystal Defects,* Longman, London.
Kinloch, A.J. and Young, R.J. (1983), *Fracture Behaviour of Polymers*, Applied Science, London.
Treloar, L.R.G. (1975), *The Physics of Rubber Elasticity*, 3rd Edn, Clarendon Press, Oxford.
Ward, I.M. (1983), *Mechanical Properties of Solid Polymers*, 2nd Edn, Wiley-Interscience, London.
Ward, I.M. (Ed.), (1986), *Developments in Oriented Polymers-2*, Elsevier Applied Science, London.
Williams, J.G. (1973), *Stress Analysis of Polymers,* Longman, London.

Problems

5.1 Mechanical models are often used to describe the viscoelastic behaviour of polymers. The Maxwell model which consists of a spring and dashpot in series predicts that the variation of strain, e, with time, t, for a viscoelastic material can be described by an equation of the form

$$\frac{de}{dt} = \frac{1}{E}\frac{d\sigma}{dt} + \frac{\sigma}{\eta} \qquad \begin{array}{l} E = \text{modulus of spring} \\ \eta = \text{viscosity of dashpot} \end{array}$$

where σ is the applied stress. The Voigt model which consists of a spring and dashpot in parallel leads to an equation of the form

$$\frac{de}{dt} = \frac{\sigma}{\eta} - \frac{Ee}{\eta}$$

(i) Derive equations describing the variation of strain with time for both models subjected to a constant stressing-rate (i.e. $d\sigma/dt$ = const.).

(ii) Derive equations describing the variation of stress with time when both models are subjected to a constant strain-rate (de/dt = const.).

5.2 A viscoelastic polymer which can be assumed to obey the Boltzmann superposition principle is subjected to the following loading history. At time, $t = 0$, a tensile stress of 10 MN m^{-2} is applied and maintained for 100 s. The stress is then removed instantaneously. If the creep compliance of the material is given by

$$J(t) = J_0(1 - \exp(-t/\tau_0))$$

where $J_0 = 2\,m^2\,GN^{-1}$ and $\tau_0 = 200$ s, what is the net creep strain after (a) 100 s and (b) 200 s?

5.3 A ring-shaped seal, made from a viscoelastic material, is used to seal a joint between two rigid pipes. When incorporated in the joint the seal is held at a fixed compressive strain of 0.2. Assuming that the seal can be treated as a Maxwell model, determine the time before the seal begins to leak under an internal fluid pressure of 0.3 MN m^{-2}. It can be assumed that the relaxation time τ_0 of the material is 300 days and the short-term (instantaneous) modulus of the material is 3 MN m^{-2}.

5.4 A viscoelastic polymer is subjected to an oscillating sinusoidal load applied at an angular frequency, ω. Assuming that the variation of the strain, e, and stress, σ, with time, t, can be represented by equations of the form

$$e = e_0 \sin \omega t$$

$$\sigma = \sigma_0 \sin(\omega t + \delta)$$

where δ is the phase lag between the stress and strain, show that the energy dissipated per cycle of the deformation ΔU can be given by

$$\Delta U = \sigma_0 e_0 \pi \sin \delta$$

5.5 Show that the integral in Equation (5.136) can be evaluated to give

$$\Delta S = -\frac{1}{2}Nk(\lambda_1^2 + \lambda_2^2 + \lambda_3^2 - 3)$$

5.6 The isothermal reversible work of deformation w per unit volume of an elastomer is given by the statistical theory of elastomer deformation as

$$w = \tfrac{1}{2}G(\lambda_1^2 + \lambda_2^2 + \lambda_3^2 - 3)$$

where λ_1, λ_2, and λ_3 are the principal extension ratios. Using this equation deduce the stress–strain relations for the general two-dimensional strain of a sheet of material and hence show that the stress–strain relation for an equal two-dimensional extension of λ is

$$\sigma_t = G(\lambda^2 - 1/\lambda^4)$$

where σ_t is the true stress acting on planes normal to the plane of stretch.

5.7 The tensile stress–strain curve of a polymer may be represented by $\sigma_t = ae^b$ where σ_t is the true stress and e is the nominal strain, and a and b are constants. Show that necking will occur when

$$e = b/(1-b)$$

Determine also the tensile yield stress of the material in terms of a and b.

5.8 If a polymer obeys a pressure-dependent von Mises yield criterion of the form

$$A[\sigma_1 + \sigma_2 + \sigma_3] + B[(\sigma_1 - \sigma_2)^2 + (\sigma_2 - \sigma_3)^2 + (\sigma_3 - \sigma_1)^2] = 1$$

where σ_1, σ_2 and σ_3 are the principal stresses and A and B are constants, show that the material will yield under the action of hydrostatic stress alone and calculate the hydrostatic stress required to cause yield, in terms of the yield stresses of the material in tension and compression.

5.9 Polystyrene is capable of undergoing either crazing or shear yielding during deformation depending upon the state of stress imposed. It is found that it undergoes shear yielding at a stress of 73 MN m^{-2} in uniaxial tension and at 92 MN m^{-2} in uniaxial compression at a particular strain-rate and temperature. Under the same conditions it is found to undergo crazing at a stress of 47 MN m^{-2} in uniaxial tension and under a biaxial tensile stress of 45 MN m^{-2}. Assuming that polystyrene obeys a pressure-dependent von Mises yield criterion and a critical strain crazing criterion, determine the conditions of stress at which crazing and shear yielding will take place simultaneously for plane stress deformation. (Poisson's ratio for polystyrene may be taken as 0.33.)

5.10 A sample of an oriented semi-crystalline polymer is deformed in tension in an X-ray diffractometer. The position of the (002) peak is found to change as the stress on the sample is increased as indicated below

Stress/MN m^{-2}	(Bragg angle)/degrees
0	37.483
40	37.477
80	37.471
120	37.466
160	37.460
200	37.454

Determine the Young's modulus of the crystals in the polymer in the chain-direction assuming that the stress on the crystals is equal to the stress applied to the specimen as a whole. (The radiation used was CuKα $\lambda = 1.542$ Å.)

5.11 It is thought that polyethylene is capable of undergoing twinning through a kinking of the molecules in a similar way to which twinning takes place in polydiacetylene single crystals. The most likely planes on which the kinking can occur are thought to be (200), (020) and (110). Draw molecular sketches of each of these twins viewed perpendicular to the plane of shear and hence or otherwise determine

(a) the twinning plane in each case;
(b) the angle through which the molecules bend;
(c) the direction of shear for each twin;
(d) the twin involving the lowest shear.

(Polyethylene is orthorhombic with $a = 7.40$ Å, $b = 4.93$ Å, $c = 2.537$ Å.)

5.12 The fracture strength of an un-notched parallel-sided sheet of a brittle polymer of thickness 5 mm and breadth 25 mm is found to be 85 MN m^{-2}. If the critical stress intensity factor, K_c, for the polymer is determined from a separate experiment on a notched specimen to be 1.25 MN m$^{-3/2}$ calculate the inherent flaw size for the polymer. (The value of K_c for a single-edge notched sheet specimen may be determined from the appropriate expression in Table 5.5.)

5.13 The following data were obtained from an experiment upon polystyrene at room temperature where the fracture stress, σ_f was measured as a function of crack length, a.

σ_f/MPa	a/mm
17.2	10
17.5	9
17.9	8
18.4	7
19.0	6
20.0	5
22.0	4
25.0	3
31.0	2
40.0	1
41.0	0.5
42.0	0.1

Determine for polystyrene
(a) $\mathcal{G}_c$, (b) K_c and (c) the inherent flaw size.
(Assume the Young's modulus is 3 GPa.)

5.14 It is found that the rate of growth of a crack in a non-crystallizing elastomer $(\mathrm{d}a/\mathrm{d}t)$ can be given by an equation of the form

$$\mathrm{d}a/\mathrm{d}t = q\mathcal{T}^n$$

where $\mathcal{T}$ is the tearing energy, and q and n are constants for a given set of testing conditions. Determine the time-to-failure, t_f, for a sheet of elastomer held at constant extension and containing a small edge-crack of initial length a_0, given that for the sheet

$$\mathcal{T} = KaW$$

where K is a constant and W is the stored energy density in the material at the strain imposed.

Answers to problems

Chapter 2

2.1 2nd order reaction, catalyst used, rate constant $k = 2.43 \times 10^{-4}$ dm^3 mol^{-1} s^{-1}. After 1 hour $p = 0.731$ and after 5 hours $p = 0.931$.

2.2 Percentage conversion $= 99.0\%$ (i.e. $p = 0.990$).

2.3 $\bar{x}_n = 75.9$, $\bar{M}_n = 8.58$ kg mol^{-1}, and $\bar{M}_w = 17.07$ kg mol^{-1}.

2.4 Flory statistical theory gives $p_c = 0.816$ whereas Carothers theory gives $p_c = 0.917$.

2.5 Mixture before reaction: $\bar{M}_n = 15\,000$ g mol^{-1}, $\bar{M}_w = 40\,000$ g mol^{-1}. The product from the reaction has: $\bar{M}_n = 15\,000$ g mol^{-1}, $M_w = 30\,000$ g mol^{-1}.

2.6 Combination.

2.7 Half-life $= 56.63$ h. The initiator concentration will have decreased by a factor of 0.988 (i.e. to $0.988[I]_0$).

2.8 (i): (a) no effect, (b) increases by a factor of 2; (ii): (a) increases by a factor of 4, (b) increases by a factor of 2; (iii): (a) increase by a factor of 4, (b) decrease by a factor of 2 [neglecting effects of chain transfer].

2.9 $R_p = 5.87 \times 10^{-6}$ mol dm^{-3} s^{-1} and $\bar{M}_n = 349.7$ kg mol^{-1}.

2.10 R_p is increased by a factor of 2.63, and $\bar{x}_n$ is decreased by a factor of 0.699.

2.11 $\bar{M}_n = 5\,470$ g mol^{-1}.

2.12 $R_p = 0.165$ mol dm^{-3} s^{-1} and $\bar{M}_n = 1\,560$ kg mol^{-1}

2.13 The polystyrene end-blocks have $\bar{M}_n = 15.0$ kg mol^{-1} and the central polyisoprene block has $\bar{M}_n = 100.0$ kg mol^{-1}.

2.15 (i) The mol% styrene repeat units in the copolymers formed are: (a) 42.7, (b) 51.0, (c) 59.7, (d) 57.4, (e) 50.5, (f) 94.6, (g) 98.2; (ii) the mol% styrene required in comonomer mixtures are: (a) 57.2, (b) 48.5, (c) 32.9, (d) 24.0, (e) any value in the range 0.1–99.9, (f) 3.3, (g) 1.3; (iii) the mol% azeotropic compositions with respect to styrene are: (b) 52.9, (c) 76.6, (d) 61.5, (e) 50.5.

2.16 $r_S = 0.40$ and $r_{AN} = 0.04$.

Chapter 3

3.1 (a) $\bar{M}_n = 46\,667$ g mol^{-1} and $\bar{M}_w = 78\,571$ g mol^{-1}; (b) $\bar{M}_n = 20\,930$ g mol^{-1} and $\bar{M}_w = 46\,667$ g mol^{-1}; (c) $\bar{M}_n = 43\,384$ g mol^{-1} and $\bar{M}_w = 86\,950$ g mol^{-1}.

3.2 $\bar{M}_n = 1.5$ kg mol^{-1}.

3.3 $\chi = 0.426$.

3.5 (a) 1 320 nm, (b) 1 078 nm, (c) 20.17 nm.

3.6 $\bar{M}_n = 143.6$ kg mol^{-1} and $\chi = 0.463$.

3.7 $\bar{M}_n = 2.0$ kg mol^{-1}

3.8 $\bar{M}_w = 735.8$ kg mol^{-1}, $\langle s^2 \rangle_z^{1/2} = 48.4$ nm, and $A_2 = 4.19 \times 10^{-4}$ m^3 mol kg^{-2}.

3.9 Sample B: $[\eta] = 26.16$ cm^3 g^{-1}; sample E: $[\eta] = 127.28$ cm^3 g^{-1}; $K = 8.15 \times 10^{-2}$ cm^3 g^{-1} (g mol^{-1})$^{-0.50}$ with $a = 0.50$.

3.10 $\chi = 1.66$.

3.11 $M = 119.15$ kg mol^{-1}.

Chapter 4

4.1 The planes giving rise to the reflections are (110), (200) and (020).

4.2 (a) Chain repeat, $c = 2.54$ Å
(b) Higher order lines cannot be obtained since $\sin\varphi \leqslant 1$ (Fig. 4.3). Second order lines could be obtained if X-rays with wavelength $\lambda < 1.27$ Å were used.
(c) The specimen would have to be tilted by 17.65°.

4.3 Linear polymer, $x_c = 0.802$
Branched polymer, $x_c = 0.420$
Branched molecules do not crystallize as easily as linear ones.

4.4 Equilibrium melting temperature, $T_m^\circ = 349$ K.
The fold surface energy, $\gamma_e = 0.048$ J m^{-2}

4.5 The equilibrium melting temperature, $T_m^\circ = 79.5$°C.

4.6 The glass transition temperature, $T_g = 89°C$.

4.7 Assuming the rule of mixtures the copolymer composition should be 82% (by weight) n-butyl acrylate and 18% (by weight) styrene. The material would only be transparent at room temperature since $\bar{n}$ will vary differently with temperature for each component.

4.8 (a) Polyethylene (flexible molecule) lower than polypropylene (side group hinders rotation)

(b) Poly(methyl acrylate) lower than poly(vinyl acetate)(bulky side group further from main chain).

(c) Poly(but-1-ene) (long bulky side group) higher than poly(but-2-ene) (shorter side groups).

(d) Poly(ethylene oxide) (flexible molecule) lower than poly(vinyl alcohol) (—OH side groups).

(e) Poly(ethyl acrylate) (long flexible side group) lower than poly(methyl methacrylate) (shorter bulky side groups).

NB All pairs are isomers.

4.9 The copolymer has $T_g = 246$ K.

4.10 The average number of branches per molecule $= 3.15$

Chapter 5

5.1 (i) Maxwell model: $\dfrac{de}{dt} = \dfrac{R}{E} + \dfrac{Rt}{\eta}$

Voigt Model: $e = \dfrac{R}{E}\left\{t - \tau_o\left[1 - \exp\left[-\dfrac{t}{\tau_o}\right]\right]\right\}$

where constant stressing rate, $R = \dfrac{d\sigma}{dt}$

(ii) Maxwell model: $\sigma = S\eta\left[1 - \exp\left[-\dfrac{t}{\tau_o}\right]\right]$

Voigt model: $\sigma = ESt + S\eta$
where constant straining rate, $S = \dfrac{de}{dt}$

5.2 (a) Creep strain after 100 s, $e(100) = 0.00787$
(b) Creep strain after 200 s, $e(200) = 0.00477$

5.3 Time before seal leaks $= 208$ days.

5.4 $\Delta U = \sigma_o e_o \pi \sin\delta$

5.5 $\Delta S = -\tfrac{1}{2}Nk(\lambda_1^2 + \lambda_2^2 + \lambda_3^2 - 3)$

5.6 $\sigma_t = G\left[\lambda^2 - \dfrac{1}{\lambda^4}\right]$

5.7   Necking occurs when $e = b/(1 - b)$
Tensile stress at yield, $\sigma_t = ab^b(1 - b)^{-b}$

5.8   The hydrostatic pressure to cause yielding,

$$P_c = \frac{\sigma_{yc}\,\sigma_{yt}}{3(\sigma_{yc} - \sigma_{yt})}$$

5.9 $\sigma_1 = 59\,\text{MN m}^{-2}$ and $\sigma_2 = -30\,\text{MN m}^{-2}$
or $\sigma_1 = -30\,\text{MN m}^{-2}$ and $\sigma_2 = 59\,\text{MN m}^{-2}$

5.10 The chain direction Youngs modulus is approximately $300\,\text{GN m}^{-2}$

5.11 See Bevis, M. (1978) *Colloid and Polymer Science*, **256** p. 234

Molecular Plane	(a) Twinning Plane	(b) Angle of Bend	(c) Direction of Shear
(200)	(202)	37.8°	[20$\bar{2}$]
(020)	(022)	54.5°	[02$\bar{2}$]
(110)	(112)	34.4°	~[11$\bar{2}$]*

* irrational

(d) Twin with lowest shear is on (112)

5.12 Inherent flaw size, $a_o = 55\,\mu\text{m}$

5.13 (a) Fracture energy, $\mathcal{G}_c = 1.3\,\text{kJ m}^{-2}$
(b) Fracture toughness, $K_c = 2.0\,\text{MN m}^{-3/2}$
(c) Inherent flaw size, $a_o = 1\,\text{mm}$.

5.14 The time to failure is given by

$$t_f = \frac{1}{qK^nW^n(n-1)a_o{}^{n-1}}$$

Index